43 iwe 620
lbf 514
01. Expl.

Ausgeschieden im Jahr 2025

INSTANTANEOUS POWER THEORY AND APPLICATIONS TO POWER CONDITIONING

IEEE Press
445 Hoes Lane
Piscataway, NJ 08854

IEEE Press Editorial Board
Mohamed E. El-Hawary, *Editor in Chief*

R. Abari	T. G. Croda	R. J. Herrick
S. Basu	S. Farshchi	M. S. Newman
A. Chatterjee	S. V. Kartalopoulos	N. Schulz
T. Chen	B. M. Hammerli	

Kenneth Moore, *Director of IEEE Book and Information Services (BIS)*
Catherine Faduska, *Senior Acquisitions Editor*
Jeanne Audino, *Project Editor*

INSTANTANEOUS POWER THEORY AND APPLICATIONS TO POWER CONDITIONING

Hirofumi Akagi
Professor of Electrical Engineering
TIT—Tokyo Institute of Technology, Japan

Edson Hirokazu Watanabe
Professor of Electrical Engineering
UFRJ—Federal University of Rio de Janeiro, Brazil

Mauricio Aredes
Associate Professor of Electrical Engineering
UFRJ—Federal University of Rio de Janeiro, Brazil

IEEE PRESS SERIES ON POWER ENGINEERING

Mohamed E. El-Hawary, *Series Editor*

IEEE PRESS

WILEY-INTERSCIENCE
A JOHN WILEY & SONS, INC., PUBLICATION

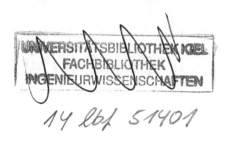

Copyright © 2007 by the Institute of Electrical and Electronics Engineers, Inc. All rights reserved.

Published by John Wiley & Sons, Inc., Hoboken, New Jersey.
Published simultaneously in Canada.

No part of this publication may be reproduced, stored in a retrieval system, or transmitted in any form or by any means, electronic, mechanical, photocopying, recording, scanning, or otherwise, except as permitted under Section 107 or 108 of the 1976 United States Copyright Act, without either the prior written permission of the Publisher, or authorization through payment of the appropriate per-copy fee to the Copyright Clearance Center, Inc., 222 Rosewood Drive, Danvers, MA 01923, (978) 750-8400, fax (978) 750-4470, or on the web at www.copyright.com. Requests to the Publisher for permission should be addressed to the Permissions Department, John Wiley & Sons, Inc., 111 River Street, Hoboken, NJ 07030, (201) 748-6011, fax (201) 748-6008, or online at http://www.wiley.com/go/permission.

Limit of Liability/Disclaimer of Warranty: While the publisher and author have used their best efforts in preparing this book, they make no representations or warranties with respect to the accuracy or completeness of the contents of this book and specifically disclaim any implied warranties of merchantability or fitness for a particular purpose. No warranty may be created or extended by sales representatives or written sales materials. The advice and strategies contained herein may not be suitable for your situation. You should consult with a professional where appropriate. Neither the publisher nor author shall be liable for any loss of profit or any other commercial damages, including but not limited to special, incidental, consequential, or other damages.

For general information on our other products and services or for technical support, please contact our Customer Care Department within the United States at (800) 762-2974, outside the United States at (317) 572-3993 or fax (317) 572-4002.

Wiley also publishes its books in a variety of electronic formats. Some content that appears in print may not be available in electronic format. For information about Wiley products, visit our web site at www.wiley.com.

Library of Congress Cataloging-in-Publication Data is available.

ISBN 978-0-470-10761-4

10 9 8 7 6 5 4 3 2 1

*This book is dedicated
to all the scientists and engineers who have participated in
the development of Instantaneous Power Theory and
Applications to Power Conditioning*

and

*to our families
Nobuko, Chieko, and Yukiko,
Yukiko, Edson Hiroshi, and Beatriz Yumi,
Marilia, Mariah, and Maynara.*

CONTENTS

Preface	xiii
1. Introduction	**1**
1.1. Concepts and Evolution of Electric Power Theory	2
1.2. Applications of the *p-q* Theory to Power Electronics Equipment	4
1.3. Harmonic Voltages in Power Systems	5
1.4. Identified and Unidentified Harmonic-Producing Loads	7
1.5. Harmonic Current and Voltage Sources	8
1.6. Basic Principles of Harmonic Compensation	11
1.7. Basic Principles of Power Flow Control	14
References	17
2. Electric Power Definitions: Background	**19**
2.1. Power Definitions Under Sinusoidal Conditions	20
2.2. Voltage and Current Phasors and the Complex Impedance	22
2.3. Complex Power and Power Factor	24
2.4. Concepts of Power Under Non-Sinusoidal Conditions— Conventional Approaches	25
2.4.1. Power Definitions by Budeanu	25
2.4.1.A. Power Tetrahedron and Distortion Factor	28
2.4.2. Power Definitions by Fryze	30
2.5. Electric Power in Three-Phase Systems	31
2.5.1. Classifications of Three-Phase Systems	31
2.5.2. Power in Balanced Three-Phase Systems	34
2.5.3. Power in Three-Phase Unbalanced Systems	36

2.6. Summary		37
References		38

3 The Instantaneous Power Theory — 41

3.1. Basis of the p-q Theory — 42
    3.1.1. Historical Background of the p-q Theory — 42
    3.1.2. The Clarke Transformation — 43
        3.1.2.A. Calculation of Voltage and Current Vectors when Zero-Sequence Components are Excluded — 45
    3.1.3. Three-Phase Instantaneous Active Power in Terms of Clarke Components — 47
    3.1.4. The Instantaneous Powers of the p-q Theory — 48
3.2. The p-q Theory in Three-Phase, Three-Wire Systems — 49
    3.2.1. Comparisons with the Conventional Theory — 53
        3.2.1.A. Example #1—Sinusoidal Voltages and Currents — 53
        3.2.1.B. Example #2—Balanced Voltages and Capacitive Loads — 54
        3.2.1.C. Example #3—Sinusoidal Balanced Voltage and Nonlinear Load — 55
    3.2.2. Use of the p-q Theory for Shunt Current Compensation — 59
        3.2.2.A. Examples of Appearance of Hidden Currents — 64
            3.2.2.A.1 Presence of the Fifth Harmonic in Load Current — 64
            3.2.2.A.2 Presence of the Seventh Harmonic in Load Current — 67
    3.2.3. The Dual p-q Theory — 68
3.3. The p-q Theory in Three-Phase, Four-Wire Systems — 71
    3.3.1. The Zero-Sequence Power in a Three-Phase Sinusoidal Voltage Source — 72
    3.3.2. Presence of Negative-Sequence Components — 74
    3.3.3. General Case-Including Distortions and Imbalances in the Voltages and in the Currents — 75
    3.3.4. Physical Meanings of the Instantaneous Real, Imaginary, and Zero-Sequence Powers — 79
    3.3.5. Avoiding the Clarke Transformation in the p-q Theory — 80
    3.3.6. Modified p-q Theory — 82
3.4. Instantaneous abc Theory — 87
    3.4.1. Active and Nonactive Current Calculation by Means of a Minimization Method — 89
    3.4.2. Generalized Fryze Currents Minimization Method — 94
3.5. Comparisons between the p-q Theory and the abc Theory — 98
    3.5.1. Selection of Power Components to be Compensated — 101
3.6. Summary — 102
References — 104

4 Shunt Active Filters — 109

- 4.1. General Description of Shunt Active Filters — 111
    - 4.1.1. PWM Converters for Shunt Active Filters — 112
    - 4.1.2. Active Filter Controllers — 113
- 4.2. Three-Phase, Three-Wire Shunt Active Filters — 116
    - 4.2.1. Active Filters for Constant Power Compensation — 118
    - 4.2.2. Active Filters for Sinusoidal Current Control — 134
        - 4.2.2.A. Positive-Sequence Voltage Detector — 138
            - 4.2.2.A.1 Main Circuit of the Voltage Detector — 138
            - 4.2.2.A.2 Phase-Locked-Loop (PLL) Circuit — 141
        - 4.2.2.B. Simulation Results — 145
    - 4.2.3. Active Filters for Current Minimization — 145
    - 4.2.4. Active Filters for Harmonic Damping — 150
        - 4.2.4.A. Shunt Active Filter Based on Voltage Detection — 151
        - 4.2.4.B. Active Filter Controller Based on Voltage Detection — 152
        - 4.2.4.C. An Application Case of Active Filter for Harmonic Damping — 157
            - 4.2.4.C.1 The Power Distribution Line for the Test Case — 158
            - 4.2.4.C.2 The Active Filter for Damping of Harmonic Propagation — 159
            - 4.2.4.C.3 Experimental Results — 160
            - 4.2.4.C.4 Adjust of the Active Filter Gain — 168
    - 4.2.5. A Digital Controller — 173
        - 4.2.5.A. System Configuration of the Digital Controller — 174
            - 4.2.5.A.1 Operating Principle of PLL and PWM Units — 175
            - 4.2.5.A.2 Sampling Operation in the A/D Unit — 177
        - 4.2.5.B. Current Control Methods — 178
            - 4.2.5.B.1 Modeling of Digital Current Control — 178
            - 4.2.5.B.2 Proportional Control — 179
            - 4.2.5.B.3 Deadbeat Control — 180
            - 4.2.5.B.4 Frequency Response of Current Control — 181
- 4.3. Three-Phase, Four-Wire Shunt Active Filters — 182
    - 4.3.1. Converter Topologies for Three-Phase, Four-Wire Systems — 183
    - 4.3.2. Dynamic Hysteresis-Band Current Controller — 184
    - 4.3.3. Active Filter Dc Voltage Regulator — 186
    - 4.3.4. Optimal Power Flow Conditions — 187
    - 4.3.5. Constant Instantaneous Power Control Strategy — 189
    - 4.3.6. Sinusoidal Current Control Strategy — 192
    - 4.3.7. Performance Analysis and Parameter Optimization — 195
        - 4.3.7.A. Influence of the System Parameters — 195
        - 4.3.7.B. Dynamic Response of the Shunt Active Filter — 196

4.3.7.C. Economical Aspects	201
4.3.7.D. Experimental Results	203
4.4. Shunt Selective Harmonic Compensation	208
4.5. Summary	216
References	217

5 Hybrid and Series Active Filters — 221

5.1. Basic Series Active Filter	221
5.2. Combined Series Active Filter and Shunt Passive Filter	223
5.2.1. Example of An Experimental System	226
5.2.1.A. Compensation Principle	226
5.2.1.A.1 Source Harmonic Current $\dot{I}_{Sh}$	228
5.2.1.A.2 Output Voltage of Series Active Filter: $\dot{V}_C$	229
5.2.1.A.3 Shunt Passive Filter Harmonic Voltage: $\dot{V}_{Fh}$	229
5.2.1.B. Filtering Characteristics	230
5.2.1.B.1 Harmonic Current Flowing From the Load to the Source	230
5.2.1.B.2 Harmonic Current Flowing from the Source to the Shunt Passive Filter	231
5.2.1.C. Control Circuit	231
5.2.1.D. Filter to Suppress Switching Ripples	233
5.2.1.E. Experimental Results	234
5.2.2. Some Remarks about the Hybrid Filters	237
5.3. Series Active Filter Integrated with a Double-Series Diode Rectifier	238
5.3.1. The First-Generation Control Circuit	241
5.3.1.A. Circuit Configuration and Delay Time	241
5.3.1.B. Stability of the Active Filter	242
5.3.2. The Second-Generation Control Circuit	244
5.3.3. Stability Analysis and Characteristics Comparison	246
5.3.3.A. Transfer Function of the Control Circuits	246
5.3.3.B. Characteristics Comparisons	247
5.3.4. Design of a Switching-Ripple Filter	248
5.3.4.A. Design Principle	248
5.3.4.B. Effect on the System Stability	250
5.3.4.C. Experimental Testing	251
5.3.5. Experimental Results	252
5.4. Comparisons Between Hybrid and Pure Active Filters	253
5.4.1. Low-Voltage Transformerless Hybrid Active Filter	255
5.4.2. Low-Voltage Transformerless Pure Shunt Active Filter	258
5.4.3. Comparisons Through Simulation Results	259
5.5. Conclusions	261
References	262

6 Combined Series and Shunt Power Conditioners — 265
6.1. The Unified Power Flow Controller (UPFC) — 267
    6.1.1. FACTS and UPFC Principles — 268
        6.1.1.A. Voltage Regulation Principle — 269
        6.1.1.B. Power Flow Control Principle — 270
    6.1.2. A Controller Design for the UPFC — 274
    6.1.3. UPFC Approach Using a Shunt Multipulse Converter — 281
        6.1.3.A. Six-Pulse Converter — 282
        6.1.3.B. Quasi 24-Pulse Converter — 286
        6.1.3.C. Control of Active and Reactive Power in Multipulse Converters — 288
        6.1.3.D. Shunt Multipulse Converter Controller — 290
6.2. The Unified Power Quality Conditioner (UPQC) — 293
    6.2.1. General Description of the UPQC — 294
    6.2.2. A Three-Phase, Four-Wire UPQC — 297
        6.2.2.A. Power Circuit of the UPQC — 297
        6.2.2.B. The UPQC Controller — 299
            6.2.2.B.1 PWM Voltage Control with Minor Feedback Control Loop — 300
            6.2.2.B.2 Series Active Filter Controller — 301
            6.2.2.B.3 Integration of the Series and Shunt Active Filter Controllers — 305
            6.2.2.B.4 General Aspects — 307
        6.2.2.C. Analysis of the UPQC Dynamic — 308
            6.2.2.C.1 Optimizing the Power System Parameters — 309
            6.2.2.C.2 Optimizing the Parameters in the Control Systems — 311
            6.2.2.C.3 Simulation Results — 312
            6.2.2.C.4 Experimental Results — 320
    6.2.3. The UPQC Combined with Passive Filters (Hybrid UPQC) — 326
        6.2.3.A. Controller of the Hybrid UPQC — 331
        6.2.3.B. Experimental Results — 337
6.3. The Universal Active Power Line Conditioner (UPLC) — 343
    6.3.1. General Description of the UPLC — 344
    6.3.2. The Controller of the UPLC — 347
        6.3.2.A. Controller for the Configuration #2 of UPLC — 355
    6.3.3. Performance of the UPLC — 355
        6.3.3.A. Normalized System Parameters — 355
        6.3.3.B. Simulation Results of Configuration #1 of UPLC — 360
        6.3.3.C. Simulation Results of Configuration #2 of UPLC — 368
    6.3.4. General Aspects — 370
6.4. Summary — 371
References — 371

Index — 375

PREFACE

The concept of "instantaneous active and reactive power" was first presented in 1982 in Japan. Since then, many scientists and engineers have made significant contributions to its modifications in three-phase, four-wire circuits, its expansions to more than three-phase circuits, and its applications to power electronics equipment. However, neither a monograph nor book on this subject has been available in the market. Filling this gap was the main motivation for writing this book. The instantaneous power theory, or simply "the p-q theory," makes clear the physical meaning of what the instantaneous active and reactive power is in a three-phase circuit. Moreover, it provides a clear insight into how energy flows from a source to a load, or circulates between phases, in a three-phase circuit.

At the beginning of writing this book, we decided to try to present the basic concepts of the theory as didactically as possible. Hence, the book was structured to present in Chapter 1 the problems related to nonlinear loads and harmonics. Chapter 2 describes the background of electrical power definitions based on conventional theories. Then, Chapter 3 deals with the instantaneous power theory. In this chapter, special attention is paid to the effort to offer abundant materials intended to make the reader understand the theory, particularly for designing controllers for active filters for power conditioning. Part of Chapter 3 is dedicated to presenting alternative sets of instantaneous power definitions. One of the alternatives, called the "modified p-q theory," expands the original imaginary power definition to an instantaneous imaginary power vector with three components. Another approach, called in this book the "*abc* theory," uses the *abc*-phase voltages and currents directly to define the active and nonactive current components. Comparisons in difference and physical meaning between these theories conclude Chapter 3.

Chapter 4 is exclusively dedicated to shunt active filters with different filter structures, showing clearly whether energy storage elements such as capacitors and inductors are necessary or not, and how much they are theoretically dispensable to the active filters. This consideration of the energy storage elements is one of the strongest points in the instantaneous power theory. Chapter 5 addresses series active filters, including hybrid configurations of active and passive filters. The hybrid configurations may provide an economical solution to harmonic filtering, particularly in medium-voltage, adjustable-speed motor drives.

Chapter 6 presents combined series and shunt power conditioners, including the unified power quality conditioner (UPQC), and the unified power flow controller (UPFC) that is a FACTS (flexible ac transmission system) device. Finally, it leads to the universal active power line conditioner (UPLC) that integrates the UPQC with the UPFC in terms of functionality.

Pioneering applications of the p-q theory to power conditioning are illustrated throughout the book, which helps the reader to understand the substantial nature of the instantaneous power theory, along with distinct differences from conventional theories.

The authors would like to acknowledge the encouragement and support received from many colleagues in various forms. The first author greatly appreciates his former colleagues, Prof. A. Nabae, the late Prof. I. Takahashi, and Mr. Y. Kanazawa at the Nagaoka University of Technology, where the p-q theory was born in 1982 and research on its applications to pure and hybrid active filters was initiated to spur many scientists and engineers to do further research on theory and practice.

The long distance between the homelands of the authors was not a serious problem because research-supporting agencies like CNPq (Conselho Nacional de Desenvolvimento Científico e Tecnológico) and JSPS (the Japan Society for the Promotion of Science) gave financial support for the authors to travel to Brazil or to Japan when a conference was held in one of these countries. Thus, the authors were able to meet and discuss face to face the details of the book, which would not be easily done over the Internet. The support received from the Fundação de Amparo à Pesquisa do Estado do Rio de Janeiro (FAPERJ) is also acknowledged.

Finally, our special thanks go to the following former graduate students and visiting scientists who worked with the authors: S. Ogasawara, K. Fujita, T. Tanaka, S. Atoh, F. Z. Peng, Y. Tsukamoto, H. Fujita, S. Ikeda, T. Yamasaki, H. Kim, K. Wada, P. Jintakosonwit, S. Srianthumrong, Y. Tamai, R. Inzunza, and G. Casaravilla, who are now working in industry or academia. They patiently helped to develop many good ideas, to design and build experimental systems, and to obtain experimental results. Without their enthusiastic support, the authors could not have published this book.

H. AKAGI, E. H. WATANABE, AND M. AREDES

Tokyo/Rio de Janeiro, August, 2006

CHAPTER 1

INTRODUCTION

The instantaneous active and reactive power theory, or the so-called "p-q Theory," was introduced by Akagi, Kanazawa, and Nabae in 1983. Since then, it has been extended by the authors of this book, as well as other research scientists. This book deals with the theory in a complete form for the first time, including comparisons with other sets of instantaneous power definitions. The usefulness of the p-q Theory is confirmed in the following chapters dealing with applications in controllers of compensators that are generically classified here as active power line conditioners.

The term "power conditioning" used in this book has much broader meaning than the term "harmonic filtering." In other words, the power conditioning is not confined to harmonic filtering, but contains harmonic damping, harmonic isolation, harmonic termination, reactive-power control for power factor correction, power flow control, and voltage regulation, load balancing, voltage-flicker reduction, and/or their combinations. Active power line conditioners are based on leading-edge power electronics technology that includes power conversion circuits, power semiconductor devices, analog/digital signal processing, voltage/current sensors, and control theory.

Concepts and evolution of electric power theory are briefly described below. Then, the need for a consistent set of power definitions is emphasized to deal with electric systems under nonsinusoidal conditions. Problems with harmonic pollution in alternating current systems (ac systems) are classified, including a list of the principal harmonic-producing loads. Basic principles of harmonic compensation are introduced. Finally, this chapter describes the fundaments of power flow control. All these topics are the subjects of scope, and will be discussed deeply in the following chapters of the book.

Instantaneous Power Theory and Applications to Power Conditioning. By Akagi, Watanabe, & Aredes
Copyright © 2007 the Institute of Electrical and Electronics Engineers, Inc.

1.1. CONCEPTS AND EVOLUTION OF ELECTRIC POWER THEORY

One of main points in the development of alternating current (ac) transmission and distribution power systems at the end of the 19th century was based on sinusoidal voltage at constant-frequency generation. Sinusoidal voltage with constant frequency has made easier the design of transformers and transmission lines, including very long distance lines. If the voltage were not sinusoidal, complications would appear in the design of transformers, machines, and transmission lines. These complications would not allow, certainly, such a development as the generalized "electrification of the human society." Today, there are very few communities in the world without ac power systems with "constant" voltage and frequency.

With the emergence of sinusoidal voltage sources, the electric power network could be made more efficient if the load current were in phase with the source voltage. Therefore, the concept of reactive power was defined to represent the quantity of electric power due to the load current that is not in phase with the source voltage. The average of this reactive power during one period of the line frequency is zero. In other words, this power does not contribute to energy transfer from the source to the load. At the same time, the concepts of apparent power and power factor were created. Apparent power gives the idea of how much power can be delivered or consumed if the voltage and current are sinusoidal and perfectly in phase. The power factor gives a relation between the average power actually delivered or consumed in a circuit and the apparent power at the same point. Naturally, the higher the power factor, the better the circuit utilization. As a consequence, the power factor is more efficient not only electrically but also economically. Therefore, electric power utilities have specified lower limits for the power factor. Loads operated at low power factor pay an extra charge for not using the circuit efficiently.

For a long time, one of the main concerns related to electric equipment was power factor correction, which could be done by using capacitor banks or, in some cases, reactors. For all situations, the load acted as a linear circuit drawing a sinusoidal current from a sinusoidal voltage source. Hence, the conventional power theory based on active-, reactive- and apparent-power definitions was sufficient for design and analysis of power systems. Nevertheless, some papers were published in the 1920s, showing that the conventional concept of reactive and apparent power loses its usefulness in nonsinusoidal cases [1,2]. Then, two important approaches to power definitions under nonsinusoidal conditions were introduced by Budeanu [3,4], in 1927 and Fryze [5] in 1932. Fryze defined power in the time domain, whereas Budeanu did it in the frequency domain. At that time, nonlinear loads were negligible, and little attention was paid to this matter for a long time.

Since power electronics was introduced in the late 1960s, nonlinear loads that consume nonsinusoidal current have increased significantly. In some cases, they represent a very high percentage of the total loads. Today, it is common to find a house without linear loads such as conventional incandescent lamps. In most cases, these lamps have been replaced by electronically controlled fluorescent lamps. In industrial applications, an induction motor that can be considered as a linear load in a steady state is now equipped with a rectifier and inverter for the purpose of achieving adjustable-speed control. The induction motor together with its drive is no longer a linear load. Unfortunately, the previous power definitions under nonsinusoidal currents were dubious, thus leading to misinterpretations in some cases. Chapter 2 presents a review of some theories dealing with nonsinusoidal conditions.

As pointed out above, the problems related to nonlinear loads have significantly increased with the proliferation of power electronics equipment. The modern equipment behaves as a nonlinear load drawing a significant amount of harmonic current from the power network. Hence, power systems in some cases have to be analyzed under nonsinusoidal conditions. This makes it imperative to establish a consistent set of power definitions that are also valid during transients and under nonsinusoidal conditions.

The power theories presented by Budeanu [3,4] and Fryze [5] had basic concerns related to the calculation of average power or root-mean-square values (rms values) of voltage and current. The development of power electronics technology has brought new boundary conditions to the power theories. Exactly speaking, the new conditions have not emerged from the research of power electronics engineers. They have resulted from the proliferation of power converters using power semiconductor devices such as diodes, thyristors, insulated-gate bipolar transistors (IGBTs), gate-turn-off (GTO) thyristors, and so on. Although these power converters have a quick response in controlling their voltages or currents, they may draw reactive power as well as harmonic current from power networks. This has made it clear that conventional power theories based on average or rms values of voltages and currents are not applicable to the analysis and design of power converters and power networks. This problem has become more serious and clear during comprehensive analysis and design of active filters intended for reactive-power compensation as well as harmonic compensation.

From the end of the 1960s to the beginning of the 1970s, Erlicki and Emanuel-Eigeles [6], Sasaki and Machida [7], and Fukao, Iida, and Miyairi [8] published their pioneer papers presenting what can be considered as a basic principle of controlled reactive-power compensation. For instance, Erlicki and Emanuel-Eigeles [6] presented some basic ideas like "compensation of distortive power is unknown to date. . . ." They also determined that "a non-linear resistor behaves like a reactive-power generator while having no energy-storing elements," and presented the very first approach to active power-factor control. Fukao, Iida and Miyairi [8] stated that "by connecting a reactive-power source in parallel with the load, and by controlling it in such a way as to supply reactive power to the load, the power network will only supply active power to the load. Therefore, ideal power transmission would be possible."

Gyugyi and Pelly [9] presented the idea that reactive power could be compensated by a naturally commutated cycloconverter without energy storage elements. This idea was explained from a physical point of view. However, no specific mathematical proof was presented. In 1976, Harashima, Inaba, and Tsuboi [10] presented, probably for the first time, the term "instantaneous reactive power" for a single-phase circuit. That same year, Gyugyi and Strycula [11] used the term "active ac power filters" for the first time. A few years later, in 1981, Takahashi, Fujiwara, and Nabae published two papers [12,13] giving a hint of the emergence of the instantaneous power theory or "*p-q* Theory." In fact, the formulation they reached can be considered a subset of the *p-q* Theory that forms the main scope of this book. However, the physical meaning of the variables introduced to the subset was not explained by them.

The *p-q* Theory in its first version was published in the Japanese language in 1982 [14] in a local conference, and later in *Transactions of the Institute of Electrical Engineers of Japan* [15]. With a minor time lag, a paper was published in English in an international conference in 1983 [16], showing the possibility of compensating for instantaneous reactive power without energy storage elements. Then, a more complete paper including experiental verifications was published in the *IEEE Transactions on Industry Applications* in 1984 [17].

The *p-q* Theory defines a set of instantaneous powers in the time domain. Since no restrictions are imposed on voltage or current behaviors, it is applicable to three-phase systems with or without neutral conductors, as well as to generic voltage and current waveforms. Thus, it is valid not only in steady states, but also during transient states. Contrary to other traditional power theories treating a three-phase system as three single-phase circuits, the *p-q* Theory deals with all the three phases at the same time, as a unity system. Therefore, this theory always considers three-phase systems together, not as a superposition or sum of three single-phase circuits. It was defined by using the $\alpha\beta0$-transformation, also known as the Clarke Transformation [18], which consists of a real matrix that transforms three-phase voltages and currents into the $\alpha\beta0$-stationary reference frames. As will be seen in this book, the *p-q* Theory provides a very efficient and flexible basis for designing control strategies and implementing them in the form of controllers for power conditioners based on power electronics devices.

There are other approaches to power definitions in the time domain. Chapter 3 is dedicated to the time-domain analysis of power in three-phase circuits, and it is especially dedicated to the *p-q* Theory.

1.2. APPLICATIONS OF THE *p-q* THEORY TO POWER ELECTRONICS EQUIPMENT

The proliferation of nonlinear loads has spurred interest in research on new power theories, thus leading to the *p-q* Theory. This theory can be used for the design of power electronics devices, especially those intended for reactive-power compensation. Various problems related to harmonic pollution in power systems have been investigated and discussed for a long time. These are listed as follows:

- *Overheating of transformers and electrical motors.* Harmonic components in voltage and/or current induce high-frequency magnetic flux in their magnetic core, thus resulting in high losses and overheating of these electrical machines. It is common to oversize transformers and motors by 5 to 10% to overcome this problem [19].

- *Overheating of capacitors for power-factor correction.* If the combination of line reactance and the capacitor for power-factor correction has a resonance at the same frequency as a harmonic current generated by a nonlinear load, an overcurrent may flow through the capacitor. This may overheat it, possibly causing damage.
- *Voltage waveform distortion.* Harmonic current may cause voltage waveform distortion that can interfere with the operation of other electronic devices. This is a common fact when rectifiers are used. Rectifier currents distort the voltage waveforms with notches that may change the zero-crossing points of the voltages. This can confuse the rectifier control circuit itself, or the control circuit of other equipment.
- *Voltage flicker.* In some cases, the harmonic spectra generated by nonlinear loads have frequency components below the line frequency. These undesirable frequency components, especially in a frequency range of 8 to 30 Hz, may cause a flicker effect in incandescent lamps, which is a very uncomfortable effect for people's eyes. The arc furnace is one of the main contributors to this kind of problem.
- *Interference with communication systems.* Harmonic currents generated by nonlinear loads may interfere with communication systems like telephones, radios, television sets, and so on.

In the past, these problems were isolated and few. Nowadays, with the increased number of nonlinear loads present in the electric grid, they are much more common. On the other hand, the need for highly efficient and reliable systems has forced researchers to find solutions to these problems. In many cases, harmonic pollution cannot be tolerated.

As will be shown in Chapter 3, the *p-q* Theory that defines the instantaneous real and imaginary powers is a flexible tool, not only for harmonic compensation, but also for reactive-power compensation. For instance, FACTS (Flexible AC Transmission System) equipment that was introduced in [20] can be better understood if one has a good knowledge of the *p-q* Theory.

All these problems have encouraged the authors to write this book, which contributes to harmonic elimination from power systems. Moreover, an interesting application example of the *p-q* Theory will be presented, dealing with power flow control in a transmission line. This theory can also be used to control grid-connected converters like those used in solar energy systems [21], and can also be extended to other distributed power generation systems such as wind energy systems and fuel cells.

1.3. HARMONIC VOLTAGES IN POWER SYSTEMS

Tables 1-1 and 1-2 show the maximum and minimum values of total harmonic distortion (THD) in voltage and dominant voltage harmonics in a typical power system

Table 1-1. THD in voltage and 5th harmonic voltage in a high-voltage power transmission system

	Over 187 kV		154–22 kV	
	THD	5th harmonic	THD	5th harmonic
Maximum	2.8 %	2.8 %	3.3 %	3.2 %
Minimum	1.1 %	1.0 %	1.4 %	1.3 %

in Japan, measured in October 2001 [22]. The individual harmonic voltages and the resulting THD in high-voltage power transmission systems tend to be less than those in the 6.6-kV power distribution system. The primary reason is that the expansion and interconnection of high-voltage power transmission systems has made the systems stiffer with an increase of short-circuit capacity. For the distribution system, the maximum value of 5th harmonic voltage in a commercial area that was investigated exceeded its allowable level of 3%, considering Japanese guidelines, while the maximum THD value in voltage was marginally lower than its allowable level of 5%.

According to [23], the maximum value of 5th harmonic voltage in the downtown area of a 6.6-kV power distribution system in Japan exceeds 7% under light-load conditions at night. They also have pointed out another significant phenomenon. The 5th harmonic voltage increases on the 6.6-kV bus at the secondary of the power transformer installed in a substation, whereas it decreases on the 77-kV bus at the primary, under light-load conditions at night. These observations based on the actual measurement suggest that the increase of 5th harmonic voltage on the 6.6-kV bus at night is due to "harmonic propagation" as a result of series and/or parallel harmonic resonance between line inductors and shunt capacitors for power-factor correction installed on the distribution system. This implies that not only harmonic compensation, but also harmonic damping, is a viable and effective way to solve harmonic pollution in power distribution systems. Hence, electric power utilities should have responsibility for actively damping harmonic propagation throughout power distribution systems. Individual consumers and end-users are responsible for keeping the current harmonics produced by their own equipment within specified limits. Both problems of harmonic elimination and harmonic damping are exhaustively discussed and analyzed in this book.

Table 1-2. THD in voltage and 5th harmonic voltage in a medium-voltage power distribution system

	6 kV			
	Residential area		Commercial area	
	THD	5th harmonic	THD	5th harmonic
Maximum	3.5 %	3.4 %	4.6 %	4.3 %
Minimum	3.0 %	2.9 %	2.1 %	1.2 %

1.4. IDENTIFIED AND UNIDENTIFIED HARMONIC-PRODUCING LOADS

Nonlinear loads drawing nonsinusoidal currents from three-phase, sinusoidal, balanced voltages are classified as identified and unidentified loads. High-power diode or thyristor rectifiers, cycloconverters, and arc furnaces are typically characterized as identified harmonic-producing loads, because electric power utilities identify the individual nonlinear loads installed by high-power consumers on power distribution systems in many cases. Each of these loads generates a large amount of harmonic current. The utilities can determine the point of common coupling (PCC) of high-power consumers who install their own harmonic-producing loads on power distribution systems. Moreover, they can determine the amount of harmonic current injected from an individual consumer.

A "single" low-power diode rectifier produces a negligible amount of harmonic current compared with the system total current. However, multiple low-power diode rectifiers can inject a significant amount of harmonics into the power distribution system. A low-power diode rectifier used as a utility interface in an electric appliance is typically considered as an unidentified harmonic-producing load.

So far, less attention has been paid to unidentified loads than identified loads. Harmonic regulations or guidelines such as IEEE 519-1992 are currently applied, with penalties on a voluntary basis, to keep current and voltage harmonic levels in check. The final goal of the regulations or guidelines is to promote better practices in both power systems and equipment design at minimum social cost.

Table 1-3 shows an analogy in unidentified and identified sources between harmonic pollution and air pollution. The efforts by researchers and engineers in the automobile industry to comply with the Clean Air Act Amendments of 1970 led to success in suppressing CO, HC, and NOx contained in automobile exhaust. As a result, the reduction achieved was 90% when gasoline-fueled passenger cars in the 1990s are compared with the same class of cars at the beginning of the 1970s. Moreover, state-of-the-art technology has brought much more reductions in the exhaust of modern gasoline-fueled passenger cars. It is interesting that the development of the automobile industry, along with the proliferation of cars, has made it

Table 1-3. Analogy between harmonic pollution and air pollution

Sources	Harmonic pollution	Air pollution
Unidentified	• TV sets and personal computers • Inverter-based home appliances such as adjustable-speed heat pumps for air conditioning • Adjustable-speed motor drives	• Gasoline-fueled vehicles • Diesel-fueled vehicles
Identified	• Bulk diode/thyristor rectifiers • Cycloconverters • Arc furnaces	• Chemical plants • Coal and oil steam power plants

possible to absorb the increased cost related to the reduction of harmful components in exhaust emitted by gasoline-fueled vehicles [24].

Harmonic regulations or guidelines are effective in overcoming "harmonic pollution." Customers pay for the cost of high performance, high efficiency, energy savings, reliability, and compactness brought by power electronics technology. However, they are unwilling to pay for the cost of solving the harmonic pollution generated by power electronics equipment unless regulations or guidelines are enacted. It is expected that the continuous efforts by power electronics researchers and engineers will make it possible to absorb the increased cost for solving the harmonic pollution.

1.5. HARMONIC CURRENT AND VOLTAGE SOURCES

In most cases, either a harmonic current source or a harmonic voltage source can represent a harmonic-producing load, from a practical point of view.

Figure 1-1(a) shows a three-phase diode rectifier with an inductive load. Note that a dc inductor L_{dc} is directly connected in series to the dc side of the diode rectifier. Here, L_{ac} is the ac inductor existing downstream of the point of common coupling (PCC). This inductor may be connected to the ac side of the diode rectifier, and/or the leakage inductance of a transformer may be installed at the ac side of the rectifier for voltage matching and/or electrical isolation. Note that this transformer is disregarded in Fig. 1-1(a). On the other hand, L_S represents a simplified equivalent inductance of the power system existing upstream of the PCC. In most cases, the inductance value of L_S is much smaller than that of L_{ac}. In other words, the short-circuit capacity upstream of the PCC is much larger than that downstream of the PCC, so that L_S is very small and might be neglected.

Since the inductance value of L_{dc} is larger than that of the ac inductor L_{ac} in many cases, the waveform of i_{Sh} is independent of L_{ac}. This allows us to treat the rectifier as a harmonic *current* source when attention is paid to harmonic voltage and current, as shown in Fig. 1-1(b). Strictly speaking, the current waveform is slightly influenced by the ac inductance because of the so-called "current overlap effect" as long as L_{ac} is not equal to zero. Note that the system inductor L_S plays an essential role in producing the harmonic voltage at the PCC, whereas the ac inductor L_{ac} makes no contribution.

Figure 1-1(c) shows a simplified circuit derived from Fig. 1-1(b), where i_{Shn} is not an instantaneous current but an rms current at the *n*th-order harmonic frequency and, moreover, L_{ac} is hidden in the harmonic current source. The product of the system reactance X_{Sn} (at *n*th order of harmonic frequency) and the harmonic current i_{Shn} gives the rms value of the *n*th-order harmonic voltage $\dot{V}_{Shn}$ appearing at the PCC.

Figure 1-2(a) shows a three-phase diode rectifier with a capacitive load. Unlike Fig. 1-1(a), a dc capacitor C_{dc} is directly connected in parallel to the dc side of the diode rectifier. As a result, the diode rectifier, seen from the ac side, can be characterized as a harmonic voltage source v_h, as shown in Fig. 1-2(b). Note that it would

1.5. HARMONIC CURRENT AND VOLTAGE SOURCES

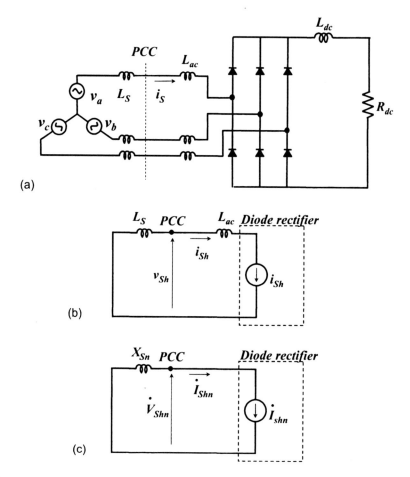

Figure 1-1. A three-phase diode rectifier with an inductive load. (a) Power circuit, (b) equivalent circuit for harmonic voltage and current on a per-phase basis, (c) simplified circuit.

be difficult to describe the waveform of v_h because i_S is a discontinuous waveform and, moreover, the conducting interval of each diode depends on L_{ac} among other circuit parameters. The reason why the rectifier can be considered as harmonic voltage source is that the harmonic impedance downstream of the ac terminals of the rectifier is much smaller than that upstream of the ac terminals. This means that the supply harmonic current i_{Sh} is strongly influenced by the ac inductance L_{ac}. Unless the ac inductance exists, the rectifier would draw a distorted pulse current with an extremely high peak value from the ac mains because only the system inductor L_S, which is too small in many cases, would limit the harmonic current. Invoking the so-called "equivalent transformation between a voltage source and a current source" allows us to obtain Fig. 1-2(c) from Fig. 1-2(b). Here, I_{Shn} and V_{hn} are an

10 INTRODUCTION

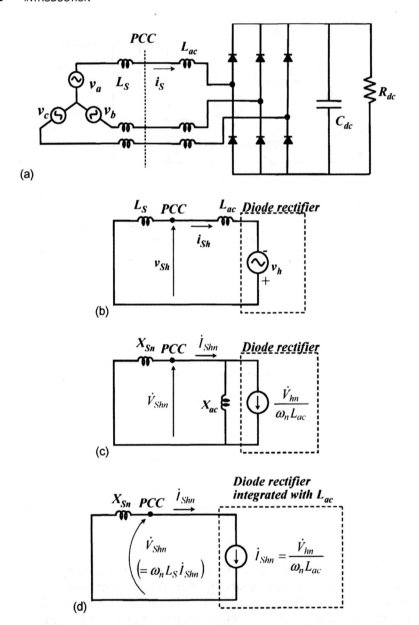

Figure 1-2. A three-phase diode rectifier with a capacitive load. (a) Power circuit, (b) equivalent circuit for harmonic voltage and current on a per-phase basis, (c) equivalent transformation between a voltage source and current source, (d) simplified circuit with the assumption that $L_S \ll L_{ac}$.

rms current and an rms voltage at the n-th order harmonic frequency, respectively. Assuming that the inductance value of L_{ac} is much larger than that of L_S allows us to eliminate L_{ac} from Fig. 1-2(c). This results in a simplified circuit, as shown in Fig. 1-2(d). It is quite interesting that the diode rectifier including the ac inductance L_{ac} is not a harmonic voltage source but a harmonic current source as long as it is seen downstream of the PCC, although the diode rectifier itself is characterized as a harmonic voltage source.

1.6. BASIC PRINCIPLES OF HARMONIC COMPENSATION

Figure 1-3 shows a basic circuit configuration of a shunt active filter in a three-phase, three-wire system. This is one of the most fundamental active filters intended for harmonic-current compensation of a nonlinear load. For the sake of simplicity, no system inductor exists upstream of the point of common coupling. This shunt active filter equipped with a current minor loop is controlled to draw the compensating current i_C from the ac power source, so that it cancels the harmonic current contained in the load current i_L.

Figure 1-4 depicts voltage and current waveforms of the ac power source v_a, the source current i_{Sa}, the load current i_{La}, and the compensating current i_{Ca} in the a-phase, under the following assumptions. The smoothing reactor L_{dc} in the dc side of the rectifier is large enough to keep constant the dc current, the active filter operates as an ideal controllable current source, and the ac inductor L_{ac} is equal to zero. The

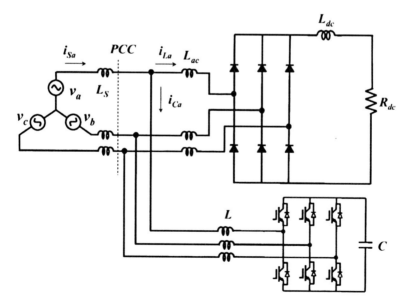

Figure 1-3. System configuration of a stand-alone shunt active filter.

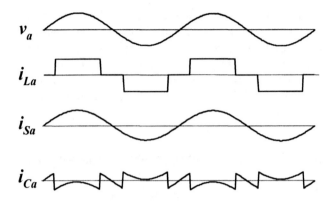

Figure 1-4. Waveforms of voltage and current in Fig. 1-3.

shunt active filter should be applied to a nonlinear load that can be considered as a harmonic current source, such as a diode/thyristor rectifier with an inductive load, an arc furnace, and so on. At present, a voltage-source PWM converter is generally preferred as the power circuit of the active filter, instead of a current-source PWM converter. A main reason is that the insulated-gate bipolar transistor (IGBT), which is one of the most popular power switching devices, is integrated with a free-wheeling diode, so that such an IGBT is much more cost-effective in constructing the voltage-source PWM converter than the current-source PWM converter. Another reason is that the dc capacitor indispensable for the voltage-source PWM converter is more compact and less heavy than the dc inductor for the current-source PWM inverter.

In addition to harmonic-current compensation, the shunt active filter has the capability of reactive-power compensation. However, adding this function to the shunt active filter brings an increase in current rating to the voltage-source PWM converter. Chapter 4 describes the shunt active filter in detail.

Figure 1-5 shows a basic circuit configuration of a series active filter in three-phase, three-wire systems. The series active filter consists of either a three-phase voltage-source PWM converter or three single-phase voltage-source PWM converters, and it is connected in series with the power lines through either a three-phase transformer or three single-phase transformers. Unlike the shunt active filter, the series active filter acts as a controllable *voltage* source and, therefore, the voltage-source PWM converter has no current minor loop. This makes the series active filter suitable for compensation of a harmonic voltage source such as a three-phase diode rectifier with a capacitive load.

Figure 1-6 depicts voltage and current waveforms of the mains voltage v_a, the supply current i_{Sa}, the load voltage v_{La} (the voltage at the ac side of the rectifier), and the compensating voltage v_{Ca} in the a-phase. Here, the following assumptions are made: the capacitor C_{dc} in the dc side of the rectifier is larger enough to keep constant dc voltage, the active filter operates as an ideal controllable voltage source, and the ac inductor L_{ac} is equal to zero. In a real system, the series active filter would not work properly if no ac inductor existed. In other words, an ac inductor is

1.6. BASIC PRINCIPLES OF HARMONIC COMPENSATION

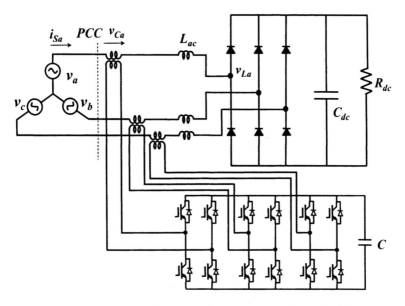

Figure 1-5. System configuration of a stand-alone series active filter.

essential to achieve proper operation, even if the series active filter is assumed to be ideal. The reason is that the ac inductor plays an important role in supporting voltage difference between source voltage v_S and the sum of the compensating voltage v_C and the load voltage v_L. Since the supply current is identical to the input current of the rectifier, it is changed from a distorted discontinuous waveform to a sinusoidal, continuous waveform as a result of achieving harmonic-voltage compensation. This means that the conducting interval of each diode becomes 180°.

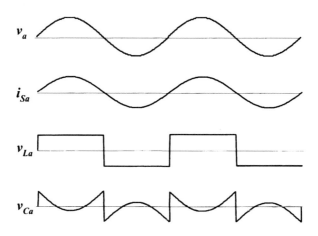

Figure 1-6. Waveforms of currents and voltages in Fig. 1-5.

When the series active filter is intended only for harmonic compensation, the compensating voltage v_C has no fundamental-frequency component. The dc voltage of the diode rectifier can be regulated by intentionally controlling the fundamental-frequency component of the compensating voltage.

Table 1.4 summarizes comparisons between the shunt and series active filters. Note that the series active filter has a "dual" relationship in each item with respect to the shunt active filter. Chapter 5 presents discussions of the series active filter.

Considering that the distribution system is based on the concept that a voltage source is delivered to the final user, it is common to consider it as an almost ideal voltage source. Therefore, in most cases where the source impedance is relatively small, the voltage waveform is considered purely sinusoidal even when the load is nonlinear and the current is distorted. In this case, the basic compensation principle is based on a shunt active filter, as shown in Fig. 1-3.

1.7. BASIC PRINCIPLE OF POWER FLOW CONTROL

This book is mainly dedicated to the problems related to power quality, for which harmonic elimination and damping are two of the major concerns. In fact, these points are more closely related to the Custom Power concept introduced in [25], which is normally applied to low-voltage systems or distribution systems. Chapters 4, 5, and 6 will present shunt, series, and combined shunt–series active filters. Moreover, it will be seen that the use of the *p-q* Theory makes it easy to perform reactive-power compensation. On the other hand, the concept of FACTS (Flexible AC Transmission Systems), as introduced in [20], has the main objective of improving controllability of power transmission systems that are operated at higher voltage and power level than power distribution systems can be. This improvement of controllability actually means a fast, continuous, and dynamic reactive-power control, as well as power-flow control or active-power control. Reactive-power control is used for power-factor correction, voltage regulation, and stability im-

Table 1.4. Comparisons between shunt and series active filters

	Shunt Active Filter	Series Active Filter
Circuit configuration	Fig. 1-3	Fig. 1-5
Power circuit	Voltage-source PWM converter with current minor loop	Voltage-source PWM converter without current minor loop
Operation	Current source: i_C	Voltage source: v_C
Suitable nonlinear load	Diode / thyristor rectifier with inductive load	Diode rectifier with capacitive load
Additional function	Reactive-power compensation	ac voltage regulation

provement. Power-flow control is important only in enhancing the power transfer capability of a given transmission line. The enhanced transmission capability leads to avoiding construction of new parallel lines. Naturally, this is an important solution to mitigating an environmental impact issue that is a world-wide concern nowadays.

Many concepts dealt with in this book can be extended to FACTS applications. Some of the FACTS applications can be imported to harmonic-eliminating devices to improve their performance. If a lossless transmission line with the same voltage magnitude at both terminals is considered, the power flow through it is given by

$$P = \frac{V^2}{X_L} \sin \delta \qquad (1.1)$$

where V is the terminal line voltage, X_L is the line series reactance, and δ is the phase-angle displacement between the sending-end and receiving-end voltages. It is clear under the constant-voltage condition that power flow can be controlled by adjusting the line reactance X_L or the phase angle δ.

Figure 1-7 shows a simplified transmission system. Let the sending (source) and receiving (load) terminal voltages be v_S and v_L, respectively. The transmission-line series inductance is represented by L_L, and v_C is the compensating voltage of the series compensator based on power electronics. If v_C is a voltage in quadrature with respect to the line current I, and has a magnitude proportional to this current, this compensator works like a capacitor or an inductor, which can be used to increase or to decrease the transmitted power. In this case, the voltage source v_C does not generate or absorb any active power. However, in a more general case, the magnitude of v_C is limited only by the voltage rating of the series compensator, although the voltage source has no limitation in phase angle.

Figure 1-8 shows the phasor diagram for the system in Fig. 1-7, with focus on a generic compensating voltage v_C. It also shows the locus of the phasor associated with v_C. Since the line current and the compensating voltage are not in quadrature, the series compensator has to absorb or generate active power. When the controllable series voltage source is in fact implemented by a voltage-fed PWM converter,

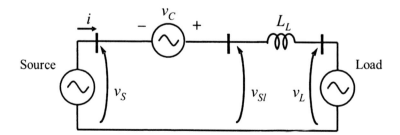

Figure 1-7. Simplified transmission system controlled by a series compensator based on power electronics.

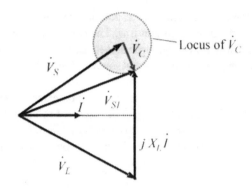

Figure 1-8. Phasor diagram for the voltages in Fig. 1-7.

an energy source or sink is necessary at the dc side of the series converter. In this sense, the concept of the Unified Power Flow Controller (UPFC) was proposed in [20], where a shunt converter is coupled to the series one through the common dc link, keeping energy balanced between the two converters. Figure 1-9 shows the basic conceptual diagram of the UPFC.

The UPFC consists of series and shunt compensators based on voltage-fed converters. As the series compensator needs an energy source or sink in its dc side, both converters are connected back to back, so that power can be exchanged between the two converters. The shunt converter is normally operated as a STATCOM (Static Synchronous Compensator) [20] for the purpose of controlling reactive power at the ac side, or regulating the ac bus voltage. It also can adjust the active power flowing into, or out of, the dc link. The series converter is operated to control the active power flowing through the transmission line. These concepts are explained in Chapter 6, and a controller based on the *p-q* Theory is also presented. By joining the concepts behind the UPFC and the concepts of shunt and series ac-

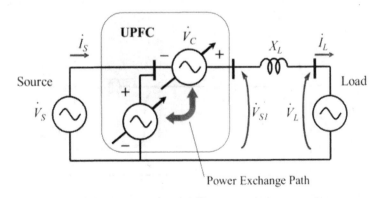

Figure 1-9. Basic conceptual block diagram of a Unified Power Flow Controller (UPFC).

tive filtering, a new device, referred to as Universal Active Power Line Conditioner (UPLC), is also introduced in Chapter 6.

REFERENCES

[1] W. V. Lyon, "Reactive Power and Unbalanced Circuits," *Electrical World*, vol. 75, no. 25, 1920, pp. 1417–1420.

[2] F. Buchholz, "Die Drehstrom-Scheinleistung bei ungleichmäßiger Belastung der drei Zweige," *Licht und Kraft, Zeitschrift für elekt. Energie-Nutzung*, no. 2, Jan. 1922, pp. 9–11.

[3] C. I. Budeanu, "Puissances reactives et fictives," *Instytut Romain de l'Energie*, pub. no. 2, Bucharest, 1927.

[4] C. I. Budeanu, "The Different Options and Conceptions Regarding Active Power in Non-sinusoidal Systems," *Instytut Romain de l'Energie*, pub. no. 4, Bucharest, 1927.

[5] S. Fryze, "Wirk-, Blind- und Scheinleistung in elektrischen Stromkreisen mit nicht-sinusförmigem Verlauf von Strom und Spannung," *ETZ-Arch. Elektrotech.*, vol. 53, 1932, pp. 596–599, 625–627, 700–702.

[6] M. S. Erlicki and A. Emanuel-Eigeles, "New Aspects of Power Factor Improvements Part I—Theoretical Basis," *IEEE Transactions on Industry and General Applications*, vol. IGA-4, 1968, July/August, pp. 441–446.

[7] H. Sasaki and T. Machida, "A New Method to Eliminate AC Harmonic by Magnetic Compensation—Consideration on Basic Design," *IEEE Transactions on Power Apparatus and Systems*, vol. 90, no. 5, 1970, pp. 2009–2019.

[8] T. Fukao, H. Iida, and S. Miyairi, "Improvements of the Power Factor of Distorted Waveforms by Thyristor Based Switching Filter," *Transactions of the IEE-Japan*, Part B, vol. 92, no. 6, 1972, pp. 342–349 (in Japanese).

[9] L. Gyugyi and B. R. Pelly, *Static Power Frequency Changers: Theory, Performance and Application*, Wiley, 1976.

[10] F. Harashima, H. Inaba, and K. Tsuboi, "A Closed-loop Control System for the Reduction of Reactive Power Required by Electronic Converters," *IEEE Transactions on IECI*, vol. 23, no. 2, May 1976, pp. 162–166.

[11] L. Gyugyi and E. C. Strycula, "Active ac Power Filters," in *Proceedings of IEEE Industry Application Annual Meeting*, vol. 19-C, 1976, pp. 529–535.

[12] I. Takahashi and A. Nabae, "Universal Reactive Power Compensator," in *Proceedings of IEEE—Industry Application Society Annual Meeting Conference Record*, 1980, pp. 858–863.

[13] I. Takahashi, K. Fujiwara, and A. Nabae, "Distorted Current Compensation System Using Thyristor Based Line Commutated Converters," *Transactions of the IEE-Japan*, Part B, vol. 101, no.3, 1981, pp. 121–128 (in Japanese).

[14] H. Akagi, Y. Kanazawa, and A. Nabae, "Principles and Compensation Effectiveness of Instantaneous Reactive Power Compensator Devices," in *Meeting of the Power Semiconductor Converters Researchers—IEE-Japan*, SPC-82-16, 1982 (in Japanese).

[15] H. Akagi, Y. Kanazawa, and A. Nabae, "Generalized Theory of Instantaneous Reactive Power and Its Applications," *Transactions of the IEE-Japan*, Part B, vol. 103, no.7, 1983, pp. 483–490 (in Japanese).

[16] H. Akagi, Y. Kanazawa, and A. Nabae, "Generalized Theory of the Instantaneous Reactive Power in Three-Phase Circuits," in *IPEC'83—International Power Electronics Conference,* Tokyo, Japan, 1983, pp. 1375–1386.

[17] H. Akagi, Y. Kanazawa, and A. Nabae, "Instantaneous Reactive Power Compensator Comprising Switching Devices Without Energy Storage Components," *IEEE Transactions on Industry Applications,* vol. IA-20, no. 3, 1984, pp. 625–630.

[18] E. Clarke, *Circuit Analysis of A-C Power Systems, Vol. I—Symmetrical and Related Components,* Wiley, 1943.

[19] N. Mohan, T. Undeland, and W. P. Robbins, *Power Electronics—Converters, Applications, and Design,* Wiley, 2003.

[20] N. Hingorani and L. Giugyi, *Understanding Facts: Concepts and Technology of Flexible AC Transmission Systems,* IEEE Press, 1999.

[21] P. G. Barbosa, E. H. Watanabe, G. B. Rolim, and R. Hanitsch, "Novel control strategy for grid-connected dc-ac converters with load power factor correction," *IEE Proceedings—Generation, Transmission and Distribution,* vol. 145, issue 5, September, 1998, pp. 487–493.

[22] "Investigation into Execution of Harmonic Guidelines for Household and Office Electric Appliances," *IEEJ SC77A Domestic Committee Report,* May 2002, pp. 7–9 (in Japanese).

[23] K. Oku, O. Nakamura, and K. Uemura, "Measurement and analysis of harmonics in power distribution systems," *IEEJ Transactions,* vol. 114-B, no. 3, 1994, pp. 234–241 (in Japanese).

[24] H. Akagi, "New trends in active filters for power conditioning," *IEEE Transactions on Industry Applications,* vol. 32, no. 6, 1996, pp. 1312–1322.

[25] N. G. Hingorani, "Introducing Custom Power," *IEEE Spectrum,* June, 1995, pp. 41–48.

CHAPTER 2

ELECTRIC POWER DEFINITIONS: BACKGROUND

There are several basic concepts that must be established before the analysis of electric power systems involving power electronic devices can begin. The calculation of electric system variables, for instance, voltage, current, power factor, and active and reactive power, under nonsinusoidal conditions is perhaps the cornerstone of this analysis. The concepts and definitions of electric power for sinusoidal ac systems are well established and accepted worldwide. However, under nonsinusoidal conditions, several and different power definitions are still in use. For instance, the conventional concepts of reactive and apparent power lose their usefulness in nonsinusoidal cases [1,2]. This problem has existed for many years and is still with us. Unfortunately, no agreement on a universally applicable power theory has been achieved yet.

At the beginning, two important approaches to power definitions under nonsinusoidal conditions were introduced by Budeanu in 1927 [4,5], and by Fryze in 1932 [3]. Budeanu worked in the frequency domain, whereas Fryze defined the power in the time domain. Unfortunately, those power definitions are dubious, and may lead to misinterpretation in some cases. No other relevant contributions were made until the 1970s, because power systems were satisfactorily well represented as balanced and sinusoidal ac sources and loads. However, the problems related to nonlinear loads became increasingly significant at the beginning of advances in power electronics devices. These modern devices behave as nonlinear loads and most of them draw a significant amount of harmonic current from the power system. An increasing number of nonlinear loads are being connected to the network. Hence, this has spurred analysis of power systems under nonsinusoidal conditions. It is imperative to establish a consistent set of power definitions that are valid also under transient and nonsinusoidal conditions.

Many different approaches to power definitions can be found in the literature. They are based on the frequency or time domains. Although system engineers used to deal with power systems in the frequency domain, the authors believe that power definitions in the time domain are more appropriate for the analysis of power systems if they are operating under nonsinusoidal conditions.

Power definitions in the frequency domain will be summarized in this chapter, just to give a background for the time-domain approach to be presented in the next chapter. The electric power definitions in the frequency domain will be presented just to make it evident why they should be complemented to become more coherent and useful in the analysis and design of modern power systems involving nonlinear loads.

2.1. POWER DEFINITIONS UNDER SINUSOIDAL CONDITIONS

The definitions of electric power for single-phase, sinusoidal systems have been well established. Nowadays, there are no divergences between the results obtained in the time and frequency domains.

An ideal single-phase system with a sinusoidal voltage source and a linear (resistive–inductive) load has voltage and current that are analytically represented by

$$v(t) = \sqrt{2}V \sin(\omega t) \qquad i(t) = \sqrt{2}I \sin(\omega t - \phi) \qquad (2.1)$$

where V and I represent the root-mean-square (rms) values of the voltage and current, respectively, and ω is the angular line frequency. The instantaneous (active) power is given by the product of the instantaneous voltage and current, that is,

$$p(t) = v(t)i(t) = 2VI \sin(\omega t) \sin(\omega t - \phi)$$
$$= VI \cos \phi - VI \cos(2\omega t - \phi) \qquad (2.2)$$

Equation (2.2) shows that the instantaneous power of the single-phase system is not constant. It has an oscillating component at twice the line frequency added to a dc level (average value) given by $VI \cos \phi$. Decomposing the oscillating component and rearranging (2.2) yields the following equation with two terms, which derives the traditional concept of active and reactive power:

$$p(t) = \underbrace{VI \cos \phi [1 - \cos(2\omega t)]}_{(I)} - \underbrace{VI \sin \phi \sin(2\omega t)}_{(II)} \qquad (2.3)$$

The decomposition given by (2.3) shows two parts of the instantaneous power that can be interpreted as:

Part I has an average value equal to $VI \cos \phi$ and has an oscillating component on it, pulsing at twice the line frequency. This part never becomes negative (provided that $-90° \leq \phi \leq 90°$) and, therefore, represents an unidirectional power flow from the source to the load.

Part II has a pure oscillating component at the double frequency (2ω), and has a peak value equal to $VI \sin \phi$. Clearly, it has a zero average value.

Conventionally the instantaneous power given in (2.3) is represented by three "constant" powers: the *active power*, the *reactive power* and the *apparent power*. These powers are discussed below.

Active power P. The average value of part I is defined as the *active (average) power* P:

$$P = VI \cos \phi \qquad (2.4)$$

The unit for the measurement of the active power in the International System is the Watt (W).

Reactive power Q. The conventional *reactive power* Q is just defined as the peak value of part II.

$$Q = VI \sin \phi \qquad (2.5)$$

The unit for the measurement of the reactive power in the International System is the var (volt-ampere reactive).

A signal for the displacement angle ϕ should be adopted to characterize the inductive or capacitive nature of the load. In this book, the most common convention will be adopted, assigning *positive* values for the reactive power of *inductive* loads. Capacitive loads have reactive power with the negative sign.

Several authors refer to the reactive power as "*the portion of power that does not realize work,*" or "*oscillating power*" [6]. The reactive power as conventionally defined represents a power component with zero average value. However, this assertion is not complete, as will be shown later. This physical meaning for the reactive power was established when the basic reactive devices were only capacitors and inductors. At that time, there were no power electronics devices. These devices are capable of creating reactive power without energy storage elements. This subject will be discussed in the next chapter.

Now, the instantaneous power $p(t)$ can be rewritten as

$$p(t) = \underbrace{P[1 - \cos(2\omega t)]}_{(I)} - \underbrace{Q \sin(2\omega t)}_{(II)} \qquad (2.6)$$

Figure 2-1 illustrates the above power components for a given voltage and current. In this figure, the ac current lags the ac voltage by a displacement angle ϕ, which is equal to $\pi/3$. From (2.6) and Fig. 2-1, it is easy to understand that the energy flow (instantaneous power) in the ac single-phase system is not unidirectional and not constant. During the time interval corresponding to area "A," the

22 ELECTRIC POWER DEFINITIONS: BACKGROUND

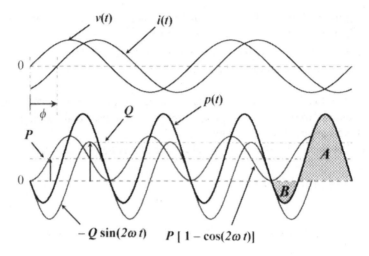

Figure 2-1. Conventional concepts of active and reactive power.

source is delivering energy to the load, whereas during the time interval corresponding to area "*B*," a percentage of this amount of energy is being returned back to the source.

Another power quantity that is commonly used to define the power rating of electrical equipment is the *apparent power S*.

Apparent power *S* is defined as

$$S = V \cdot I \tag{2.7}$$

The unit for the apparent power in the International System is VA (Volt-Ampere). This power is usually understood to represent the "*maximum reachable active power at unity power factor*" [7]. The definition of power factor is given in the next section.

2.2. VOLTAGE AND CURRENT PHASORS AND COMPLEX IMPEDANCE

Sometimes, the analysis of power systems can be significantly simplified by using the phasor notation instead of using sinusoidal voltages and currents as time functions given in (2.1). Therefore, a short review of phasor definitions will be presented here.

A sinusoidal time function $f(t)$, with a given angular frequency of ω, can be represented as the imaginary part of a complex number as follows:

$$f(t) = \sqrt{2}A \sin(\omega t + \phi) = \mathrm{Im}\{\dot{F} \cdot e^{j\omega t}\} \tag{2.8}$$

Here, $\dot{F}$ is a complex number with a magnitude of $\sqrt{2}A$ and phase ϕ, and is defined as the phasor related to the sinusoidal function $f(t)$. Thus,

$$\dot{F} = \sqrt{2}A \angle \phi \tag{2.9}$$

The representation of a sinusoidal waveform by a phasor is only possible if it is purely sinusoidal with an angular frequency of ω, and "frozen" for a complete cycle. Therefore, phasor representation, although practical, is valid only for steady-state conditions.

A voltage phasor $\dot{V}$ and a current phasor $\dot{I}$ can be represented by complex numbers in its polar or Cartesian notation:

$$\begin{vmatrix} \dot{V} = V \angle \theta_V = V_\mathcal{R} + jV_\mathcal{J} \\ V_\mathcal{R} = V \cos \theta_V \quad V_\mathcal{J} = V \sin \theta_V \end{vmatrix} \begin{vmatrix} \dot{I} = I \angle \theta_I = I_\mathcal{R} + jI_\mathcal{J} \\ I_\mathcal{R} = I \cos \theta_I \quad I_\mathcal{J} = I \sin \theta_I \end{vmatrix} \tag{2.10}$$

where V and I are the rms values of the sinusoidal voltage and current time functions, and θ_V and θ_I are the phase angles at a given time instant. Positive phase angles are conventionally measured in the counterclockwise direction. Figure 2-2 shows the sinusoidal voltage and current time functions and the corresponding phasor representation at an instant when $\omega t + \theta_V = 0$. The displacement angle ϕ between $\dot{V}$ and $\dot{I}$ is given by $\phi = \theta_V - \theta_I$.

If the voltages and currents of an ac power system are sinusoidal, it is possible to define the concept of impedance. For instance, on a series RLC impedance branch, the ratio between the terminal voltage phasor and the current phasor is equal to a complex number, defined as *complex impedance* $\mathbf{Z}$, given by:

$$\mathbf{Z} = \frac{\dot{V}}{\dot{I}} = \frac{V \angle \theta_V}{I \angle \theta_I} = \frac{V}{I} \angle (\theta_V - \theta_I) = V \angle \theta_Z = R + j\left(\omega L - \frac{1}{\omega C}\right) \tag{2.11}$$

It is known that the current through an inductive load lags its terminal voltage, and the current through a capacitive load leads its terminal voltage. Therefore, the above convention of complex impedance produces a positive impedance angle (positive reactance) for inductive loads.

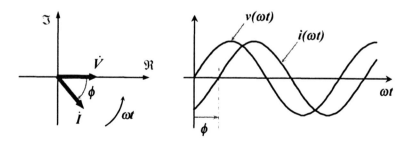

Figure 2-2. Relation between phasors and sinusoidal time functions.

2.3. COMPLEX POWER AND POWER FACTOR

A complex power, S, can be defined as the product of the voltage and current phasors. In order to be coherent with the sign convention of the reactive power given in (2.5) that is related to a "positive reactive power for inductive loads," the following definition of complex power should use the conjugate value ($\dot{I}^* = I \angle -\theta_I$) of the current phasor.

Complex power S is defined as

$$\mathbf{S} = \dot{V}\dot{I}^* = (V \angle \theta_V)(I \angle -\theta_I) = \underbrace{VI \cos(\theta_V - \theta_I)}_{\phi}{}^P + j\underbrace{VI \sin(\theta_V - \theta_I)}_{\phi}{}^Q \quad (2.12)$$

Note that the absolute value of the complex power is equal to the apparent power defined in (2.7), that is,

$$|\mathbf{S}| = \sqrt{[VI \cos(\theta_V - \theta_I)]^2 + [VI \sin(\theta_V - \theta_I)]^2} = S = VI \quad (2.13)$$

The term $\cos \phi$ is equal to the ratio between the active power P and the apparent power S, and is referred to as *power factor* λ. In the modern literature, it is common to find the symbol "PF" to denote power factor.

Power factor λ (PF) is defined as

$$\lambda = \mathrm{PF} = \cos \phi = \frac{P}{S} \quad (2.14)$$

The concept of complex power and power factor can be graphically represented in the well-known triangle of powers, as shown in Fig. 2-3. This figure, together with the above set of power and power factor definitions can be summarized as follows. If the load is not purely resistive, the reactive power Q is not zero, and the ac-

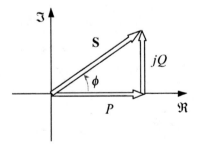

Figure 2-3. Graphical representation of the complex power—the triangle of powers.

tive power P is smaller than the apparent power S. Thus, the power factor λ is smaller than unity. These are the traditional meanings of the above electric powers defined under pure sinusoidal conditions [6]. They are widely used in industry to characterize electric equipment like transformers, machines, and so on. Unfortunately, these concepts of power are not valid, or lead to misinterpretations, under nonsinusoidal conditions [16]. This point will be deeply discussed in the next sections of this chapter.

2.4. CONCEPTS OF POWER UNDER NONSINUSOIDAL CONDITIONS—CONVENTIONAL APPROACHES

The concepts of power under nonsinusoidal conditions are not unique; they are divergent and lead to different results in some aspects. Two distinct sets of power definitions are commonly used; one is established in the frequency domain and the other in the time domain. They are presented below, to highlight their inconsistencies and to show why they are inadequate for use in controllers of power-line conditioners.

2.4.1. Power Definitions by Budeanu

A set of power definitions established by Budeanu [4,5] in 1927 is still very important for the analysis of power systems in the frequency domain. He introduced definitions that are valid for generic waveforms of voltage and current. However, since they are defined in the frequency domain, they can be applied only in steady-state analysis. In other words, they are limited to periodic waveforms of voltage and current.

If a single-phase ac circuit with a generic load and a source is in steady state, its voltage and current waveforms can be decomposed in Fourier series. Then, the corresponding phasor for each harmonic component can be determined, and the following definitions of powers can be derived.

Apparent power S is defined as

$$S = VI \qquad (2.15)$$

The apparent power in (2.15) is, in principle, identical to that given in (2.7). The difference is that V and I are the rms values of generic, periodic voltage and current waveforms, which are calculated as

$$V = \sqrt{\frac{1}{T}\int_0^T v^2(t)dt} = \sqrt{\sum_{n=1}^{\infty} V_n^2} \qquad I = \sqrt{\frac{1}{T}\int_0^T i^2(t)dt} = \sqrt{\sum_{n=1}^{\infty} I_n^2} \qquad (2.16)$$

Here, V_n and I_n correspond to the rms value of the nth order harmonic components of the Fourier series, and T is the period of the fundamental component. No direct current (dc) component is being considered in this analysis. The displacement angle of each pair of nth order harmonic voltage and current components is represented by ϕ_n. Budeanu defined the active and reactive powers as follows.

Active power P:

$$P = \sum_{n=1}^{\infty} P_n = \sum_{n=1}^{\infty} V_n I_n \cos \varphi_n \qquad (2.17)$$

Reactive power Q:

$$Q = \sum_{n=1}^{\infty} Q_n = \sum_{n=1}^{\infty} V_n I_n \sin \varphi_n \qquad (2.18)$$

The definitions of the apparent power and the reactive power seem to come from the necessity of quantifying in ac systems *"the portion of power that does not realize work."* In other words, it might be interesting to have indexes for quantifying the quality of the power being supplied by a power-generating system. However, under nonsinusoidal conditions, both reactive power and apparent power cannot characterize satisfactorily the issues of power quality or the efficiency of the transmission system. For instance, the above-defined reactive power does not include cross products between voltage and current harmonics at different frequencies. Note that neither the active power in (2.17), nor the reactive power in (2.18), includes the products of harmonic components at different frequencies. Furthermore, (2.18) comprises the algebraic sum of "harmonic reactive power" components that are positive or negative or even cancel each other, depending on the several harmonic displacement angles ϕ_n.

The loss of power quality under nonsinusoidal conditions is better characterized by another power definition, the *distortion power D,* that was introduced by Budeanu. This distortion power complements the above set of power definitions.

Distortion power D is defined as

$$D^2 = S^2 - P^2 - Q^2 \qquad (2.19)$$

The powers defined in (2.15) to (2.19) are well known and widely used in the analysis of circuit systems operating under nonsinusoidal conditions. However, only the active power P, as defined in (2.17), has a clear physical meaning, not only in the sinusoidal case, but also under nonsinusoidal conditions [8,9]. The active power represents the *average* value of the instantaneous active power. In other words, it represents the average ratio of energy transfer between two electric subsystems. In contrast, the reactive power and the apparent power as introduced by Budeanu [4,5] are just mathematical formulations, as an extension of the definitions for the sinusoidal case, without clear physical meanings. Another drawback is that a common instrument for power measurement based on the power definitions in the frequency domain cannot indicate easily a loss of power quality in practical cases. This point is clarified in the following example.

2.4. CONCEPTS OF POWER UNDER NONSINUSOIDAL CONDITIONS

Example #1
Consider an ideal sinusoidal voltage source $v(t) = \sin(2\pi 50 t)$ [pu] supplying:

a) $i(t) = \sin(2\pi 50 t - \pi/4)$ [pu] for a linear load
b) $i(t) = \sin(2\pi 50 t - \pi/4) + 0.1\sin(14\pi 50 t - \pi/4)$ [pu] for a nonlinear load

The fundamental components of voltage and current in both cases are identical. Therefore, the active and reactive powers calculated from (2.17) and (2.18) are the same in both cases:

$$P = \sum_{n=1}^{\infty} V_n I_n \cos \varphi_n = \frac{1}{\sqrt{2}} \frac{1}{\sqrt{2}} \cos\left(\frac{\pi}{4}\right) = 0.3536$$

$$Q = \sum_{n=1}^{\infty} V_n I_n \sin \varphi_n = \frac{1}{\sqrt{2}} \frac{1}{\sqrt{2}} \sin\left(\frac{\pi}{4}\right) = 0.3536$$

Case (a) is purely sinusoidal, whereas case (b) is not. The rms value of the voltage (V) and the current (I) in case (a) is equal to $1/\sqrt{2} = 0.7071$, whereas the rms of the current in case (b) is

$$I = \sqrt{\frac{1}{T}\int_0^T i^2(t)dt} = \sqrt{\sum_{n=1}^{\infty} I_n^2} = \sqrt{\left(\frac{1}{\sqrt{2}}\right)^2 + \left(\frac{0.1}{\sqrt{2}}\right)^2} = 0.7106$$

Although a 10% harmonic current with respect to the fundamental current was added in case (b), an increase of only 0.5% was verified in the rms current with respect to that in case (a). This causes an increase of only 0.5% in the apparent power S, which would be very difficult to be detected by conventional instruments of power measurement. This example is the evidence that instruments for power measurements based on rms values of voltage and current in the frequency domain may be inadequate to deal with power systems under nonsinusoidal conditions. Note that case (b) is undesirable in terms of harmonics and power quality. Moreover, the measurement of a distorted current is not so easy. Figure 2-4 shows

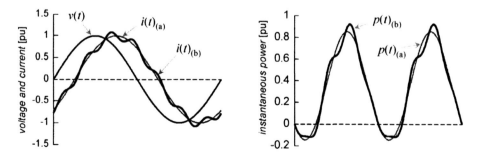

Figure 2-4. Instantaneous voltage, current, and power of Example #1.

waveforms of the voltage and currents considered above, and the produced instantaneous powers. The seventh harmonic current added in case (b) causes an increase of 8% in the peak value of the instantaneous power. The peak value is 0.8535 pu in case (a), whereas it is 0.9219 pu in case (b). In summary, this harmonic current induces an increase of 0.5% in the apparent power, 8% in the peak power (16 times), and produces no effect on the active or reactive powers defined in (2.17) and (2.18).

2.4.1.A. Power Tetrahedron and Distortion Factor

Due to the presence of the distortion power D under nonsinusoidal conditions, graphical power representation is given on a three-dimensional reference frame, instead of the power triangle described in Fig. 2-3. Figure 2-5 shows the new graphical power representation that is well known as a power tetrahedron.

Under nonsinusoidal conditions, the apparent power defined in (2.7) and (2.15) differs from the complex power defined in (2.12). As a result, the equivalence between them is no longer valid, as shown in (2.13). A new complex power S_{PQ} can be defined from (2.17) and (2.18) that corresponds to the new active and reactive power defined by Budeanu. This new complex power S_{PQ} is defined as

$$S_{PQ} = P + jQ = \sum_{n=1}^{\infty} P_n + j\sum_{n=1}^{\infty} Q_n = \sum_{n=1}^{\infty} V_n I_n \cos \varphi_n + j\sum_{n=1}^{\infty} V_n I_n \sin \varphi_n \quad (2.20)$$

The relation between the apparent power S and the complex power S_{PQ} is given by

$$S = VI = \sqrt{P^2 + Q^2 + D^2} = \sqrt{|S_{PQ}|^2 + D^2} \quad (2.21)$$

The power factor λ is defined in (2.14) as the ratio of the active power with respect to the apparent power. Now, λ is equal to $\cos \theta$ in the power tetrahedron of Fig. 2-5.

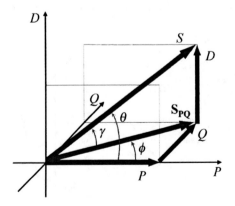

Figure 2-5. The power tetrahedron.

As can be seen in (2.21), the absolute value of the complex power ($|S_{PQ}| = [P^2 + Q^2]^{1/2}$) is different from the apparent power S under nonsinusoidal conditions. The ratio between P and $|S_{PQ}|$ is defined as the *displacement factor* (cos ϕ), although it is the power factor under sinusoidal conditions. Another term, the *distortion factor* (cos γ), is defined as the ratio between the length of S_{PQ} and the apparent power S. These factors are listed below.

Power factor λ:

$$\lambda = \cos \theta = \frac{P}{S} \tag{2.22}$$

Displacement factor cos ϕ:

$$\cos \phi = \frac{P}{|S_{PQ}|} \tag{2.23}$$

Distortion factor cos γ:

$$\cos \gamma = \frac{|S_{PQ}|}{S} \tag{2.24}$$

The following relation is valid:

$$\lambda = \cos \theta = \frac{P}{S} = \cos \phi \cdot \cos \gamma \tag{2.25}$$

Remarks

The reactive power Q in (2.18) and the harmonic power D in (2.19) are mathematical formulations that may lead to false interpretations, particularly when these concepts are extended to the analysis of three-phase circuits. The above equations treat electric circuits under nonsinusoidal conditions as a sum of several *independent* circuits excited at different frequencies. The calculated powers do not provide any consistent basis for designing passive filters or for controlling active power line conditioners.

The apparent power S has at least four different definitions as it appears in a widely known dictionary [14]. One of them considers it as a number that gives the basic rating of electrical equipment. However, it is not a unique definition and, therefore, the definition given in (2.15) is just one of them, named "the rms volt-ampere" [9].

It is more difficult to find reasons for applicability of the apparent power if a three-phase, four-wire, nonsinusoidal power source is connected to generic loads [16]. These problems are reported in several works, and many researchers have tried to solve them [17,18,19,20,21]. More recently, an IEEE Trial-Use Standard [15] was published for definitions used for measurement of electric power quantities under sinusoidal, nonsinusoidal, balanced, or unbalanced conditions. It lists the mathematical expressions that were used in the past, as well as new expressions, and explains the features of the new definitions.

2.4.2. Power Definitions by Fryze

In the early 1930s, Fryze proposed a set of power definitions based on rms values of voltage and current [3]. The basic equations according to the Fryze's approach are given below.

Active power P_w:

$$P_w = \frac{1}{T}\int_0^T p(t)\,dt = \frac{1}{T}\int_0^T v(t)i(t)\,dt = V_w I = VI_w \tag{2.26}$$

where V and I are the voltage and current rms values and V_w and I_w are the active voltage and active current defined below. The rms values of voltage and current are calculated as given in (2.16). Together with the active power P_w, these rms values form the basis of the Fryze's approach. From them, all other parameters can be defined and calculated as follows.

Apparent power P_S:

$$P_S = VI \tag{2.27}$$

Active power factor λ:

$$\lambda = \frac{P_w}{P_S} = \frac{P_w}{VI} \tag{2.28}$$

Reactive power P_q:

$$P_q = \sqrt{P_S^2 - P_w^2} = V_q I = VI_q \tag{2.29}$$

where V_q and I_q are the reactive voltage and current as defined below.

Reactive power factor λ_q:

$$\lambda_q = \sqrt{1 - \lambda^2} \tag{2.30}$$

Active voltage V_w and **active current** I_w:

$$V_w = \lambda \cdot V \qquad I_w = \lambda \cdot I \tag{2.31}$$

Reactive voltage V_q and **reactive current** I_q:

$$V_q = \lambda_q \cdot V \qquad I_q = \lambda_q \cdot I \tag{2.32}$$

Fryze defined reactive power as comprising all the portions of voltage and current, which does not contribute to the active power P_w. Note that the active power P_w is defined as the average value of the instantaneous active power. This concept of active and reactive power is well accepted nowadays. For instance, Czarnecki has improved this approach, going into detail by dividing reactive power P_q into four subparts according to their respective origins in electric circuits [11,12,13].

It is possible to demonstrate that there is no difference between the *active* power and the *apparent* power defined by Fryze in time domain and Budeanu in frequency domain. It is easy to confirm that the active power calculated from (2.17) is always the same as from (2.26). Both apparent powers from (2.15) and from (2.27) are also the same. However, the reactive power given in (2.18) by Budeanu is different from that in (2.29) by Fryze.

Fryze verified that the active power factor λ reaches its maximum ($\lambda = 1$) if and only if the instantaneous current is proportional to the instantaneous voltage, otherwise $\lambda < 1$ [3]. However, under nonsinusoidal conditions, the fact of having the current proportional to the voltage does not ensure an optimal power flow from the electromechanical energy conversion point of view, as will be shown later. If the concepts defined above are applied to the analysis of three-phase systems, they may lead to cases in which the three-phase instantaneous active power contains an oscillating component even if the three-phase voltage and current are proportional (unity power factor $\lambda = 1$). All these remarks will be clarified in the next chapter, which will present the instantaneous active and reactive power theory.

The above set of power definitions does not need any decomposition of generic voltage or current waveform in Fourier series, although it still requires the calculation of rms values of voltage and current. Hence, it is not valid during transient phenomena.

2.5. ELECTRIC POWER IN THREE-PHASE SYSTEMS

It is common to find three-phase circuits being analyzed as a sum of three *separate* single-phase circuits. This is a crude simplification, especially in cases involving power electronic devices or nonlinear loads. The total active, reactive, and apparent powers in three-phase circuits have been calculated just as three times the powers in a single-phase circuit, or the sum of the powers in the three single-phase, separated circuits. It is also common to find in the literature works that assign the same physical meaning or mathematical interpretation for the active, reactive, and apparent power in both single-phase and three-phase systems. The authors do not agree with these ideas because three-phase systems present some properties that are not observed in single-phase systems. These properties will be presented in this section.

2.5.1. Classifications of Three-Phase Systems

Three-phase systems can be grounded or not. If a three-phase system is grounded at more than one point under normal operation (no-fault or short-circuited operation), the ground can provide an additional path for current circulation. A three-phase system can also have a fourth conductor or the so-called "neutral wire or conductor." In both cases, the system is classified as a three-phase, four-wire system. If no ground is present or there is only one grounded node in the whole subnetwork, the system is classified as a three-phase, three-wire system or simply as a three-phase system.

So far, only single-phase systems have been considered and classified as systems "under sinusoidal conditions" or "under nonsinusoidal conditions." Hereafter, the words "distorted" system will be also used to refer to "a system under nonsinusoidal conditions." Beside these considerations, three-phase systems under sinusoidal conditions have a particular characteristic regarding the amplitude and phase angle of each phase voltage (or line current). If the amplitudes are equal and the displacement angles between the phases are equal to $2\pi/3$, the three-phase system is said to be "balanced" or "symmetrical." Otherwise, the three-phase system is "unbalanced" or "unsymmetrical." The terms balanced and unbalanced will be used in this book. The above classification is illustrated in Fig. 2-6, where three examples of three-phase voltages are given. This means that a three-phase system consisting only of a positive-sequence component or only of a negative-sequence component is a balanced system. However, only the positive-sequence component will be considered in this book as the main component.

Figure 2-6(a) and Fig. 2-6(b) show examples of three-phase balanced and unbalanced voltages, respectively. Figure 2-6(c) shows three-phase distorted and unbalanced voltages that are obtained by superposing harmonic components on the unbalanced voltages given in Fig. 2-6(b). Thus, beside the harmonics components, the voltages have also unbalanced fundamental components.

There are two kinds of imbalances, which can be better understood when the Symmetrical Components Theory [22] is applied to the three-phase phasors of voltage or current.

At first, the three-phase unbalanced phasors $(\dot{V}_a, \dot{V}_b, \dot{V}_c)$ are transformed into three other phasors: the positive-sequence phasor $\dot{V}_+$, the negative-sequence phasor $\dot{V}_-$, and the zero-sequence phasor $\dot{V}_0$. They are calculated as

$$\begin{bmatrix} \dot{V}_0 \\ \dot{V}_+ \\ \dot{V}_- \end{bmatrix} = \frac{1}{3} \begin{bmatrix} 1 & 1 & 1 \\ 1 & \alpha & \alpha^2 \\ 1 & \alpha^2 & \alpha \end{bmatrix} \begin{bmatrix} \dot{V}_a \\ \dot{V}_b \\ \dot{V}_c \end{bmatrix} \quad (2.33)$$

The constant α is a complex number that acts as a 120°-phase shift operator, that is,

$$\alpha = 1 \angle 120° = e^{j(2\pi/3)} = -\frac{1}{2} + j\frac{\sqrt{3}}{2} \quad (2.34)$$

The inverse transformation of (2.33) is given by

$$\begin{bmatrix} \dot{V}_a \\ \dot{V}_b \\ \dot{V}_c \end{bmatrix} = \begin{bmatrix} 1 & 1 & 1 \\ 1 & \alpha^2 & \alpha \\ 1 & \alpha & \alpha^2 \end{bmatrix} \begin{bmatrix} \dot{V}_0 \\ \dot{V}_+ \\ \dot{V}_- \end{bmatrix} \quad (2.35)$$

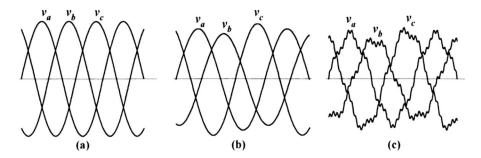

Figure 2-6. (a) Balanced voltages, (b) unbalanced voltages, and (c) distorted and unbalanced voltages.

Note that the zero-sequence phasor contributes to the phase voltages without 120° phase shifting, whereas the positive-sequence phasor contributes to the voltage in the form of an *abc* sequence and the negative-sequence phasor in the form of an *acb* sequence. The above idea of decomposition of unsymmetrical phasors in the positive-sequence phasor V_+, the negative-sequence phasor V_-, and the zero-sequence phasor V_0 can be better visualized by referring to Fig. 2-7.

Imbalances at the fundamental frequency can be caused by negative-sequence or zero-sequence components. However, it is important to note that only an imbalance from a negative-sequence component can appear in a three-phase grounded or ungrounded system. Imbalance from zero-sequence component only appears in a three-phase, four-wire (grounded) system, which induces the current flowing through the neutral wire. Contrarily, the sum of the three instantaneous phase voltages, as well as the sum of the three instantaneous line currents, is always equal to zero if they consist only of positive-sequence components and/or negative-sequence components. Hence, positive-sequence and negative-sequence components can be present in three-phase circuits with or without the neutral conductor.

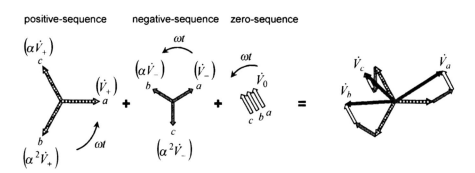

Figure 2-7. Positive-sequence, negative-sequence, and zero-sequence components of an unbalanced three-phase system.

Figure 2-7 shows phasors in the same angular frequency ω, corresponding to the decomposition of an unbalanced three-phase system into symmetrical components. The above analysis is valid for an arbitrarily chosen value of ω. For instance, three-phase, generic, periodic voltages and currents can be decomposed in Fourier series. From these series, the third (3ω), fifth (5ω), seventh (7ω), and subsequent harmonics in a three-phase generic voltage or current can be separated in groups of abc harmonics that are at a given frequency. For each abc harmonic group, the above decomposition of phasors into symmetrical components can be applied.

2.5.2. Power in Balanced Three-Phase Systems

In three-phase balanced systems, the total (three-phase) active, reactive, and apparent powers have been calculated as three times the single-phase powers defined above. Although many authors assign the same physical meaning or mathematical interpretation for the active, reactive, and apparent power in both single-phase and three-phase systems, at least one remark should be made concerning the reactive power Q. The reactive power does not describe the same phenomenon in three-phase and in single-phase circuits [23,24]. This is confirmed below.

Three-phase voltage and line current that contain only the positive-sequence fundamental component (sinusoidal and balanced system) are given by

$$\begin{cases} v_a(t) = \sqrt{2}V_+ \sin(\omega t + \phi_{V+}) \\ v_b(t) = \sqrt{2}V_+ \sin\left(\omega t + \phi_{V+} - \frac{2\pi}{3}\right) \\ v_c(t) = \sqrt{2}V_+ \sin\left(\omega t + \phi_{V+} + \frac{2\pi}{3}\right) \end{cases} \quad (2.36)$$

$$\begin{cases} i_a(t) = \sqrt{2}I_+ \sin(\omega t + \phi_{I+}) \\ i_b(t) = \sqrt{2}I_+ \sin\left(\omega t + \phi_{I+} - \frac{2\pi}{3}\right) \\ i_c(t) = \sqrt{2}I_+ \sin\left(\omega t + \phi_{I+} + \frac{2\pi}{3}\right) \end{cases} \quad (2.37)$$

The symbols V_+, I_+, ϕ_{V+}, and ϕ_{I+} with subscript "+" are used to emphasize the nature of the positive-sequence component.

For a three-phase system with or without a neutral conductor, the three-phase instantaneous active power $p_{3\phi}(t)$ describes the total instantaneous energy flow per time unit being transferred between two subsystems, and it is given by

$$p_{3\phi}(t) = v_a(t)i_a(t) + v_b(t)i_b(t) + v_c(t)i_c(t) = p_a(t) + p_b(t) + p_c(t) \quad (2.38)$$

2.5. ELECTRIC POWERS IN THREE-PHASE SYSTEMS

Substituting (2.36) and (2.37) in (2.38) results in

$$p_{3\phi}(t) = V_+I_+\left[\cos(\phi_{V+} - \phi_{I+}) - \cos(2\omega t + \phi_{V+} + \phi_{I+}) +\right.$$

$$+ \cos(\phi_{V+} - \phi_{I+}) - \cos\left(2\omega t + \phi_{V+} + \phi_{I+} + \frac{2\pi}{3}\right) + \qquad (2.39)$$

$$\left. + \cos(\phi_{V+} - \phi_{I+}) - \cos\left(2\omega t + \phi_{V+} + \phi_{I+} - \frac{2\pi}{3}\right)\right]$$

$$p_{3\phi}(t) = 3V_+I_+\cos(\phi_{V+} - \phi_{I+}) = 3P$$

The sum of the three time-dependent terms in (2.39) is always equal to zero. Hence, the instantaneous active three-phase power $p_{3\phi}(t)$ is constant, that is, it is time-independent. In contrast, the single-phase power defined in (2.2) contains a time dependent term. This term was decomposed into two terms in (2.3), from which the active power P and the reactive power Q were defined. Here, the three-phase instantaneous active power is constant and equal to $3P$ (three times the single-phase active power). For consistency with the definitions in single-phase systems, the three-phase active (average) power $P_{3\phi}$, is defined as

$$P_{3\phi} = 3P = 3V_+I_+\cos(\phi_{V+} - \phi_{I+}) \qquad (2.40)$$

For three-phase *balanced* systems, the three-phase apparent power can be defined from the voltage and current phasors, similar to the definition of a single-phase system under sinusoidal conditions. Because the three-phase circuit is balanced, the idea that a "three-phase circuit can be considered as three single-phase circuits" may be adopted. In this case, the idea, "what happens in a given phase also will happen in the next phase, 1/3 of the fundamental period later, and so on" is also used. Hence, the three-phase apparent power $S_{3\phi}$ is defined as

$$S_{3\phi} = 3S = 3V_+I_+ \qquad (2.41)$$

Furthermore, a three-phase complex power $\mathbf{S_{3\phi}}$ is defined as three times the single-phase complex power defined in (2.12). A definition of three-phase reactive power $Q_{3\phi}$ can be derived from the imaginary part of the definition of this three-phase complex power as follows.

Three-phase complex power $\mathbf{S_{3\phi}}$:

$$\mathbf{S_{3\phi}} = 3\dot{V}_+\dot{I}_+^* = 3V_+ \angle \phi_{V+} \, I_+ \angle -\phi_{I+}$$

$$= 3V_+I_+\underbrace{\cos(\phi_{V+} - \phi_{I+})}_{P_{3\phi}} + j3V_+I_+\underbrace{\sin(\phi_{V+} - \phi_{I+})}_{Q_{3\phi}} \qquad (2.42)$$

Three-phase reactive power $Q_{3\phi}$:

$$Q_{3\phi} = 3Q = 3V_+I_+ \sin(\phi_{V+} - \phi_{I+}) \tag{2.43}$$

The reactive power $Q_{3\phi}$ in (2.42) or in (2.43) is just a mathematical definition without any precise physical meaning. In other words, the three-phase circuit includes no oscillating power components in the three-phase instantaneous active power $p_{3\phi}(t)$ given in (2.39), in contrast to the single-phase instantaneous active power given in (2.3). Thus, there is no portion of power that can be related to a unidirectional oscillating power from which the active (average) power is defined by $P = VI \cos \phi$, or an "oscillating power component that does not realize work," defined by $Q = VI \sin \phi$.

For example, a three-phase ideal generator supplying a balanced, ideal capacitor bank produces no mechanical torque if losses are neglected. In this case, each phase voltage phasor is orthogonal to the corresponding line current phasor, and $P_{3\phi}$ is equal to zero. A generator supplying no active power means, ideally, that there is no mechanical torque. On the other hand, a single-phase ideal generator supplying an ideal capacitor has an oscillating mechanical torque due to the presence of the oscillating portion of active power, as described in (2.3). Therefore, it is not correct to think that the three-phase reactive power represents an oscillating energy between the source and the load if all the three phases of the system are considered together, not as three single-phase circuits, but as a three-phase circuit. This is the reason why a three-phase system should not be considered as a sum of three separate single-phase systems. Nevertheless, the above single-phase approach to power definitions is useful in terms of simplifying the analysis of power systems that can be approximated to balanced and sinusoidal models.

2.5.3. Power in Three-Phase Unbalanced Systems

The traditional concepts of apparent power and reactive power are in contradiction if applied to unbalanced and/or distorted three-phase systems. Both approaches of Budeanu in Section 2.4.1 and of Fryze in Section 2.4.2 are not adequate in unbalanced/distorted three-phase systems.

Based on rms values of voltage and current, two definitions of three-phase apparent power are commonly used by some authors [7,9,13,25,26]:

i) "Per phase" calculation:

$$S_{3\phi} = \sum_k S_k = \sum_k V_k I_k, \quad k = (a, b, c) \tag{2.44}$$

ii) "aggregate rms value" calculation:

$$S_\Sigma = \sqrt{\sum_k V_k^2} \sqrt{\sum_k I_k^2}, \quad k = (a, b, c) \tag{2.45}$$

Here, V_a, V_b, V_c and I_a, I_b, I_c are the rms values of the phase voltages (line-to-neutral voltages) and line currents, as calculated in (2.16).

It is possible to demonstrate that for a balanced and sinusoidal case, the apparent powers from (2.44) and (2.45) are equivalent. However, under nonsinusoidal or unbalanced conditions, the result always holds that $S_\Sigma \leq S_{3\phi}$. The powers S_Σ and $S_{3\phi}$ are mathematical definitions without any clear physical meaning [11,16]. However, some authors use one of them to have the sense of "maximum reachable active power at unity power factor" [7].

In (2.45), the concept of *aggregate voltage and current* is used. Schering [16] pointed out that Buchholz [2][25] introduced the concept of *aggregate* voltage and current ("Kollektivstrom" and "Kollektivspannung" in the original work written in German) in 1919 to define an apparent power for generic loads. The (rms) aggregate voltage and current are given by

$$V_\Sigma = \sqrt{V_a^2 + V_b^2 + V_c^2} \quad \text{and} \quad I_\Sigma = \sqrt{I_a^2 + I_b^2 + I_c^2} \qquad (2.46)$$

The aggregate voltage in (2.46), which is calculated from the rms values of the phase voltages V_a, V_b, and V_c, gives the rms value of the line voltage if the system is sinusoidal and balanced. The same situation happens with the aggregate current that is calculated from the rms values of line currents I_a, I_b, and I_c. Moreover, the concept of the aggregate voltage can also be useful if instantaneous values of phase voltages are used. An example follows:

$$\left. \begin{array}{l} v_a(t) = \sqrt{2}V_+ \sin(\omega t + \phi_{V+}) \\ v_b(t) = \sqrt{2}V_+ \sin\left(\omega t + \phi_{V+} - \dfrac{2\pi}{3}\right) \\ v_c(t) = \sqrt{2}V_+ \sin\left(\omega t + \phi_{V+} + \dfrac{2\pi}{3}\right) \end{array} \right\} \quad v_\Sigma = \sqrt{v_a^2(t) + v_b^2(t) + v_c^2(t)} = \sqrt{3}V_+ \qquad (2.47)$$

Under three-phase unbalanced/distorted systems, the instantaneous aggregate voltage has an oscillating portion that superposes the dc value given in (2.47).

It is not possible to establish a consistent set of power definitions (active power, reactive power, power factor, etc.) from the apparent power given in (2.44) or (2.45). This would lead to a subsequent definition of reactive or harmonic powers that would seem to be mathematical definitions without any clear physical meaning. Therefore, the *instantaneous active power*—a universal concept—will be chosen in the next chapter as the fundamental equation for power definitions in three-phase systems.

2.6. SUMMARY

This chapter presented an overview of some power theories, starting from the classical references dating from the 1920s. It was interesting to show that long ago some researchers like Fryze and Budeanu had studied the problem. Most of the concepts presented here are not only important for knowing the past work but also to better understand the new concepts that will be presented in the next chapter.

Although this chapter starts with the power definitions under sinusoidal conditions that are quite common situations and well known to electrical engineers, they are necessary to create a solid base for the rest of the theory. Voltage and current phasors, as well as the resulting complex impedance, were presented just as a review of these concepts, necessary to define the complex power and power factor. Concepts of power under nonsinusoidal conditions for conventional approaches were analyzed, starting with the concepts presented by Budeanu in the frequency domain. The power tetrahedron and the distortion factor were introduced to better fix the concepts of power under nonsinusoidal conditions. Then, power definitions by Fryze in the time domain and the resulting electric power in three-phase systems were presented, after classifications of three-phase systems with imbalances and/or distortions. Power in three-phase balanced systems and power in three-phase unbalanced and distorted systems were presented, along with detailed discussions on the concept of apparent power in both systems.

REFERENCES

[1] W. V. Lyon, "Reactive Power and Unbalanced Circuits," *Electrical World,* vol. 75, no. 25, 1920, pp. 1417–1420.

[2] F. Buchholz, "Die Drehstrom-Scheinleistung bei ungleichmäßiger Belastung der drei Zweige," *Licht und Kraft, Zeitschrift für elekt. Energie-Nutzung,* no. 2, Jan. 1922, pp. 9–11.

[3] S. Fryze, "Wirk-, Blind- und Scheinleistung in elektrischen Stromkreisen mit nicht-sinusförmigem Verlauf von Strom und Spannung," *ETZ-Arch. Elektrotech.,* vol. 53, 1932, pp. 596–599, 625–627, 700–702.

[4] C. I. Budeanu, "Puissances reactives et fictives," *Instytut Romain de l'Energie,* pub. no. 2, Bucharest, 1927.

[5] C. I. Budeanu, "The Different Options and Conceptions Regarding Active Power in Nonsinusoidal Systems", *Instytut Romain de l'Energie,* pub. no. 4, Bucharest, 1927.

[6] R. M. Kerchner and G. F. Corcoran, *Alternating-Current Circuits,* Wiley, 1938.

[7] A. E. Emanuel, "Apparent and Reactive Powers in Three-Phase Systems: In Search of a Physical Meaning and a Better Resolution," *ETEP—Eur. Trans. Elect. Power Eng.,* vol. 3, no. 1, Jan./Feb. 1993, pp. 7–14.

[8] L. S. Czarnecki, "What Is Wrong with the Budeanu Concept of Reactive and Distortion Power and Why It Should Be Abandoned," *IEEE Trans. Instr. Meas.,* vol. IM-36, no. 3, 1987, pp. 834–837.

[9] P. Filipski and R. Arseneau, "Definition and Measurement of Apparent Power under Distorted Waveform Conditions," *IEEE Tutorial Course on Non-sinusoidal Situations,* 90EH0327-7, 1990, pp. 37–42.

[10] M. Depenbrock, D. A. Marshall, and J. D. van Wyk, "Formulating Requirements for a Universally Applicable Power Theory as Control Algorithm in Power Compensators," *ETEP—Eur. Trans. Elect. Power Eng.,* vol. 4, no. 6, Nov./Dec. 1994, pp. 445–455.

REFERENCES

[11] L. S. Czarnecki, "Comparison of Power Definitions for Circuits with Non-sinusoidal Waveforms," *IEEE Tutorial Course on Non-sinusoidal Situations*, 90EH0327-7, 1990, pp. 43–50.

[12] L. S. Czarnecki, "Misinterpretations of Some Power Properties of Electric Circuits," *IEEE Trans. Power Delivery*, vol. 9, no. 4, Oct. 1994, pp. 1760–1769.

[13] L. S. Czarnecki, "Power Related Phenomena in Three-Phase Unbalanced Systems," *IEEE Trans. Power Delivery*, vol. 10, no. 3, July 1995, pp. 1168–1176.

[14] *IEEE Standard Dictionary of Electrical and Electronics Terms*, Third Edition, IEEE/Wiley, 1984.

[15] *IEEE Trial-Use Standard Definitions for the Measurement of Electric Power Quantities Under Sinusoidal, Nonsinusoidal, Balanced, or Unbalanced Conditions*, IEEE Std 1459-2000, Jan. 2000.

[16] H. Schering, "Die Definition der Schein- und Blindleistung sowie des Leistungsfaktors bei Mehrphasenstrom," *Elektrotechnische Zeitschrift*, vol. 27., July 1924, pp. 710–712.

[17] V. N. Nedelcu, "Die einheitlische Leistungstheorie der unsymmetrischen und mehrwelligen Mehrphasensysteme," *ETZ-Arch. Elektrotech.* vol. 84, no. 5, 1963, pp. 153–157.

[18] Z. Nowomiejski, "Generalized Theory of Electric Power," *Arch. für Elektrotech.*, vol. 63, 1981, pp. 177–182.

[19] A. E. Emanuel, "Energetical Factors in Power Systems with Nonlinear Loads," *Arch. für Elektrotech.*, vol. 59, 1977, pp. 183–189.

[20] R. Arseneau and P. Filipski, "Application of a Three Phase Nonsinusoidal Calibration System for Testing Energy and Demand Meters under Simulated Field Conditions," *IEEE Trans. Power Delivery*, vol. 3, July 1988, pp. 874–879.

[21] E. W. Kimbark, *Direct Current Transmission*, vol. 1, Wiley, 1971.

[22] C. L. Fortescue, "Method of Symmetrical Co-ordinates Applied to the Solution of Polyphase Networks," *A.I.E.E. Trans.*, vol. 37, June 1918, pp. 1027–1140.

[23] E. H. Watanabe, R. M. Stephan, and M. Aredes, "New Concepts of Instantaneous Active and Reactive Powers in Electrical Systems with Generic Loads," *IEEE Trans. Power Delivery*, vol. 8, no. 2, Apr. 1993, pp. 697–703.

[24] M. Aredes and E. H. Watanabe, "New Control Algorithms for Series and Shunt Three-Phase Four-Wire Active Power Filters," *IEEE Trans. Power Delivery*, vol. 10, no. 3, July 1995, pp. 1649–1656.

[25] F. Buchholz, "Die Darstellung der Begriffe Scheinleistung und Scheinarbeit bei Mehrphasenstrom," *Elektro-J. 9*, 1929, pp. 15–21.

[26] M. Depenbrock, "The FBD-Method, a Generally Applicable Tool for Analysing Power Relations," *IEEE Trans. Power Systems*, vol. 8, no. 2, May 1993, pp. 381–387.

[27] J. H. R. Enslin and J. D. van Wyk, "Measurement and Compensation of Fictitious Power under Nonsinusoidal Voltage and Current Conditions," *IEEE Trans. Instr. Meas.*, vol. IM-37, no. 3, 1988, pp. 403–408.

[28] M. A. Slonim and J. D. van Wyk, "Power Components in Systems with Sinusoidal and Nonsinusoidal Voltage and/or Current Conditions," *IEE Proc.*, vol. 135, Pt. B, no. 2, Mar. 1988, pp. 76–84.

[29] D. A. Marshall, J. D. van Wyk, "An Evaluation of the Real-Time Compensation of Fic-

titious Power in Electric Energy Networks," *IEEE Trans. Power Delivery,* vol. 6, no. 4, Oct. 1991, pp. 1774–1780.

[30] P. Tenti, "The Relation Between Global Performance Indexes: Power Factor and FBD Factor," *ETEP—European Transactions on Electrical Power Engineering,* vol. 4, no. 6, Nov./Dec. 1994, pp. 505–507.

CHAPTER 3

THE INSTANTANEOUS POWER THEORY

As discussed in the previous chapter, research on the calculation and physical interpretation of energy flow in an electric circuit dates back to the 1920s. It is amazing to find excellent early works dealing with most of the important aspects of this energy flow. However, the basic concern was related to the average power or rms value of voltage and current. Although the concept presented by Fryze [1] uses rms values for power analysis, it treats a three-phase circuit as a unit, not a superposition of three single-phase circuits. The development of power electronics devices and their associated converters has brought new boundary conditions to the energy flow problem. This is not exactly because the problem is new, but because these converters behave as nonlinear loads and represent a significant amount of power compared with other traditional linear loads. The speed response of these converters and the way they generate reactive power and harmonic components have made it clear that conventional approaches to the analysis of power are not sufficient in terms of taking average or rms values of variables. Therefore, time-domain analysis has evolved as a new manner to analyze and understand the physical nature of the energy flow in a nonlinear circuit. This chapter is dedicated to the time-domain analysis of power in a three-phase electric circuit.

The theories that deal with instantaneous power can be mainly classified into the following two groups. The first one is developed based on the transformation from the *abc* phases to three-orthogonal axes, and the other is done directly on the *abc* phases. The first one is what will be called the *p-q* Theory that is based on the *abc* to $\alpha\beta 0$ transformation. The second one has no specific name. Because it deals directly with the *abc* phases, it will be called the *abc* Theory in this book. Finally, comparisons between the two theories will be presented.

3.1. BASIS OF THE p-q THEORY

The *p-q* Theory is based on a set of instantaneous powers defined in the time domain. No restrictions are imposed on the voltage or current waveforms, and it can be applied to three-phase systems with or without a neutral wire for three-phase generic voltage and current waveforms. Thus, it is valid not only in the steady state, but also in the transient state. As will be seen in the following chapters, this theory is very efficient and flexible in designing controllers for power conditioners based on power electronics devices.

Other traditional concepts of power are characterized by treating a three-phase system as three single-phase circuits. The *p-q* Theory first transforms voltages and currents from the *abc* to $\alpha\beta 0$ coordinates, and then defines instantaneous power on these coordinates. Hence, this theory always considers the three-phase system as a unit, not a superposition or sum of three single-phase circuits.

3.1.1. Historical Background of the p-q Theory

The *p-q* Theory in its first version was published in the Japanese language in July 1982 in a local conference and later in the journal *Transactions of the IEE-Japan* [3]. With a minor time lag it was published in 1983 in an international conference [4], and, in 1984, in the *IEEE Transactions on Industry Applications*, including experimental verification [5]. The development of this theory was based on various previous works written by power electronics specialists interested in reactive-power compensation. From the end of 1960s to the beginning of 1970s, some papers related to what can be considered as a basic principle of reactive-power compensation were published [6,7,8]. The authors of [6] presented some basic ideas like "... compensation of distortive power are unknown to date...." They also assured that "a nonlinear resistor behaves like a reactive-power generator while having no energy-storing elements," and presented the very first approach to power-factor correction. Fukao and his coauthors in [8], the authors said, "... by connecting a reactive-power source in parallel with the load and by controlling it in such a way as to supply reactive power to the load, the power network will only supply active power. Therefore, an ideal power transmission would be possible."

Gyugyi and Pelly in [9] presented the idea that reactive power could be compensated by a naturally commuted cycloconverter without energy storage elements. Generation of reactive power without energy storage elements was also investigated in [10]. This idea was explained from a physical point of view, but no specific mathematical proof was presented. In 1976, Harashima and his coauthors presented in [11], probably for the first time, the term "instantaneous reactive power" for a single-phase circuit. In that same year, Gyugyi and Strycula [12] used the term "active ac power filters" for the first time. In 1981, Takahashi and his coauthors in [13] and [14] gave a hint to the emergence of the *p-q* Theory. The formulation they reached is in fact a subset of the *p-q* Theory. However, no physical meaning of the variables introduced in the two papers was explained. This theory will be explained in the next sections.

The *p-q* Theory uses the $\alpha\beta 0$ transformation, also known as the Clarke transformation [15], which consists of a real matrix that transforms three-phase voltages and currents into the $\alpha\beta 0$ stationary reference frames. Therefore, the presentation of the *p-q* Theory will start with this transformation, followed by the theory itself, its physical meanings and interpretations, and comparisons with conventional power theories.

3.1.2. The Clarke Transformation

The $\alpha\beta 0$ transformation or the Clarke transformation [15] maps the three-phase instantaneous voltages in the *abc* phases, v_a, v_b, and v_c, into the instantaneous voltages on the $\alpha\beta 0$-axes v_α, v_β, and v_0. The Clarke Transformation and its inverse transformation of three-phase generic voltages are given by

$$\begin{bmatrix} v_0 \\ v_\alpha \\ v_\beta \end{bmatrix} = \sqrt{\frac{2}{3}} \begin{bmatrix} \frac{1}{\sqrt{2}} & \frac{1}{\sqrt{2}} & \frac{1}{\sqrt{2}} \\ 1 & -\frac{1}{2} & -\frac{1}{2} \\ 0 & \frac{\sqrt{3}}{2} & -\frac{\sqrt{3}}{2} \end{bmatrix} \begin{bmatrix} v_a \\ v_b \\ v_c \end{bmatrix} \quad (3.1)$$

$$\begin{bmatrix} v_a \\ v_b \\ v_c \end{bmatrix} = \sqrt{\frac{2}{3}} \begin{bmatrix} \frac{1}{\sqrt{2}} & 1 & 0 \\ \frac{1}{\sqrt{2}} & -\frac{1}{2} & \frac{\sqrt{3}}{2} \\ \frac{1}{\sqrt{2}} & -\frac{1}{2} & -\frac{\sqrt{3}}{2} \end{bmatrix} \begin{bmatrix} v_0 \\ v_\alpha \\ v_\beta \end{bmatrix} \quad (3.2)$$

Similarly, three-phase generic instantaneous line currents, i_a, i_b, and i_c, can be transformed on the $\alpha\beta 0$ axes by

$$\begin{bmatrix} i_0 \\ i_\alpha \\ i_\beta \end{bmatrix} = \sqrt{\frac{2}{3}} \begin{bmatrix} \frac{1}{\sqrt{2}} & \frac{1}{\sqrt{2}} & \frac{1}{\sqrt{2}} \\ 1 & -\frac{1}{2} & -\frac{1}{2} \\ 0 & \frac{\sqrt{3}}{2} & -\frac{\sqrt{3}}{2} \end{bmatrix} \begin{bmatrix} i_a \\ i_b \\ i_c \end{bmatrix} \quad (3.3)$$

and its inverse transformation is

$$\begin{bmatrix} i_a \\ i_b \\ i_c \end{bmatrix} = \sqrt{\frac{2}{3}} \begin{bmatrix} \frac{1}{\sqrt{2}} & 1 & 0 \\ \frac{1}{\sqrt{2}} & -\frac{1}{2} & \frac{\sqrt{3}}{2} \\ \frac{1}{\sqrt{2}} & -\frac{1}{2} & -\frac{\sqrt{3}}{2} \end{bmatrix} \begin{bmatrix} i_0 \\ i_\alpha \\ i_\beta \end{bmatrix} \qquad (3.4)$$

One advantage of applying the $\alpha\beta 0$ transformation is to separate zero-sequence components from the *abc*-phase components. The α and β axes make no contribution to zero-sequence components. No zero-sequence current exists in a three-phase, three-wire system, so that i_0 can be eliminated from the above equations, thus resulting in simplification. If the three-phase voltages are balanced in a four-wire system, no zero-sequence voltage is present, so that v_0 can be eliminated. However, when zero-sequence voltage and current components are present, the complete transformation has to be considered.

If v_0 can be eliminated from the transformation matrixes, the Clarke transformation and its inverse transformation become

$$\begin{bmatrix} v_\alpha \\ v_\beta \end{bmatrix} = \sqrt{\frac{2}{3}} \begin{bmatrix} 1 & -\frac{1}{2} & -\frac{1}{2} \\ 0 & \frac{\sqrt{3}}{2} & -\frac{\sqrt{3}}{2} \end{bmatrix} \begin{bmatrix} v_a \\ v_b \\ v_c \end{bmatrix} \qquad (3.5)$$

and

$$\begin{bmatrix} v_a \\ v_b \\ v_c \end{bmatrix} = \sqrt{\frac{2}{3}} \begin{bmatrix} 1 & 0 \\ -\frac{1}{2} & \frac{\sqrt{3}}{2} \\ -\frac{1}{2} & -\frac{\sqrt{3}}{2} \end{bmatrix} \begin{bmatrix} v_\alpha \\ v_\beta \end{bmatrix} \qquad (3.6)$$

Similar equations hold in the line currents.

The real matrices in (3.5) and (3.6) suggest an axis transformation as shown in Fig. 3-1. They are stationary axes and should not be confused with the concepts of voltage or current phasors, like those illustrated in Chapter 2 (Fig. 2-2). Here, instantaneous values of phase voltages and line currents referred to the *abc* stationary axes are transformed into the $\alpha\beta$ stationary axes, or vice-versa. The *a*, *b*, and *c* axes are spatially shifted by $2\pi/3$ rad from each other while the α and β axes are orthogonal, and the α axis is parallel to the *a* axis. The direction of the β axis was chosen in such a way that if voltage or current spatial vectors on the *abc* coordinates rotate

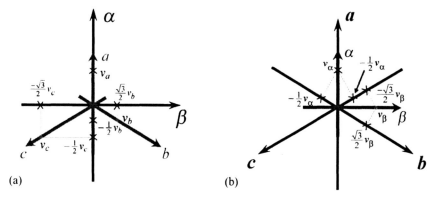

Figure 3-1. Graphical representations. (a) The *abc* to $\alpha\beta$ transformation (Clarke transformation). (b) Inverse $\alpha\beta$ to *abc* transformation (inverse Clarke transformation).

in the *abc* sequence, they would rotate in the $\alpha\beta$ sequence on the $\alpha\beta$ coordinates. This idea is well explained in the following section.

3.1.2.A. Calculation of Voltage and Current Vectors When Zero-Sequence Components are Excluded

If v_0 can be neglected, an instantaneous voltage vector is defined from the instantaneous α- and β-voltage components, that is,

$$\mathbf{e} = v_\alpha + jv_\beta \tag{3.7}$$

Similarly, if i_0 can be neglected, the instantaneous current vector is defined as

$$\mathbf{i} = i_\alpha + ji_\beta \tag{3.8}$$

The above instantaneous vectors can be represented in a complex plane, where the real axis is the α axis, and the imaginary axis is the β axis of the Clarke transformation. It should be noted that the vectors defined above are functions of time, because they consist of the Clarke components of the instantaneous phase voltages and line currents in a three-phase system. Therefore, they should not be misinterpreted as phasors. The usefulness of this vector definition is illustrated in the following example. Later, it will be used also to define a new concept of instantaneous complex apparent power.

Consider the following sinusoidal balanced phase voltages and line currents of a three-phase linear circuit.

$$\begin{cases} v_a(t) = \sqrt{2}V\cos(\omega t + \phi_V) \\ v_b(t) = \sqrt{2}V\cos\left(\omega t + \phi_V - \frac{2\pi}{3}\right) \\ v_c(t) = \sqrt{2}V\cos\left(\omega t + \phi_V + \frac{2\pi}{3}\right) \end{cases} \quad \begin{cases} i_a(t) = \sqrt{2}I\cos(\omega t + \phi_I) \\ i_b(t) = \sqrt{2}I\cos\left(\omega t + \phi_I - \frac{2\pi}{3}\right) \\ i_c(t) = \sqrt{2}I\cos\left(\omega t + \phi_I + \frac{2\pi}{3}\right) \end{cases} \tag{3.9}$$

The angles ϕ_V and ϕ_I are the voltage and current phases, respectively, with respect to a given reference.

The above voltages and currents consist of a single symmetrical component in the fundamental positive sequence (see Fig. 2-7). Thus, they are sinusoidal and balanced. These voltages and currents can be transformed to the $\alpha\beta$ reference frames by using (3.5) and its similar current equations. The above three-phase voltages and currents transformed into the α–β reference frames are given by:

$$\begin{cases} v_\alpha = \sqrt{3}V\cos(\omega t + \phi_V) \\ v_\beta = \sqrt{3}V\sin(\omega t + \phi_V) \end{cases} \text{and} \quad \begin{cases} i_\alpha = \sqrt{3}I\cos(\omega t + \phi_I) \\ i_\beta = \sqrt{3}I\sin(\omega t + \phi_I) \end{cases}$$

Now, a voltage vector **e** and current vector **i** are derived as follows:

$$\mathbf{e} = v_\alpha + jv_\beta \Rightarrow \begin{vmatrix} \mathbf{e} = \sqrt{3}V[\cos(\omega t + \phi_V) + j\sin(\omega t + \phi_V)] \\ \mathbf{e} = \sqrt{3}V\, e^{j(\omega t + \phi_V)} \end{vmatrix} \quad (3.10)$$

and

$$\mathbf{i} = i_\alpha + ji_\beta \Rightarrow \begin{vmatrix} \mathbf{i} = \sqrt{3}I[\cos(\omega t + \phi_I) + j\sin(\omega t + \phi_I)] \\ \mathbf{i} = \sqrt{3}I\, e^{j(\omega t + \phi_I)} \end{vmatrix} \quad (3.11)$$

Therefore, in the case of a three-phase balanced sinusoidal system the voltage and current vectors have constant amplitudes and rotate in the clockwise direction, or in the $\alpha \rightarrow \beta$ sequence, at the angular frequency ω, as suggested in Fig. 3-2.

The same result is found if the instantaneous *abc* voltages and currents are taken directly to compose vectors on a complex plane. To confirm the above result obtained by using the Clarke Transformation, the relative positions of the *a*, *b*, and *c* axes and the α and β axes must be kept as suggested in Fig. 3-1. The *a* axis must be coincident with the α axis that is the real axis of the complex plane. The β axis is the imaginary axis shifted by $\pi/2$ from the real α axis. Each time function of the *abc*-phase voltages is multiplied by a proper unitary complex phase shifter to lay in the

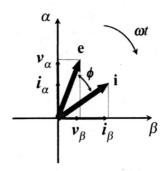

Figure 3-2. Vector representation of voltages and currents on the α-β reference frames.

corresponding axis direction. Thus, the complex voltage vector can be composed by

$$e_{abc} = v_a e^{j0} + v_b e^{j2\pi/3} + v_c e^{-j2\pi/3} \qquad (3.12)$$

By replacing the time functions for v_a, v_b, and v_c given in (3.9) and making some manipulation, the following equation is obtained:

$$e_{abc} = \frac{3\sqrt{2}}{2} V[\cos(\omega t + \phi_V) + j \sin(\omega t + \phi_V)] = \frac{3\sqrt{2}}{2} V e^{j(\omega t + \phi_V)} \qquad (3.13)$$

Note that the voltage vector e_{abc} in (3.13) has a constant amplitude, and rotates in the clockwise direction at the angular frequency ω, like e defined in (3.10). Therefore, if **e** turns clockwise, in the $\alpha \rightarrow \beta$ sequence, at the synchronous angular speed, e_{abc} does the same, passing sequentially through $a \rightarrow b \rightarrow c$ axes at the same angular speed. By comparison, the following relation can be written:

$$\mathbf{e} = \sqrt{\frac{2}{3}} \mathbf{e}_{abc} \qquad (3.14)$$

The vector representation of three-phase instantaneous voltages and currents has been used increasingly in the field of power electronics. For instance, it has been used in vector control of ac motor drives, in space-vector pulse-width-modulation (PWM) of power converters, as well as in control of power conditioners.

3.1.3. Three-Phase Instantaneous Active Power in Terms of Clarke Components

The Clarke transformation and its inverse transformation as used in (3.1) to (3.4) have the property of being invariant in power. This feature is very suitable when the focus is put on the analysis of instantaneous power in three-phase systems.

All traditional power definitions summarized in the last chapter require as a precondition that the system be in the steady state. The three-phase instantaneous active power has a clear and universally accepted physical meaning, and is also valid during transient states.

> For a three-phase system with or without a neutral conductor in the steady state or during transients, the three-phase instantaneous active power $p_{3\phi}(t)$ describes the total instantaneous energy flow per second between two subsystems.

The three-phase instantaneous active power, $p_{3\phi}(t)$, is calculated from the instantaneous phase voltages and line currents as

$$\begin{array}{c} p_{3\phi}(t) = v_a(t)i_a(t) + v_b(t)i_b(t) + v_c(t)i_c(t) \\ \Updownarrow \\ p_{3\phi} = v_a i_a + v_b i_b + v_c i_c \end{array} \qquad (3.15)$$

where v_a, v_b, and v_c are the instantaneous phase voltages and i_a, i_b, and i_c the instantaneous line currents, as shown in Fig. 3-3. In a system without a neutral wire, v_a, v_b, and v_c are measured from a common point of reference. Sometimes, it is called the "ground" or "fictitious star point." However, this reference point can be set arbitrarily and $p_{3\phi}$, calculated from (3.15), always results in the same value for all arbitrarily chosen reference points for voltage measurement. For instance, if the b phase is chosen as a reference point, the measured "phase voltages" and the three-phase instantaneous active power, $p_{3\phi}$, are calculated as

$$p_{3\phi} = (v_a - v_b)i_a + (v_b - v_b)i_b + (v_c - v_b)i_c = v_{ab}i_a + v_{cb}i_c \qquad (3.16)$$

This explains why it is possible to use $(n-1)$ wattmeters to measure the active power in n-wire systems.

The three-phase instantaneous active power can be calculated in terms of the $\alpha\beta 0$ components if (3.2) and (3.4) are used to replace the abc variables in (3.15).

$$p_{3\phi} = v_a i_a + v_b i_b + v_c i_c \Leftrightarrow p_{3\phi} = v_\alpha i_\alpha + v_\beta i_\beta + v_0 i_0 \qquad (3.17)$$

The property of power invariance as a result of using the Clarke transformation is shown in (3.17). The *p-q* Theory will exploit this feature.

3.1.4. The Instantaneous Powers of the *p-q* Theory

The *p-q* Theory is defined in three-phase systems with or without a neutral conductor. Three instantaneous powers—the instantaneous zero-sequence power p_0, the instantaneous real power p, and the instantaneous imaginary power q—are defined from the instantaneous phase voltages and line currents on the $\alpha\beta 0$ axes as

$$\begin{bmatrix} p_0 \\ p \\ q \end{bmatrix} = \begin{bmatrix} v_0 & 0 & 0 \\ 0 & v_\alpha & v_\beta \\ 0 & v_\beta & -v_\alpha \end{bmatrix} \begin{bmatrix} i_0 \\ i_\alpha \\ i_\beta \end{bmatrix} \qquad (3.18)$$

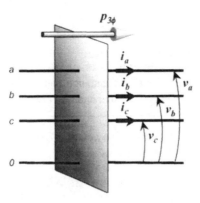

Figure 3-3. Three-phase instantaneous active power.

There are no zero-sequence current components in three-phase, three-wire systems, that is, $i_0 = 0$. In this case, only the instantaneous powers defined on the $\alpha\beta$ axes exist, because the product $v_0 i_0$ in (3.17) is always zero. Hence, in three-phase, three-wire systems, the instantaneous real power p represents the total energy flow per time unity in terms of $\alpha\beta$ components. In this case, $p_{3\phi} = p$.

The instantaneous imaginary power q has a nontraditional physical meaning that will be explained later in the next section.

3.2. THE p-q THEORY IN THREE-PHASE, THREE-WIRE SYSTEMS

Another way to introduce the p-q Theory for three-phase, three-wire systems is to use the instantaneous voltage and current vectors defined in (3.7) and (3.8). The conventional concept of the complex power defined in (2.12) uses a voltage phasor and the conjugate of a current phasor. Thus, it is valid only for a system in the steady state with a fixed line frequency. A new definition of *instantaneous complex power* is possible, using the instantaneous vectors of voltage and current. The instantaneous complex power **s** is defined as the product of the voltage vector **e** and the conjugate of the current vector **i***, given in the form of complex numbers:

$$\mathbf{s} = \mathbf{e} \cdot \mathbf{i}^* = (v_\alpha + jv_\beta)(i_\alpha - ji_\beta) = \underbrace{(v_\alpha i_\alpha + v_\beta i_\beta)}_{p} + j\underbrace{(v_\beta i_\alpha - v_\alpha i_\beta)}_{q} \quad (3.19)$$

The instantaneous real and imaginary powers defined in (3.18) are part of the instantaneous complex power, **s**, defined in (3.19). Since instantaneous voltages and currents are used, there are no restrictions in **s**, and it can be applied during steady states or during transients.

The original definition of p and q in [2]–[5] was based on the following equation:

$$\begin{bmatrix} p \\ q \end{bmatrix} = \begin{bmatrix} v_\alpha & v_\beta \\ -v_\beta & v_\alpha \end{bmatrix} \begin{bmatrix} i_\alpha \\ i_\beta \end{bmatrix} \quad (3.20)$$

Equations (3.18) and (3.20) are applicable to the analysis and design in the same manner. However, this book adopts (3.18), because a positive value of the instantaneous imaginary power q in (3.18) corresponds to the product of a positive-sequence voltage and a lagging (inductive) positive-sequence current, in agreement with the conventional concept of reactive power.

In the following explanation, the $\alpha\beta$ currents will be set as functions of voltages and the real and imaginary powers p and q. This is very suitable for better explaining the physical meaning of the powers defined in the p-q Theory. From (3.18), it is possible to write

$$\begin{bmatrix} i_\alpha \\ i_\beta \end{bmatrix} = \frac{1}{v_\alpha^2 + v_\beta^2} \begin{bmatrix} v_\alpha & v_\beta \\ v_\beta & -v_\alpha \end{bmatrix} \begin{bmatrix} p \\ q \end{bmatrix} \quad (3.21)$$

The right-hand side of (3.21) can have its terms expanded as

$$\begin{bmatrix} i_\alpha \\ i_\beta \end{bmatrix} = \frac{1}{v_\alpha^2 + v_\beta^2} \begin{bmatrix} v_\alpha & v_\beta \\ v_\beta & -v_\alpha \end{bmatrix} \begin{bmatrix} p \\ 0 \end{bmatrix} + \frac{1}{v_\alpha^2 + v_\beta^2} \begin{bmatrix} v_\alpha & v_\beta \\ v_\beta & -v_\alpha \end{bmatrix} \begin{bmatrix} 0 \\ q \end{bmatrix} \quad (3.22)$$

$$\triangleq \begin{bmatrix} i_{\alpha p} \\ i_{\beta p} \end{bmatrix} + \begin{bmatrix} i_{\alpha q} \\ i_{\beta q} \end{bmatrix}$$

The above current components can be defined as shown below.
Instantaneous active current on the α axis $i_{\alpha p}$:

$$i_{\alpha p} = \frac{v_\alpha}{v_\alpha^2 + v_\beta^2} p \quad (3.23)$$

Instantaneous reactive current on the α axis $i_{\alpha q}$:

$$i_{\alpha q} = \frac{v_\beta}{v_\alpha^2 + v_\beta^2} q \quad (3.24)$$

Instantaneous active current on the β axis $i_{\beta p}$:

$$i_{\beta p} = \frac{v_\beta}{v_\alpha^2 + v_\beta^2} p \quad (3.25)$$

Instantaneous reactive current on the β axis $i_{\beta q}$:

$$i_{\beta q} = \frac{-v_\alpha}{v_\alpha^2 + v_\beta^2} q \quad (3.26)$$

The instantaneous power on the α and β coordinates are defined as p_α and p_β, respectively, and are calculated from the instantaneous voltages and currents on the $\alpha\beta$ axes as follows:

$$\begin{bmatrix} p_\alpha \\ p_\beta \end{bmatrix} = \begin{bmatrix} v_\alpha i_\alpha \\ v_\beta i_\beta \end{bmatrix} = \begin{bmatrix} v_\alpha i_{\alpha p} \\ v_\beta i_{\beta p} \end{bmatrix} + \begin{bmatrix} v_\alpha i_{\alpha q} \\ v_\beta i_{\beta q} \end{bmatrix} \quad (3.27)$$

Note that, in the three-phase, three-wire system, the three-phase instantaneous active power in terms of Clarke components in (3.17) is equal to the instantaneous real power defined in (3.18). From (3.27) and (3.18), the real power can be given by the sum of p_α and p_β. Therefore, rewriting this sum by using (3.27) yields the following equation:

$$p = v_\alpha i_{\alpha p} + v_\beta i_{\beta p} + v_\alpha i_{\alpha q} + v_\beta i_{\beta q}$$
$$= \frac{v_\alpha^2}{v_\alpha^2 + v_\beta^2} p + \frac{v_\beta^2}{v_\alpha^2 + v_\beta^2} p + \frac{v_\alpha v_\beta}{v_\alpha^2 + v_\beta^2} q + \frac{-v_\alpha v_\beta}{v_\alpha^2 + v_\beta^2} q \quad (3.28)$$

In the above equation, there are two important points. One is that the instantaneous real power p is given only by

$$v_\alpha i_{\alpha p} + v_\beta i_{\beta p} = p_{\alpha p} + p_{\beta p} = p \quad (3.29)$$

The other is that the following relation exists for the terms dependent on q

$$v_\alpha i_{\alpha q} + v_\beta i_{\beta q} = p_{\alpha q} + p_{\beta q} = 0 \tag{3.30}$$

The above equations suggest the separation of the powers in the following types.
Instantaneous active power on the α axis $p_{\alpha p}$

$$p_{\alpha p} = v_\alpha \cdot i_{\alpha p} = \frac{v_\alpha^2}{v_\alpha^2 + v_\beta^2} p \tag{3.31}$$

Instantaneous reactive power on the α axis $p_{\alpha q}$

$$p_{\alpha q} = v_\alpha \cdot i_{\alpha q} = \frac{v_\alpha v_\beta}{v_\alpha^2 + v_\beta^2} q \tag{3.32}$$

Instantaneous active power on the β axis $p_{\beta p}$

$$p_{\beta p} = v_\beta \cdot i_{\beta p} = \frac{v_\beta^2}{v_\alpha^2 + v_\beta^2} p \tag{3.33}$$

Instantaneous reactive power on the β axis $p_{\beta q}$

$$p_{\beta q} = v_\beta \cdot i_{\beta q} = \frac{-v_\alpha v_\beta}{v_\alpha^2 + v_\beta^2} q \tag{3.34}$$

It should be noted that the watt [W] can be used as the unit of all the powers, $p_{\alpha p}$, $p_{\alpha q}$, $p_{\beta p}$, and $p_{\beta q}$, because each power is defined by the product of the instantaneous voltage on one axis and a part of the instantaneous current on the same axis.

The above equations lead to the following important conclusions:

- The instantaneous current i_α is divided into the instantaneous active component $i_{\alpha p}$ and the instantaneous reactive component $i_{\alpha q}$ as shown in (3.23) and (3.24). This same division is made for the currents on the β axis.
- The sum of the α axis instantaneous active power $p_{\alpha p}$, given in (3.31), and the β axis instantaneous active power $p_{\beta p}$, given in (3.33), corresponds to the instantaneous real power p.
- The sum of $p_{\alpha q}$ and $p_{\beta q}$ is always zero. Therefore, they neither make a contribution to the instantaneous nor average energy flow between the source and the load in a three-phase circuit. This is the reason that they were named instantaneous *reactive* power on the α and β axes. The instantaneous imaginary power q is a quantity that gives the magnitude of the powers $p_{\alpha q}$ and $p_{\beta q}$.
- Because the sum of $p_{\alpha q}$ and $p_{\beta q}$ is always zero, their compensation does not need any energy storage system, as will be shown later.

If the $\alpha\beta$ variables of the instantaneous imaginary power q as defined in (3.18) are replaced by their equivalent expressions referred to the abc axes using (3.5) and similarly for the currents, the following relation can be found:

$$q = v_\beta i_\alpha - v_\alpha i_\beta = \frac{1}{\sqrt{3}}[(v_a - v_b)i_c + (v_b - v_c)i_a + (v_c - v_a)i_b]$$
$$= \frac{1}{\sqrt{3}}(v_{ab}i_c + v_{bc}i_a + v_{ca}i_b) \quad (3.35)$$

Note that q, on the $\alpha\beta$ reference frames, is defined as the sum of products of voltages and currents on different axes. Likewise, the imaginary power q, when calculated directly from the abc phase voltages and line currents, results from the sum of products of line voltages and line currents in different phases. This expression is similar to that implemented in some instruments for measuring the three-phase reactive power. The difference is that voltage and current phasors are used in those instruments. Here, instantaneous values of voltage and current are used instead. As was shown, the imaginary power q does not contribute to the total energy flow between the source and the load, and vice-versa. The imaginary power q is a new quantity, and needs a unit to distinguish this power from the traditional reactive power. The authors propose the use of the unit "Volt-Ampere Imaginary" and the symbol "vai," making an analogy to the symbol "var" of the traditional unit "Volt-Ampere Reactive."

From now on, whenever no doubt can arise, the *instantaneous* zero-sequence power, the *instantaneous* imaginary power, and the *instantaneous* real power defined in the *p-q* Theory will be called *zero-sequence power, imaginary power,* and *real power,* respectively.

At this point, the following important remark can be written.

The imaginary power q is proportional to the quantity of energy that is being exchanged between the phases of the system. It does not contribute to the energy transfer* between the source and the load at any time.

Figure 3-4 summarizes the above explanations about the real and imaginary power. It is important to note that the conventional power theory defined reactive power as a component of the instantaneous (active) power, which has an *average* value equal to zero. Here, it is not so. The imaginary power means a sum of products of *instantaneous* three-phase voltage and current portions that do not contribute to the energy transfer between two subsystems at any time. In this book, the term *"instantaneous reactive power"* in a three-phase system is used as a synonym of the imaginary power. Therefore, they have the same physical meaning.

*The term "energy transfer" is used here in a general manner, meaning not only the energy delivered to the load, but also the energy oscillating between the source and the load.

3.2. THE p-q THEORY IN THREE-PHASE, THREE-WIRE SYSTEMS

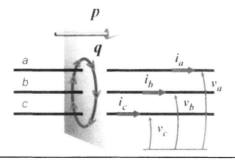

p : instantaneous total energy flow per time unit;
q : energy exchanged between the phases without transferring energy.

Figure 3-4. Physical meaning of the instantaneous real and imaginary powers.

3.2.1. Comparisons with the Conventional Theory

To better understand the meaning of p and q, some examples will be presented. The first one is for a linear circuit with sinusoidal voltages and currents, and the others are for sinusoidal voltages supplying nonlinear loads.

3.2.1.A. Example #1—Sinusoidal Voltages and Currents

Suppose a three-phase ideal voltage source supplying power to a three-phase balanced impedance. The phase voltages and line currents can be expressed as

$$\begin{cases} v_a(t) = \sqrt{2}V \sin(\omega t) \\ v_b(t) = \sqrt{2}V \sin\left(\omega t - \frac{2\pi}{3}\right) \\ v_c(t) = \sqrt{2}V \sin\left(\omega t + \frac{2\pi}{3}\right) \end{cases} \text{ and } \begin{cases} i_a(t) = \sqrt{2}I \sin(\omega t + \phi) \\ i_b(t) = \sqrt{2}I \sin\left(\omega t - \frac{2\pi}{3} + \phi\right) \\ i_c(t) = \sqrt{2}I \sin\left(\omega t + \frac{2\pi}{3} + \phi\right) \end{cases} \quad (3.36)$$

The $\alpha\beta$ transformation of the above voltages and currents are:

$$\begin{cases} v_\alpha = \sqrt{3}V \sin(\omega t) \\ v_\beta = -\sqrt{3}V \cos(\omega t) \end{cases} \text{ and } \begin{cases} i_\alpha = \sqrt{3}I \sin(\omega t + \phi) \\ i_\beta = -\sqrt{3}I \cos(\omega t + \phi) \end{cases} \quad (3.37)$$

The above two equations make it possible to calculate the real and imaginary powers:

$$\begin{cases} p = 3VI \cos\phi \\ q = -3VI \sin\phi \end{cases} \quad (3.38)$$

Both instantaneous powers are constant in this example. The real power p is equal to the conventional definition of the three-phase active power $P_{3\phi}$, whereas the imaginary power q is equal to the conventional three-phase reactive power $Q_{3\phi}$ [compare with that in (2.42)]. This example shows the correspondences between the p-q Theory and the conventional theory in the case of having sinusoidal balanced voltages and linear loads.

If the load has inductive characteristics, the imaginary power q has a positive value, and if the load is capacitive, its value is negative, in concordance with the most common definition of reactive power.

3.2.1.B. Example #2—Balanced Voltages and Capacitive Loads

To better explain the concepts behind the p-q Theory, the following two situations are examined: (i) a three-phase balanced voltage with a three-phase balanced capacitive load (capacitance C), and (ii) an unbalanced load (just one capacitor connected between two phases).

In the first case, the load is balanced under steady-state conditions. The following real and imaginary powers are obtained:

$$\begin{cases} p = 0 \\ q = -3\dfrac{V^2}{X_C} \end{cases} \quad (3.39)$$

The term X_C represents the reactance of the capacitor. As expected, in this case there is no power flowing from the source to the load. Moreover, the imaginary power is constant and coincident with the conventional three-phase reactive power.

In the second case, a capacitor (capacitance C) is connected between phases a and b. The instantaneous real and imaginary powers are given by

$$p = \frac{3V^2}{X_C} \sin\left(2\omega t + \frac{\pi}{3}\right)$$

$$q = -\frac{3V^2}{X_C}\left[1 + \cos\left(2\omega t + \frac{\pi}{3}\right)\right] \quad (3.40)$$

Each power in the above equation has no constant part, and presents an oscillating part. From the conventional power theory, it would be normal to expect only a reactive power (average imaginary power) and no real power at all. However, the results are different, and should be discussed. The reason that the real power is not zero is because the capacitor terminal voltage is varying as a sinusoidal wave and, therefore, it is being charged and discharged, justifying the energy flow given by p. In fact, if it is considered that a turbine is powering the generator, it will have to produce a positive torque when the capacitor is charging, or a negative torque when it is discharging. Of course, there will be torque ripples in its shaft. In the previous example having three balanced capacitors, one capacitor is discharging while the others are charging. In steady-state conditions, there is no total (three-phase) energy flow from the source to the capacitors.

The instantaneous imaginary power also varies with time, a nonzero average value equal to that of the previous case of the balanced three-phase capacitor bank [compare with (3.39)]. It is clear from this example that under unbalanced load conditions the *p-q* Theory presents some important insights that cannot be seen with the conventional frequency-domain theory under unbalanced load conditions.

3.2.1.C. Example #3—Sinusoidal Balanced Voltage and Nonlinear Load

Now, it will be assumed that a three-phase voltage source is balanced and purely sinusoidal, as in (3.36), and the load is a thyristor rectifier operating with a firing angle equal to 30°. The commutation angle is assumed to be null and the smoothing reactor at the dc side large enough to eliminate totally the ripple in the dc current. Figure 3-5 shows the idealized circuit.

Figure 3-6(a) shows the rectifier dc output voltage waveform v_d and Fig. 3-6(b) shows the *a*-phase voltage and the ideal current waveforms. It is well known that this three-phase typical current waveform contains, besides the fundamental, harmonics on the order $(6n \pm 1; n = 1, 2, 3 \ldots)$. The $(6n - 1)$th harmonics are of the negative-sequence type, whereas the $(6n + 1)$th harmonics are of the positive-sequence type (see Fig. 2-7). The line currents of a six-pulse thyristor rectifier operating with a firing angle equal to 30° can be represented as:

$$\begin{cases} i_a(t) = \sqrt{2}I_1 \sin\left(\omega t - \frac{\pi}{6}\right) - \sqrt{2}I_5 \sin\left(5\omega t - \frac{\pi}{6}\right) + \sqrt{2}I_7 \sin\left(7\omega t - \frac{\pi}{6}\right) - \ldots \\ i_b(t) = \sqrt{2}I_1 \sin\left(\omega t - \frac{2\pi}{3} - \frac{\pi}{6}\right) - \sqrt{2}I_5 \sin\left(5\omega t + \frac{2\pi}{3} - \frac{\pi}{6}\right) \\ \qquad + \sqrt{2}I_7 \sin\left(7\omega t - \frac{2\pi}{3} - \frac{\pi}{6}\right) - \ldots \\ i_c(t) = \sqrt{2}I_1 \sin\left(\omega t + \frac{2\pi}{3} - \frac{\pi}{6}\right) - \sqrt{2}I_5 \sin\left(5\omega t - \frac{2\pi}{3} - \frac{\pi}{6}\right) \\ \qquad + \sqrt{2}I_7 \sin\left(7\omega t + \frac{2\pi}{3} - \frac{\pi}{6}\right) - \ldots \end{cases} \quad (3.41)$$

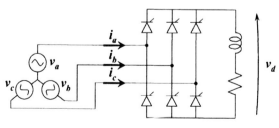

firing angle: 30°

Figure 3-5. Three-phase voltage source supplying a thyristor rectifier.

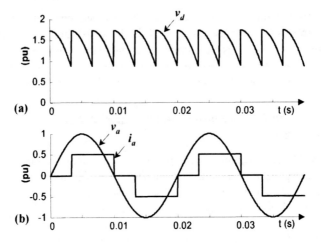

Figure 3-6. (a) Rectifier dc output voltage v_d and (b) a-phase voltage and current waveforms.

Figures 3-7(a) and (b) show the real power p and the imaginary power q, respectively. The real power was calculated using the voltages and currents at the ac side, and it is the same as if it would be calculated by the product of the dc voltage v_d and the dc current. The instantaneous input power at the ac side of the rectifier is equal to the output power at the dc side if there are no losses in the rectifier. The imaginary power is only defined for ac multiphase circuits, so it can be only calculated at the ac side.

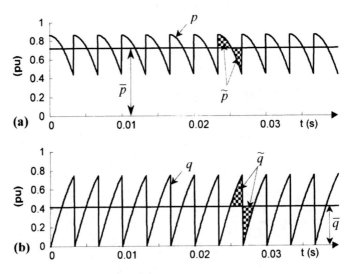

Figure 3-7. (a) Real power p and (b) imaginary power q.

These two powers have constant values and a superposition of oscillating components. Therefore, it is interesting to separate p and q into two parts:

Real power: $\quad\quad\quad\quad p = \bar{p} + \tilde{p}$

Imaginary power: $\quad q = \bar{q} + \tilde{q}$ (3.42)

$\quad\quad\quad\quad\quad\quad\quad\quad$ Average $\quad$ Oscillating
$\quad\quad\quad\quad\quad\quad\quad\quad$ powers $\quad\quad$ powers

where $\bar{p}$ and $\tilde{p}$ represent the average and oscillating parts of p, whereas $\bar{q}$ and $\tilde{q}$ represent the average and oscillating parts of q.

The real power p represents the total (three-phase) energy flow per time unity in the circuit. The average value $\bar{p}$ represents the energy flowing per time unity in one direction only. If $\bar{p}$ and $\bar{q}$ are calculated in terms of the abc components from (3.36) and (3.41), they result in

$$\begin{cases} \bar{p} = 3VI_1 \cos\dfrac{\pi}{6} \\ \bar{q} = 3VI_1 \sin\dfrac{\pi}{6} \end{cases} \quad (3.43)$$

Again, the average values of the real and imaginary power given by the p-q Theory agree with the conventional definition of three-phase active and reactive powers given in Chapter 2 (Budeanu approach). The oscillating part $\tilde{p}$ represents the oscillating energy flow per time unity, which naturally produces a zero average value, representing an amount of additional power flow in the system without effective contribution to the energy transfer from the source to the load or from the load to the source.

The imaginary power q gives the magnitude of the instantaneous reactive powers $p_{\alpha q}$ and $p_{\beta q}$, or the magnitude of the corresponding powers in the abc system. As explained before, although the powers $p_{\alpha q}$ and $p_{\beta q}$ exist in each axis, their sum is zero all the time. The average value of the imaginary power $\bar{q}$ corresponds to the conventional three-phase reactive power and does not contribute to energy transfer. The oscillating component of the imaginary power $\tilde{q}$ corresponds also to a power that is being exchanged among the three phases, without transferring any energy between source and load. In the present example, both oscillating real ($\tilde{p}$) and imaginary ($\tilde{q}$) powers are related to the presence of harmonics exclusively in the load currents. Later, general equations for $\bar{p}$, $\tilde{p}$, $\bar{q}$, and $\tilde{q}$ including harmonic voltage and current simultaneously will be presented.

Figure 3-8 shows the a-phase current component responsible for producing the average real power $\bar{p}$ and the average imaginary power $\bar{q}$. For reference, the voltages in this same phase are plotted together. These currents were calculated using (3.23) and (3.25), for $p = \bar{p}$, and (3.24) and (3.26), for $q = \bar{q}$, followed by the $\alpha\beta$ to abc transformation. The current $i_{a\bar{p}}$ is perfectly in phase with the voltage v_a, whereas $i_{a\bar{q}}$ is delayed exactly by $90°$, acting like an inductive load, as is well known for a line-commutated thyristor rectifier.

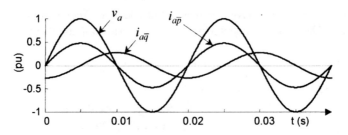

Figure 3-8. Voltage in *a* phase; currents $i_{a\tilde{p}}$ and $i_{a\tilde{q}}$.

Figure 3-9 (a), (b), and (c) show the *a*-phase current $i_{a\tilde{p}}$ responsible for producing the oscillating power $\tilde{p}$, the current $i_{a\tilde{q}}$ that produces the oscillating power $\tilde{q}$, and the sum of $i_{a\tilde{p}}$ and $i_{a\tilde{q}}$, respectively. These currents were calculated, as in the previous case, using (3.23) and (3.25) for $p = \tilde{p}$, and (3.24) and (3.26) for $q = \tilde{q}$, followed by the $\alpha\beta$ to *abc* transformation. The sum of the four components, $i_{a\bar{p}} + i_{a\bar{q}} + i_{a\tilde{p}} + i_{a\tilde{q}}$, is equal to the original square wave current shown in Fig. 3-6 (b).

The *p-q* Theory has the prominent merit of allowing complete analysis and real-time calculation of various powers and respective currents involved in a three-phase circuit. However, this is not the main point. Knowing in real time the values of undesirable currents in a circuit allow us to eliminate them. For instance, if the oscillating powers are undesirable, by compensating the currents $i_{a\tilde{p}}$ and $i_{a\tilde{q}}$ of the load and their correspondent currents in phases *b* and *c* the compensated current drawn from the network would become sinusoidal. It can be easily shown that $i_a - (i_{a\tilde{p}} + i_{a\tilde{q}})$ produces a purely sinusoidal waveform. This is one of the basic ideas of active filtering that will be presented in detail in the next chapter.

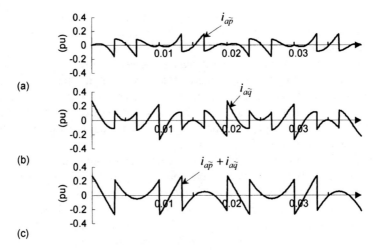

Figure 3-9. Currents $i_{a\tilde{p}}$, $i_{a\tilde{q}}$, and the sum $(i_{a\tilde{p}} + i_{a\tilde{q}})$.

3.2.2. Use of the p-q Theory for Shunt Current Compensation

One important application of the *p-q* Theory is the compensation of undesirable currents. Various examples of this kind of compensation will be presented in the next chapters. Figure 3-10 illustrates the basic idea of shunt current compensation. It shows a source (power generating system) supplying a nonlinear load that is being compensated by a shunt compensator. A kind of shunt compensator is the active filter that will be presented in detail in the next chapter. For the sake of simplicity, it is assumed that the shunt compensator behaves as a three-phase, controlled current source that can draw any set of arbitrarily chosen current references i_{Ca}^*, i_{Cb}^*, and i_{Cc}^*.

Figure 3-11 shows a general control method to be used in the controller of a shunt compensator. The calculated real power *p* of the load can be separated into its average ($\bar{p}$) and oscillating ($\tilde{p}$) parts. Likewise, the load imaginary power *q* can be separated into its average ($\bar{q}$) and oscillating ($\tilde{q}$) parts. Then, undesired portions of the real and imaginary powers of the load that should be compensated are selected. The powers to be compensated are represented by $-p_c^*$ and $-q_c^*$ in the controller shown in Fig. 3-11. The reason for including minus signals in the compensating powers is to emphasize that the compensator should draw a compensating current that produces exactly the inverse of the undesirable powers drawn by the nonlinear load. Note that the adopted current convention in Fig. 3-10 is such that the compensated current, that is, the source current, is the sum of the load current and the compensating current. Then, the inverse transformation from $\alpha\beta$ to *abc* is applied to calculate the instantaneous values of the three-phase compensating current references i_{Ca}^*, i_{Cb}^*, and i_{Cc}^*.

Figures 3-12 to 3-16 show various possible compensation results obtained from Fig. 3-10, considering the same nonlinear load as in Fig. 3-5. Each figure shows the current that should be eliminated from the load current, the compensated source

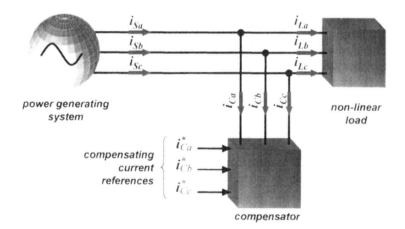

Figure 3-10. Basic principle of shunt current compensation.

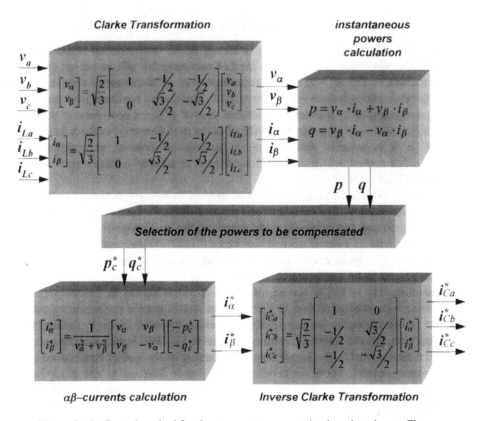

Figure 3-11. Control method for shunt current compensation based on the *p-q* Theory.

current, and p_S and q_S at the source side, along with the *a*-phase voltage. Note that the compensator must reproduce the inverse of that current to be eliminated, which is not be shown in the mentioned figures. The ideal compensated current can be calculated simply by *subtracting* the eliminated current from the load current. This result is the same if the control method shown in Fig. 3-11 is used to generate the compensating current that produces the *inverse* of the powers to be eliminated ($-p_c^*$ and $-q_c^*$). Then, this compensating current is *summed* to the load current. The powers p_S and q_S correspond to the new powers delivered from the source after compensation.

Figure 3-12 shows the compensation of the load imaginary power *q*. In this case, all current components that do not transfer energy, although they may produce losses in the network, are eliminated. The real power p_S, produced by the compensated current, is equal to that produced by the load current, whereas the imaginary power q_S is zero in the source. One interesting point is that the compensator acting as a controlled current sink draws a compensating current that depends only on *q*, so that no energy is flowing out or into this compensator. This means that, in principle, this compensator does not need any power source or energy storage system to realize

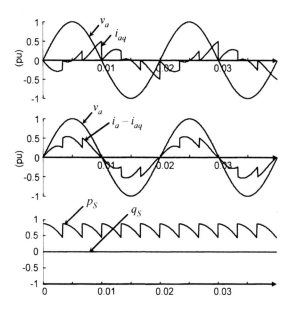

Figure 3-12. Eliminated current i_{aq}, compensated current $i_a - i_{aq}$, and the real and imaginary powers produced by the compensated current.

the compensation of q. Since the compensator is drawing currents that correspond to the load imaginary power q, including all the harmonics related to $q = \bar{q} + \tilde{q}$, this compensator must have, ideally, an infinite frequency response.

Figure 3-13 shows the same compensation as that in the previous case, except for using only the average imaginary power $\bar{q}$. In this case, the compensating current $i_{a\bar{q}}$ has no harmonic components and, therefore, the compensator draws sinusoidal currents at the line frequency. As expected, the real power p_S at the source side is equal to the real power p of the load. The imaginary power q_S has only an oscillating part as the average value $(\bar{q})$ of the load imaginary power is being compensated.

Figure 3-14 shows the case of compensation only for $\tilde{p}$. This may not be not a common situation, although it is very interesting from a theoretical point of view. Taking the current related to this power as the compensating current reference to the compensator makes constant (without ripple) the three-phase instantaneous real power that is equal to the calculated real power. The imaginary power is the same as that of the load. If the primary power source consists of a turbine generator, this kind of power compensation eliminates the torque ripple in the rotor axis, thus resulting in producing no undesirable shaft vibration. It is interesting to note that although the compensated current contains significant harmonic content, these harmonics do not influence the real power. The compensator has to supply exactly the oscillating real power $\tilde{p}$ that is being eliminated from the power source. Therefore, it must have the capability to supply and absorb energy, but with zero average value. Hence, this compensator must be coupled with an energy storage system.

62 THE INSTANTANEOUS POWER THEORY

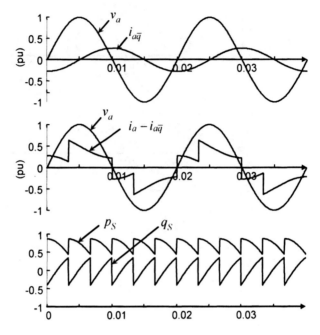

Figure 3-13. Eliminated current $i_{a\bar{q}}$, compensated current $i_a - i_{a\bar{q}}$, and the real and imaginary powers produced by the compensated current.

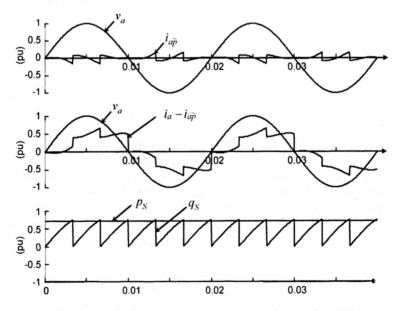

Figure 3-14. Eliminated current $i_{a\tilde{p}}$, compensated current $i_a - i_{a\tilde{p}}$, and the real and imaginary powers produced by the compensated current.

Figure 3-15 shows the case of compensation for $\tilde{p}$ and $\tilde{q}$. This means that the currents shown in Fig. 3-9 are eliminated. Now, the source current becomes purely sinusoidal. This kind of compensation is applicable when harmonic elimination is the most important issue. Although the resulting real and imaginary powers have no ripple, the current related to the average imaginary power $\bar{q}$ is still flowing out of the network, making the power factor lower than unity. In terms of conventional concepts of powers, the reactive current is not being compensated.

Figure 3-16 shows the case of compensation for $\tilde{p}$ and q. This means that all the undesirable current components of the load are being eliminated. The compensated current is sinusoidal, produces a constant real power, and does not generate any imaginary power. The nonlinear load and the compensator form an ideal, linear, purely resistive load. The source current has a minimum rms value that transfers the same energy as the original load current that produce the average real power $\bar{p}$. This is the best compensation that can be made from the power-flow point of view, because it smoothes the power drawn from the generator system. Besides, it eliminates all the harmonic currents. However, it should be pointed out that this is a particular situation in which no unbalances or distortions are present in the system voltages. Cases involving three-phase unbalanced and distorted voltages will be studied later in this chapter. It will be clear that, under nonsinusoidal system voltages, it is impossible to guarantee *simultaneously* constant real power and sinusoidal currents drawn from the network.

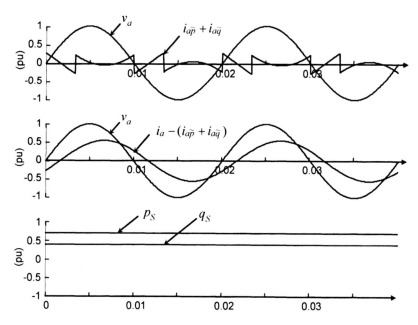

Figure 3-15. Eliminated current $i_{a\tilde{p}} + i_{a\tilde{q}}$, compensated current $i_a - (i_{a\tilde{p}} + i_{a\tilde{q}})$, and the real and imaginary powers produced by the compensated current.

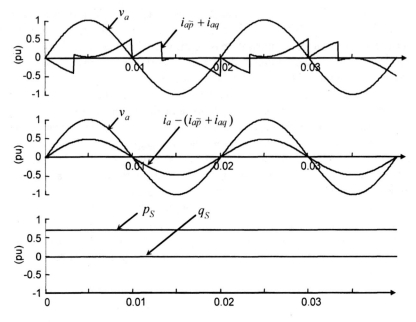

Figure 3-16. Eliminated current $i_{a\tilde{p}} + i_{aq}$, compensated current $i_a - (i_{a\tilde{p}} + i_{aq})$, and the real and imaginary powers produced by the compensated current.

3.2.2.A. Examples of Appearance of Hidden Currents

As will be shown in detail in the next chapter, there are many works showing that the *p-q* Theory is very precise for the calculation of compensating currents for shunt active filters. The compensation may be perfect if the power converter used to synthesize the compensating currents is able to generate them with high fidelity. However, one interesting phenomenon occurs when the percentage of oscillating real power $\tilde{p}$ being compensated is different from that of the oscillating imaginary power $\tilde{q}$. This phenomenon is maximized in the case when only $\tilde{p}$ or only $\tilde{q}$ is compensated. Although it was not highlighted, this situation was already verified in Fig. 3-12 and in Fig. 3-14, where the harmonic spectra of the compensated current (source current) are different from that of the original load current. In fact, when an unequal percentage of compensation of oscillating powers is used, the filtering algorithm may introduce some harmonics that are not present in the original load currents. Although some authors criticize the *p-q* Theory for this point, this phenomenon is not really a problem, and depends on what one wants to have as a result of the compensation. To simplify the analysis, a circuit with only the fifth harmonic current of the negative-sequence component will be analyzed. Then, a seventh harmonic component of a positive-sequence type is analyzed.

3.2.2.A.1. Presence of the Fifth Harmonic in Load Current.
Here, it will be assumed that the three-phase balanced voltage consists only of the fundamental

positive-sequence voltage, as given in (3.36). Besides its fundamental component, the nonlinear load current generally contains a large harmonic spectrum. However, for this analysis, only the fifth-order harmonic component will be considered. The phase angle displacements are chosen such that this current has only a negative-sequence component given by

$$\begin{cases} i_{a5}(t) = \sqrt{2}I_{-5}\sin(5\omega t + \delta_{-5}) \\ i_{b5}(t) = \sqrt{2}I_{-5}\sin\left(5\omega t + \delta_{-5} + \dfrac{2\pi}{3}\right) \\ i_{c5}(t) = \sqrt{2}I_{-5}\sin\left(5\omega t + \delta_{-5} - \dfrac{2\pi}{3}\right) \end{cases} \qquad (3.44)$$

Subscript "–" indicates a negative-sequence component and subscript "5" indicates a fifth-order harmonic frequency. The Clarke transformation of the currents in (3.44) results in

$$\begin{cases} i_{\alpha 5} = \sqrt{3}I_{-5}\sin(5\omega t + \delta_{-5}) \\ i_{\beta 5} = \sqrt{3}I_{-5}\cos(5\omega t + \delta_{-5}) \end{cases} \qquad (3.45)$$

The $\alpha\beta$ transformation of the voltage results in (3.37). From (3.37) and (3.45), the real and imaginary powers are calculated as the following oscillating components:

$$\begin{aligned} \tilde{p} &= -3V_{+1}I_{-5}\cos(6\omega t + \delta_{-5}) \\ \tilde{q} &= -3V_{+1}I_{-5}\sin(6\omega t + \delta_{-5}) \end{aligned} \qquad (3.46)$$

Note that the real and imaginary powers have only an oscillating component at six times the line frequency. Now, from these $\tilde{p}$ and $\tilde{q}$ components and using (3.23) to (3.26), the instantaneous active and reactive currents on the $\alpha\beta$ axes are calculated as follows:

$$i_{\alpha p5} = \frac{v_\alpha}{v_\alpha^2 + v_\beta^2}\tilde{p} = \frac{\sqrt{3}}{2}I_{-5}\sin(5\omega t + \delta_{-5}) - \frac{\sqrt{3}}{2}I_{-5}\sin(7\omega t + \delta_{-5}) \quad (3.47)$$

$$i_{\alpha q5} = \frac{v_\beta}{v_\alpha^2 + v_\beta^2}\tilde{q} = \frac{\sqrt{3}}{2}I_{-5}\sin(5\omega t + \delta_{-5}) + \frac{\sqrt{3}}{2}I_{-5}\sin(7\omega t + \delta_{-5}) \quad (3.48)$$

$$i_{\beta p5} = \frac{v_\beta}{v_\alpha^2 + v_\beta^2}\tilde{p} = \frac{\sqrt{3}}{2}I_{-5}\cos(5\omega t + \delta_{-5}) + \frac{\sqrt{3}}{2}I_{-5}\cos(7\omega t + \delta_{-5}) \quad (3.49)$$

$$i_{\beta q5} = \frac{-v_\alpha}{v_\alpha^2 + v_\beta^2}\tilde{q} = \frac{\sqrt{3}}{2}I_{-5}\cos(5\omega t + \delta_{-5}) - \frac{\sqrt{3}}{2}I_{-5}\cos(7\omega t + \delta_{-5}) \quad (3.50)$$

where

$$\begin{cases} i_{\alpha 5} = i_{\alpha p5} + i_{\alpha q5} \\ i_{\beta 5} = i_{\beta p5} + i_{\beta q5} \end{cases} \quad (3.51)$$

Observing the above equations makes it possible to note two interesting facts:

1. $i_{\alpha p5}$ and $i_{\alpha q5}$, as well as $i_{\beta p5}$ and $i_{\beta q5}$ contain seventh-harmonic components that were not present in the original currents.
2. In the α axis, the seventh harmonic in $i_{\alpha p5}$ is equal to the seventh harmonic in $i_{\alpha q5}$, but with the opposite signal. Therefore, they normally sum zero and do not appear in the circuit. The same is valid for the β-axis current components $i_{\beta p5}$ and $i_{\beta q5}$.

If p-q Theory is used to compensate for the currents that are dependent on $\tilde{p}$ and $\tilde{q}$, it is possible to define compensating currents using gains k_p and k_q for the above current components given in (3.51). Thus, the compensating currents would be given by

$$\begin{cases} i_{C\alpha 5} = k_p \cdot i_{\alpha p5} + k_q \cdot i_{\alpha q5} \\ i_{C\beta 5} = k_p \cdot i_{\beta p5} + k_q \cdot i_{\beta q5} \end{cases} \quad (3.52)$$

In this case, the source current would be

$$\begin{cases} i_{S\alpha 5} = i_{\alpha 5} - i_{C\alpha 5} \\ i_{S\beta 5} = i_{\beta 5} - i_{C\beta 5} \end{cases} \quad (3.53)$$

If $k_p = k_q$, the seventh-harmonic component is totally eliminated in the source current. However, if $k_p \neq k_q$, the seventh-harmonic component in $i_{\alpha p5}$ does not cancel the seventh-harmonic component in $i_{\alpha q5}$. Therefore, a harmonic component that was not present in the original currents is introduced in the source currents. For this reason, this seventh-harmonic component this type of current component will be called "hidden currents."

Part of the fifth-harmonic current component produces oscillating real power $\tilde{p}$ and is responsible for the oscillating energy flowing in a three-phase circuit. The other part of the fifth-harmonic current component does not transport energy at all, because it produces an oscillating imaginary power $\tilde{q}$.

In the filtering process, the worst situation occurs when the p-q Theory is used to compensate only for oscillating imaginary power $\tilde{q}$, or only for the oscillating real power $\tilde{p}$. In these cases, the seventh-harmonic component (the hidden current) in $i_{\alpha p5}$ and $i_{\beta p5}$ will not be cancelled by the seventh-harmonic component in $i_{\alpha q5}$ and $i_{\beta q5}$, respectively. Therefore, the hidden current will appear with maximum magnitude. It is important to note that a fifth-harmonic negative-sequence component is considered and, in this case, the hidden-current component is at a higher (seventh harmonic) frequency.

When only $\tilde{q}$ is used to filter the current, the source current after compensation will contain the hidden-current component. In principle, it is not possible to say that this is a bad or good thing. What can be said truly is that the imaginary power in the source is zero. In other words, the source currents after compensation consist only of components that contribute to the energy flow between the source and the load.

When only $\tilde{p}$ is compensated, the source current will also contain hidden currents, as in the previous case. However, all oscillating real power in the source is eliminated. Therefore, if the objective is to eliminate the oscillating energy flow in the circuit, this is the solution. This compensation technique may be interesting when dealing with motor drives or specific generation systems. In fact, this procedure is important when torque ripple in a motor or generator has to be eliminated.

On the other hand, if the objective of the filter is to eliminate partially or all of the fifth harmonic without introducing any hidden current, the oscillating real power and the oscillating imaginary power must be compensated with $k_p = k_q$. This is the case using a passive filter that has no capability to filter only the active or reactive portion of a harmonic current.

It is clear from the above analysis that the *p-q* Theory intended for current compensation brings more flexibility to the filter design.

3.2.2.A.2. Presence of the Seventh Harmonic in Load Current.

Next, a seventh-harmonic positive-sequence current component is analyzed. These currents are given by

$$\begin{cases} i_{a7}(t) = \sqrt{2}I_{+7} \sin(7\omega t + \delta_{+7}) \\ i_{b7}(t) = \sqrt{2}I_{+7} \sin\left(7\omega t + \delta_{+7} - \frac{2\pi}{3}\right) \\ i_{c7}(t) = \sqrt{2}I_{+7} \sin\left(7\omega t + \delta_{+7} + \frac{2\pi}{3}\right) \end{cases} \quad (3.54)$$

The Clarke transformation of the currents in (3.54) is given by

$$\begin{cases} i_{\alpha 7} = \sqrt{3}I_{+7} \sin(7\omega t + \delta_{+7}) \\ i_{\beta 7} = -\sqrt{3}I_{+7} \cos(7\omega t + \delta_{+7}) \end{cases} \quad (3.55)$$

Again, the $\alpha\beta$ transformation of the voltage is given in (3.37), and the real and imaginary powers are calculated as

$$\begin{aligned} \tilde{p} &= 3V_{+1}I_{+7} \cos(6\omega t + \delta_{+7}) \\ \tilde{q} &= -3V_{+1}I_{+7} \sin(6\omega t + \delta_{+7}) \end{aligned} \quad (3.56)$$

Note that the real and imaginary powers have only oscillating components at six times the system frequency. From these $\tilde{p}$ and $\tilde{q}$ components and using (3.23) to (3.26), the instantaneous active and reactive currents on the $\alpha\beta$ axes can be calculated:

$$i_{\alpha p7} = \frac{v_\alpha}{v_\alpha^2 + v_\beta^2}\tilde{p} = -\frac{\sqrt{3}}{2}I_{+7}\sin(5\omega t + \delta_{+7}) + \frac{\sqrt{3}}{2}I_{+7}\sin(7\omega t + \delta_{+7}) \quad (3.57)$$

$$i_{\alpha q7} = \frac{v_\beta}{v_\alpha^2 + v_\beta^2}\tilde{q} = \frac{\sqrt{3}}{2}I_{+7}\sin(5\omega t + \delta_{+7}) + \frac{\sqrt{3}}{2}I_{+7}\sin(7\omega t + \delta_{+7}) \quad (3.58)$$

$$i_{\beta p7} = \frac{v_\beta}{v_\alpha^2 + v_\beta^2}\tilde{p} = -\frac{\sqrt{3}}{2}I_{+7}\cos(5\omega t + \delta_{+7}) - \frac{\sqrt{3}}{2}I_{+7}\cos(7\omega t + \delta_{+7}) \quad (3.59)$$

$$i_{\beta q7} = \frac{-v_\alpha}{v_\alpha^2 + v_\beta^2}\tilde{q} = \frac{\sqrt{3}}{2}I_{+7}\cos(5\omega t + \delta_{+7}) - \frac{\sqrt{3}}{2}I_{+7}\cos(7\omega t + \delta_{+7}) \quad (3.60)$$

where

$$\begin{cases} i_{\alpha 7} = i_{\alpha p7} + i_{\alpha q7} \\ i_{\beta 7} = i_{\beta p7} + i_{\beta q7} \end{cases} \quad (3.61)$$

The above equations show that there are also hidden currents associated with positive-sequence harmonic currents. They have similar properties to those in the case of the previous fifth-order harmonic negative-sequence currents. The difference is that the frequency of the hidden currents is lower than the frequency of the original, positive-sequence harmonic current. All conclusions made for the case of the fifth-order harmonic, shown in (3.44) to (3.53), are valid for the seventh-order harmonic, shown in (3.54) to (3.61).

3.2.3. The Dual p-q Theory

The original p-q Theory was defined with the most common case of three-phase systems comprising only voltage sources in mind. However, it may be interesting to also present its dual theory that would be suitable for the cases in which three-phase current sources are present, or in which it is desirable to perform *series voltage* compensation instead of *shunt current* compensation.

In the previous section, current components were calculated as function of the $\alpha\beta$ voltages, the real power, and the imaginary power. Those equations are suitable for applications to the control method of shunt current compensation. In the *dual p-q Theory*, it is assumed that the currents, and the real and imaginary powers are known, and the voltage components should be calculated or compensated. One possible application of this dual p-q Theory is to the case of series voltage compensation that is the dual of shunt current compensation.

3.2. THE p-q THEORY IN THREE-PHASE, THREE-WIRE SYSTEMS

For simplicity, only the case of a three-phase, three-wire system will be analyzed. Therefore, no zero-sequence voltage and current components are present. For this condition, the following equation can be derived from (3.19):

$$\begin{bmatrix} p \\ q \end{bmatrix} = \begin{bmatrix} i_\alpha & i_\beta \\ -i_\beta & i_\alpha \end{bmatrix} \begin{bmatrix} v_\alpha \\ v_\beta \end{bmatrix} \quad (3.62)$$

Considering that the real and imaginary powers, as well as the currents, are known, the voltages can be calculated as function of these variables. Multiplying both sides of (3.62) by the inverse matrix of currents, the voltages are determined as functions of currents and powers by

$$\begin{bmatrix} v_\alpha \\ v_\beta \end{bmatrix} = \frac{1}{i_\alpha^2 + i_\beta^2} \begin{bmatrix} i_\alpha & -i_\beta \\ i_\beta & i_\alpha \end{bmatrix} \begin{bmatrix} p \\ q \end{bmatrix} \quad (3.63)$$

The right-hand side of (3.63) can be decomposed as

$$\begin{bmatrix} v_\alpha \\ v_\beta \end{bmatrix} = \frac{1}{i_\alpha^2 + i_\beta^2} \begin{bmatrix} i_\alpha & -i_\beta \\ i_\beta & i_\alpha \end{bmatrix} \begin{bmatrix} p \\ 0 \end{bmatrix} + \frac{1}{i_\alpha^2 + i_\beta^2} \begin{bmatrix} i_\alpha & -i_\beta \\ i_\beta & i_\alpha \end{bmatrix} \begin{bmatrix} 0 \\ q \end{bmatrix} \quad (3.64)$$

From (3.64), the following voltage components can be defined:

- Instantaneous active voltage on the α axis $v_{\alpha p}$

$$v_{\alpha p} = \frac{i_\alpha}{i_\alpha^2 + i_\beta^2} p \quad (3.65)$$

- Instantaneous reactive voltage on the α axis $v_{\alpha q}$

$$v_{\alpha q} = \frac{-i_\beta}{i_\alpha^2 + i_\beta^2} q \quad (3.66)$$

- Instantaneous active voltage on the β axis $v_{\beta p}$

$$v_{\beta p} = \frac{i_\beta}{i_\alpha^2 + i_\beta^2} p \quad (3.67)$$

- Instantaneous reactive voltage on the β axis $v_{\beta q}$

$$v_{\beta q} = \frac{i_\alpha}{i_\alpha^2 + i_\beta^2} q \quad (3.68)$$

The following equation is valid:

$$\begin{bmatrix} v_\alpha \\ v_\beta \end{bmatrix} = \begin{bmatrix} v_{\alpha p} \\ v_{\beta p} \end{bmatrix} + \begin{bmatrix} v_{\alpha q} \\ v_{\beta q} \end{bmatrix} \quad (3.69)$$

The above equations can be applied when the load or the source can be modeled as a current source.

The voltage components on the $\alpha\beta$ reference frames are derived from (3.63), whereas the current components are derived from (3.21). The voltage components defined in (3.65) to (3.68) correspond to the dual of those current components defined in (3.23) to (3.26), in the *p-q* Theory. Hence, all physical meanings associated with those current components are valid here for their dual voltage components. Furthermore, all examples of separation of load *current* components are applicable here, but now to load *voltage* components.

In Section 3.2.2, the basic principle of shunt current compensation was introduced and illustrated in Fig. 3-10. Now, a complementary principle of *series voltage* compensation is derived. Figure 3-17 illustrates this dual principle of compensation.

The shunt compensator draws a current to compensate for undesirable power components produced by the load current. The ideal series compensator of Fig. 3-17 behaves as a controlled voltage source to compensate for undesirable power components produced by the load voltage. This dual principle can determine the compensating voltage v_C^* directly from the current and the power portions to be compensated.

A general control method for calculating the compensating voltage v_C^* is illustrated in Fig. 3-18. It is the dual of that compensation method shown in Fig. 3-11 for shunt current compensation. The phase voltages at the load terminal and the line currents are measured and transformed into the $\alpha\beta$ reference frames. Then, the real and imaginary powers of the load are calculated, and the undesirable power portions are selected. From these power portions of the load powers and the line currents, the compensating voltages are calculated and inserted "instantaneously" in the power system by the series compensator. Hence, the compensated voltages v_{Sa},

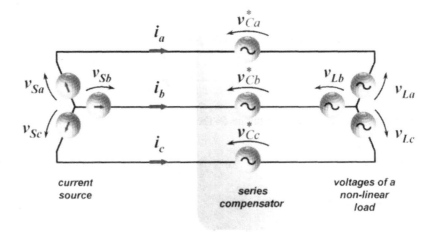

Figure 3-17. Basic principle of series voltage compensation.

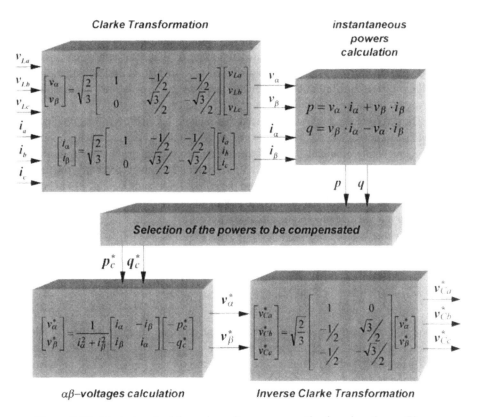

Figure 3-18. Control method for series voltage compensation based on the *p-q* Theory.

v_{Sb}, and v_{Sc} do not produce any undesirable power portions with the line currents at the load terminal.

One could prefer to design a compensator not for directly compensating portions of power but rather to guarantee, for instance, sinusoidal compensated voltage. This point and other compensation features will be explored in the next chapters. In practical cases, it is difficult to implement a control algorithm using the compensation method shown in Fig. 3-18. If this control algorithm is implemented in a controller for the series compensator, the $\alpha\beta$ voltage calculation block may realize a division by zero under noload conditions. However, this is the complement of a short-circuit situation in the case of a shunt active filter (Fig. 3-11), where the $\alpha\beta$-current calculation block would realize a division by zero.

3.3. THE *p-q* THEORY IN THREE-PHASE, FOUR-WIRE SYSTEMS

The previous section presented the *p-q* Theory for three-phase, three-wire systems. The physical meaning of the instantaneous real and imaginary power was discussed and clarified with examples. However, the presence of a fourth conductor, namely the neutral conductor, is very common in low-voltage distribution systems, in addition to the cases of grounded transmission systems. These systems are classified as three-phase, four-wire systems. The simplified transformation and equations used in the previous section are not applicable to these cases. This section presents cases involving the three instantaneous powers p, q, and p_0 of the p-q Theory.

The three-phase, four-wire systems can include both zero-sequence voltage and current as a generic case. These systems allow all the three line currents i_a, i_b, and i_c to be independent, whereas two of the three line currents are independent in three-phase, three-wire systems. Therefore, to represent the system correctly, the instantaneous zero-sequence power p_0 defined on the $\alpha\beta 0$ reference frames, p_0 has to be introduced as the third instantaneous power in addition to the instantaneous real power p and the instantaneous imaginary power q.

Mathematically, they are defined as:

$$\begin{bmatrix} p_0 \\ p \\ q \end{bmatrix} = \begin{bmatrix} v_0 & 0 & 0 \\ 0 & v_\alpha & v_\beta \\ 0 & v_\beta & -v_\alpha \end{bmatrix} \begin{bmatrix} i_0 \\ i_\alpha \\ i_\beta \end{bmatrix} \quad (3.70)$$

The real and imaginary powers have the same physical meaning as before. The difference is the additional definition of the zero-sequence power. Before explaining it, the three-phase instantaneous active power shown in (3.17) should be re-written in terms of the $\alpha\beta 0$ components:

$$p_{3\phi} = v_a i_a + v_b i_b + v_c i_c = v_\alpha i_\alpha + v_\beta i_\beta + v_0 i_0 = p + p_0 \quad (3.71)$$

This equation shows that the three-phase instantaneous active power $p_{3\phi}$ is equal to the sum of the real power p and the zero-sequence power p_0. In the case of a three-phase, three-wire circuit, the power p_0 does not exist, and so $p_{3\phi}$ is equal to p.

3.3.1. The Zero-Sequence Power in a Three-Phase Sinusoidal Voltage Source

To understand the nature of the zero-sequence power, it is considered that a three-phase sinusoidal voltage source consists of positive- and zero-sequence voltages at the angular frequency ω. The symmetrical components of this voltage source were calculated based on their phasors (see Fig. 2-2). Although this analysis is valid only for the steady-state condition, it is very elucidating. These symmetrical components were then transformed back to the time domain and rewritten as time functions (this procedure is better explained in the next section) as

$$\begin{aligned} v_a &= \sqrt{2} V_+ \sin(\omega t + \phi_{v+}) + \sqrt{2} V_0 \sin(\omega t + \phi_{v0}) \\ v_b &= \sqrt{2} V_+ \sin(\omega t - 2\pi/3 + \phi_{v+}) + \sqrt{2} V_0 \sin(\omega t + \phi_{v0}) \\ v_c &= \sqrt{2} V_+ \sin(\omega t + 2\pi/3 + \phi_{v+}) + \sqrt{2} V_0 \sin(\omega t + \phi_{v0}) \end{aligned} \quad (3.72)$$

3.3. THE p-q THEORY IN THREE-PHASE, FOUR-WIRE SYSTEMS

It is assumed that the current also has positive- and zero-sequence components, that is,

$$i_a = \sqrt{2}I_+ \sin(\omega t + \phi_{i+}) + \sqrt{2}I_0 \sin(\omega t + \phi_{i0})$$
$$i_b = \sqrt{2}I_+ \sin(\omega t - 2\pi/3 + \phi_{i+}) + \sqrt{2}I_0 \sin(\omega t + \phi_{i0}) \quad (3.73)$$
$$i_c = \sqrt{2}I_+ \sin(\omega t + 2\pi/3 + \phi_{i+}) + \sqrt{2}I_0 \sin(\omega t + \phi_{i0})$$

The subscripts + and 0 are used to define the positive- and zero-sequence components. Applying the Clarke transformation, the following voltages and currents on the $\alpha\beta 0$ reference frames are obtained:

$$v_\alpha = \sqrt{3}V_+ \sin(\omega t + \phi_{v+})$$
$$v_\beta = -\sqrt{3}V_+ \cos(\omega t + \phi_{v+}) \quad (3.74)$$
$$v_0 = \sqrt{6}V_0 \sin(\omega t + \phi_{v0})$$

and

$$i_\alpha = \sqrt{3}I_+ \sin(\omega t + \phi_{i+})$$
$$i_\beta = -\sqrt{3}I_+ \cos(\omega t + \phi_{i+}) \quad (3.75)$$
$$i_0 = \sqrt{6}I_0 \sin(\omega t + \phi_{i0})$$

Since the real and imaginary powers defined by (3.70) depend only on the positive-sequence voltage and current, these powers are similar in content and in meaning to those analyzed in the previous section. Therefore, the analysis here will be focused on the zero-sequence power p_0 that is given by

$$p_0 = 3V_0 I_0 \cos(\phi_{v0} - \phi_{i0}) - 3V_0 I_0 \cos(2\omega t + \phi_{v0} + \phi_{i0}) = \bar{p}_0 + \tilde{p}_0 \quad (3.76)$$

This power has the same characteristics as the instantaneous power in a single-phase circuit. It has an average value and an oscillating component at twice the line frequency. The average value $\bar{p}_0$ represents a unidirectional energy flow. It has the same characteristics as the conventional (average) active power. The oscillating component $\tilde{p}_0$ also transfers energy instantaneously. However, it has an average value equal to zero, because it is oscillating. The analysis shows that, in principle, the average value of the zero-sequence power helps to increase the total energy transfer, and in this sense, it can be considered as a positive point. However, even for the simplest case of a zero-sequence component in the voltage and current, the zero-sequence power p_0 cannot produce constant power $\bar{p}_0$ alone. In other words, p_0 always consists of $\tilde{p}_0$ plus $\bar{p}_0$, if $\cos(\phi_{v0} - \phi_{i0}) \neq 0$. The elimination of $\tilde{p}_0$ is accompanied by the elimination of $\bar{p}_0$ together. This is one interesting characteristic of this power, and this is one of the reasons why it is not welcome in most circuits. In summary, the zero-sequence power p_0 exists only if there are zero-sequence voltage and current. It is an instantaneous *active* power contributing to energy flow, just like in a single-phase circuit. The average zero-sequence power $\bar{p}_0$ is always associated to the oscillating component $\tilde{p}_0$. Therefore, there is no way to eliminate the oscillating component and keep the average part alone.

3.3.2. Presence of Negative-Sequence Components

For a three-phase, balanced, positive-sequence voltage, the presence of the negative-sequence components may be a serious problem. This section will analyze the case in which the voltages are sinusoidal waveforms with a frequency of ω, consisting of positive-, negative-, and zero-sequence components as given below:

$$v_a = \sqrt{2}V_+ \sin(\omega t + \phi_{v+}) + \sqrt{2}V_- \sin(\omega t + \phi_{v-}) + \sqrt{2}V_0 \sin(\omega t + \phi_{v0})$$
$$v_b = \sqrt{2}V_+ \sin(\omega t - 2\pi/3 + \phi_{v+}) + \sqrt{2}V_- \sin(\omega t + 2\pi/3 + \phi_{v-})$$
$$+ \sqrt{2}V_0 \sin(\omega t + \phi_{v0}) \quad (3.77)$$
$$v_c = \sqrt{2}V_+ \sin(\omega t + 2\pi/3 + \phi_{v+}) + \sqrt{2}V_- \sin(\omega t - 2\pi/3 + \phi_{v-})$$
$$+ \sqrt{2}V_0 \sin(\omega t + \phi_{v0})$$

and the currents consist of

$$i_a = \sqrt{2}I_+ \sin(\omega t + \phi_{i+}) + \sqrt{2}I_- \sin(\omega t + \phi_{i-}) + \sqrt{2}I_0 \sin(\omega t + \phi_{i0})$$
$$i_b = \sqrt{2}I_+ \sin(\omega t - 2\pi/3 + \phi_{i+}) + \sqrt{2}I_- \sin(\omega t + 2\pi/3 + \phi_{i-})$$
$$+ \sqrt{2}I_0 \sin(\omega t + \phi_{i0}) \quad (3.78)$$
$$i_c = \sqrt{2}I_+ \sin(\omega t + 2\pi/3 + \phi_{i+}) + \sqrt{2}I_- \sin(\omega t - 2\pi/3 + \phi_{i-})$$
$$+ \sqrt{2}I_0 \sin(\omega t + \phi_{i0})$$

Applying the Clarke transformation yields the following voltages and currents:

$$v_\alpha = \sqrt{3}V_+ \sin(\omega t + \phi_{v+}) + \sqrt{3}V_- \sin(\omega t + \phi_{v-})$$
$$v_\beta = -\sqrt{3}V_+ \cos(\omega t + \phi_{v+}) + \sqrt{3}V_- \cos(\omega t + \phi_{v-}) \quad (3.79)$$
$$v_0 = \sqrt{6}V_0 \sin(\omega t + \phi_{v0})$$

and

$$i_\alpha = \sqrt{3}I_+ \sin(\omega t + \phi_{i+}) + \sqrt{3}I_- \sin(\omega t + \phi_{i-})$$
$$i_\beta = -\sqrt{3}I_+ \cos(\omega t + \phi_{i+}) + \sqrt{3}I_- \cos(\omega t + \phi_{i-}) \quad (3.80)$$
$$i_0 = \sqrt{6}I_0 \sin(\omega t + \phi_{i0})$$

3.3. THE p-q THEORY IN THREE-PHASE, FOUR-WIRE SYSTEMS

The zero-sequence voltage and current are the same as those in the previous case; the zero-sequence power is equal to that in the previous case. Nothing changed in it due to the presence of negative-sequence components, as expected. On the other hand, the real and imaginary powers changed considerably. The powers given below are separated in their average and oscillating components:

$$\begin{cases} \bar{p} = 3V_+I_+ \cos(\phi_{v+} - \phi_{i+}) + 3V_-I_- \cos(\phi_{v-} - \phi_{i-}) \\ \bar{q} = 3V_+I_+ \sin(\phi_{v+} - \phi_{i+}) - 3V_-I_- \sin(\phi_{v-} - \phi_{i-}) \\ \tilde{p} = -3V_+I_- \cos(2\omega t + \phi_{v+} + \phi_{i-}) - 3V_-I_+ \cos(2\omega t + \phi_{v-} + \phi_{i+}) \\ \tilde{q} = -3V_+I_- \sin(2\omega t + \phi_{v+} + \phi_{i-}) + 3V_-I_+ \sin(2\omega t + \phi_{v-} + \phi_{i+}) \end{cases} \quad (3.81)$$

The following conclusions can be written from the above equations for the real and imaginary powers:

1. The positive- and negative-sequence components in voltages and currents may contribute to the average real and imaginary powers.
2. The instantaneous real and imaginary powers contain oscillating components due to the cross product of the positive-sequence voltage and the negative-sequence current, and the negative-sequence voltage and the positive-sequence current. Hence, even circuits without harmonic components may have oscillating real or imaginary powers.

3.3.3. General Case Including Distortions and Imbalances in the Voltages and in the Currents

The three-phase, four-wire system with sinusoidal positive-, negative-, and zero-sequence voltage and current components at the fundamental frequency was analyzed in the previous section. This section will discuss the generalized three-phase, four-wire system including not only those components at the fundamental frequency, but also harmonics components.

General equations relating the instantaneous powers in the p-q Theory and the theory of symmetrical components (also called *the Fortescue components* [36]) are valid in the steady state. The theory of symmetrical components based on phasors is valid only in the steady state. However, these general equations are fundamental in elucidating some important characteristics of the p-q Theory that are valid even during transients. Moreover, the general equations will be useful in understanding control methods for shunt and series active filters and other active power-line conditioners introduced in the following chapters.

For the present analysis, a three-phase, four-wire system considered in the steady state has generic, but periodic, voltages and currents. They may include the fundamental component as well as harmonic components. Further, each three-phase group of phasors in a given frequency may be unbalanced, which means that it may consist of positive-, negative-, and zero-sequence components, according to the symmetrical component theory. Generic periodic voltages and currents can be decomposed into *Fourier series* as

$$v_k(t) = \sum_{n=1}^{\infty} \sqrt{2} V_{kn} \sin(\omega_n t + \phi_{kn}) \qquad k = (a, b, c) \qquad (3.82)$$

$$i_k(t) = \sum_{n=1}^{\infty} \sqrt{2} I_{kn} \sin(\omega_n t + \phi_{kn}) \qquad k = (a, b, c) \qquad (3.83)$$

where n indicates the harmonic order. Equations (3.82) and (3.83) can be written in terms of phasors, including the fundamental ($n = 1$) and harmonic phasors, as follows:

$$\dot{V}_k = \sum_{n=1}^{\infty} V_{kn} \angle \phi_{kn} = \sum_{n=1}^{\infty} \dot{V}_{kn} \qquad k = (a, b, c) \qquad (3.84)$$

$$\dot{I}_k = \sum_{n=1}^{\infty} I_{kn} \angle \delta_{kn} = \sum_{n=1}^{\infty} \dot{I}_{kn} \qquad k = (a, b, c) \qquad (3.85)$$

Then, the symmetrical-components transformation [36] is applied to each a-b-c-harmonic group of phasors of voltages or currents to determine their positive-, negative-, and zero-sequence components, that is,

$$\begin{bmatrix} \dot{V}_{0n} \\ \dot{V}_{+n} \\ \dot{V}_{-n} \end{bmatrix} = \frac{1}{3} \begin{bmatrix} 1 & 1 & 1 \\ 1 & \alpha & \alpha^2 \\ 1 & \alpha^2 & \alpha \end{bmatrix} \begin{bmatrix} \dot{V}_{an} \\ \dot{V}_{bn} \\ \dot{V}_{cn} \end{bmatrix} \qquad (3.86)$$

The subscripts "0," "+," and "−" correspond to the zero-, positive-, and negative-sequence components, respectively. The complex number α in the transformation matrix is the 120° phase-shift operator:

$$\alpha = 1 \angle 120° = e^{j(2\pi/3)} \qquad (3.87)$$

The inverse transformation of (3.86) is given by

$$\begin{bmatrix} \dot{V}_{an} \\ \dot{V}_{bn} \\ \dot{V}_{cn} \end{bmatrix} = \begin{bmatrix} 1 & 1 & 1 \\ 1 & \alpha^2 & \alpha \\ 1 & \alpha & \alpha^2 \end{bmatrix} \begin{bmatrix} \dot{V}_{0n} \\ \dot{V}_{+n} \\ \dot{V}_{-n} \end{bmatrix} \qquad (3.88)$$

Equivalent functions of time can be derived from the phasors given by (3.88). Hence, rewriting the harmonic voltages in terms of symmetrical components in the time domain, yield the following expressions for the nth a-b-c group of harmonic voltages:

3.3. THE p-q THEORY IN THREE-PHASE, FOUR-WIRE SYSTEMS

$$\begin{cases} v_{an}(t) = \sqrt{2}V_{0n}\sin(\omega_n t + \phi_{0n}) + \sqrt{2}V_{+n}\sin(\omega_n t + \phi_{+n}) \\ \qquad + \sqrt{2}V_{-n}\sin(\omega_n t + \phi_{-n}) \\ v_{bn}(t) = \sqrt{2}V_{0n}\sin(\omega_n t + \phi_{0n}) + \sqrt{2}V_{+n}\sin[\omega_n t + \phi_{+n} - (2\pi/3)] \\ \qquad + \sqrt{2}V_{-n}\sin[\omega_n t + \phi_{-n} + (2\pi/3)] \\ v_{cn}(t) = \sqrt{2}V_{0n}\sin(\omega_n t + \phi_{0n}) + \sqrt{2}V_{+n}\sin[\omega_n t + \phi_{+n} + (2\pi/3)] \\ \qquad + \sqrt{2}V_{-n}\sin[\omega_n t + \phi_{-n} - (2\pi/3)] \end{cases} \quad (3.89)$$

Similarly, the instantaneous line currents are found to be

$$\begin{cases} i_{an}(t) = \sqrt{2}I_{0n}\sin(\omega_n t + \delta_{0n}) + \sqrt{2}I_{+n}\sin(\omega_n t + \delta_{+n}) \\ \qquad + \sqrt{2}I_{-n}\sin(\omega_n t + \delta_{-n}) \\ i_{bn}(t) = \sqrt{2}I_{0n}\sin(\omega_n t + \delta_{0n}) + \sqrt{2}I_{+n}\sin[\omega_n t + \delta_{+n} - (2\pi/3)] \\ \qquad + \sqrt{2}I_{-n}\sin[\omega_n t + \delta_{-n} + (2\pi/3)] \\ i_{cn}(t) = \sqrt{2}I_{0n}\sin(\omega_n t + \delta_{0n}) + \sqrt{2}I_{+n}\sin[\omega_n t + \delta_{+n} + (2\pi/3)] \\ \qquad + \sqrt{2}I_{-n}\sin[\omega_n t + \delta_{-n} - (2\pi/3)] \end{cases} \quad (3.90)$$

The above decomposition into symmetrical components allows the analysis of an unbalanced three-phase system as a sum of two balanced three-phase systems plus the zero-sequence component.

The harmonic voltages and currents in terms of symmetrical components, as given in (3.89) and (3.90), can replace the terms in the series given in (3.82) and (3.83), respectively. If the $\alpha\beta 0$ transformation, as defined in (3.1) and (3.3), is applied, the following expressions for the generic voltage and current transformed in the $\alpha\beta 0$-reference frames can be obtained:

$$\begin{cases} v_\alpha = \sum_{n=1}^{\infty} \sqrt{3}V_{+n}\sin(\omega_n t + \phi_{+n}) + \sum_{n=1}^{\infty} \sqrt{3}V_{-n}\sin(\omega_n t + \phi_{-n}) \\ v_\beta = \sum_{n=1}^{\infty} -\sqrt{3}V_{+n}\cos(\omega_n t + \phi_{+n}) + \sum_{n=1}^{\infty} \sqrt{3}V_{-n}\cos(\omega_n t + \phi_{-n}) \\ v_0 = \sum_{n=1}^{\infty} \sqrt{6}V_{0n}\sin(\omega_n t + \phi_{0n}) \end{cases} \quad (3.91)$$

$$\begin{cases} i_\alpha = \sum_{n=1}^{\infty} \sqrt{3}I_{+n}\sin(\omega_n t + \delta_{+n}) + \sum_{n=1}^{\infty} \sqrt{3}I_{-n}\sin(\omega_n t + \delta_{-n}) \\ i_\beta = \sum_{n=1}^{\infty} -\sqrt{3}I_{+n}\cos(\omega_n t + \delta_{+n}) + \sum_{n=1}^{\infty} \sqrt{3}I_{-n}\cos(\omega_n t + \delta_{-n}) \\ i_0 = \sum_{n=1}^{\infty} \sqrt{6}I_{0n}\sin(\omega_n t + \delta_{0n}) \end{cases} \quad (3.92)$$

78 THE INSTANTANEOUS POWER THEORY

It is possible to see that the positive- and negative-sequence components contribute to the α- and β-axis voltages and currents, whereas the 0-axis voltage and current comprise only zero-sequence components.

Further, the real power p, the imaginary power q, and the zero-sequence power p_0, as defined in (3.70), can be calculated by using the generic voltages and currents in terms of symmetrical components given by (3.91) and (3.92).

The relation between the conventional concepts of powers and the new powers defined in the p-q Theory is better visualized if the powers p, q, and p_0 are separated in their average values $\bar{p}$, $\bar{q}$, $\bar{p}_0$, and their oscillating parts $\tilde{p}$, $\tilde{q}$, $\tilde{p}_0$.

$$
\begin{array}{lrcl}
\text{Real power:} & p & = & \bar{p} + \tilde{p} \\
\text{Imaginary power:} & q & = & \bar{q} + \tilde{q} \\
\text{Zero-sequence power:} & p_0 & = & \bar{p}_0 + \tilde{p}_0
\end{array}
\qquad (3.93)
$$

$$\text{Average powers} \quad \text{Oscillating powers}$$

The resulting power expressions are as follows:

$$\bar{p}_0 = \sum_{n=1}^{\infty} 3 V_{0n} I_{0n} \cos(\phi_{0n} - \delta_{0n}) \qquad (3.94)$$

$$\bar{p} = \sum_{n=1}^{\infty} 3 V_{+n} I_{+n} \cos(\phi_{+n} - \delta_{+n}) + \sum_{n=1}^{\infty} 3 V_{-n} I_{-n} \cos(\phi_{-n} - \delta_{-n}) \qquad (3.95)$$

$$\bar{q} = \sum_{n=1}^{\infty} 3 V_{+n} I_{+n} \sin(\phi_{+n} - \delta_{+n}) + \sum_{n=1}^{\infty} -3 V_{-n} I_{-n} \sin(\phi_{-n} - \delta_{-n}) \qquad (3.96)$$

$$\tilde{p}_0 = \left\{ \sum_{\substack{m=1 \\ m \ne n}}^{\infty} \left[\sum_{n=1}^{\infty} 3 V_{0m} I_{0n} \cos((\omega_m - \omega_n)t + \phi_{0m} - \delta_{0m}) \right] + \sum_{m=1}^{\infty} \left[\sum_{n=1}^{\infty} -3 V_{0m} I_{0n} \cos((\omega_m + \omega_n)t + \phi_{0m} + \delta_{0m}) \right] \right\} \qquad (3.97)$$

$$\tilde{p} = \left\{ \sum_{\substack{m=1 \\ m \ne n}}^{\infty} \left[\sum_{n=1}^{\infty} 3 V_{+m} I_{+n} \cos((\omega_m - \omega_n)t + \phi_{+m} - \delta_{+n}) \right] + \right.$$
$$+ \sum_{\substack{m=1 \\ m \ne n}}^{\infty} \left[\sum_{n=1}^{\infty} 3 V_{-m} I_{-n} \cos((\omega_m - \omega_n)t + \phi_{-m} - \delta_{-n}) \right] +$$
$$+ \sum_{m=1}^{\infty} \left[\sum_{n=1}^{\infty} -3 V_{+m} I_{-n} \cos((\omega_m + \omega_n)t + \phi_{+m} + \delta_{-n}) \right] +$$
$$\left. + \sum_{m=1}^{\infty} \left[\sum_{n=1}^{\infty} -3 V_{-m} I_{+n} \cos((\omega_m + \omega_n)t + \phi_{-m} + \delta_{+n}) \right] \right\} \qquad (3.98)$$

$$\tilde{q} = \left\{ \sum_{\substack{m=1 \\ m \neq n}}^{\infty} \left[\sum_{n=1}^{\infty} 3V_{+m}I_{+n} \sin((\omega_m - \omega_n)t + \phi_{+m} - \delta_{+n}) \right] + \right.$$

$$+ \sum_{\substack{m=1 \\ m \neq n}}^{\infty} \left[\sum_{n=1}^{\infty} -3V_{-m}I_{-n} \sin((\omega_m - \omega_n)t + \phi_{-m} - \delta_{-n}) \right] +$$

$$+ \sum_{m=1}^{\infty} \left[\sum_{n=1}^{\infty} -3V_{+m}I_{-n} \sin((\omega_m + \omega_n)t + \phi_{+m} + \delta_{-n}) \right] +$$

$$\left. + \sum_{m=1}^{\infty} \left[\sum_{n=1}^{\infty} 3V_{-m}I_{+n} \sin((\omega_m + \omega_n)t + \phi_{-m} + \delta_{+n}) \right] \right\} \quad (3.99)$$

These generic power expressions elucidate the relations between the conventional and the instantaneous concepts of active and reactive power. For instance, it is possible to see that the well-known three-phase fundamental active power ($P = 3VI \cos \varphi$) is one term of the average real power $\bar{p}$, whereas the three-phase reactive power ($Q = 3VI \sin \varphi$) is included in the average imaginary power $\bar{q}$. All harmonics in voltage and current can contribute to the average powers $\bar{p}$ and $\bar{q}$ if they have the same frequency and have the same sequence component (positive or negative), as shown in (3.95) and (3.96). The presence of more than one harmonic frequency and/or sequence components also produce $\tilde{p}$ and $\tilde{q}$, according to (3.98) and (3.99). On the other hand, the zero-sequence power $p_0 = \bar{p}_0 + \tilde{p}_0$, that is, the sum of (3.94) and (3.97), always has the average part associated with an oscillating part $\tilde{p}_0$. Therefore, if the oscillating part is eliminated by a compensator, this compensator should be able to also deal with the average part $\bar{p}_0$ that may be present. Contrarily, Section 3.2.2 showed briefly that it is always possible to compensate for only $\tilde{p}$ or $\tilde{q}$ and leave $\bar{p}$ or $\bar{q}$ to be supplied by the source. These points will be discussed in detail in the following chapter.

3.3.4. Physical Meanings of the Instantaneous Real, Imaginary, and Zero-Sequence Powers

Before the use of the p-q Theory to develop control circuits for active power-line conditioners for current or voltage compensation, the physical meaning of all the instantaneous powers must be clearly explained. Fig. 3-19 summarizes the concepts involved in these powers.

The following conclusions are similar to all the conclusions obtained so far. However, this time they were obtained for generic voltage and current waveforms.

The instantaneous powers that the p-q Theory defines in the time domain are independent of the rms values of voltages and currents. This theory includes the conventional frequency-domain concepts of active and reactive power defined for three-phase sinusoidal balanced systems as a particular case. Therefore, the p-q Theory in the time domain is not contradictory but complementary to the conventional theories in the frequency domain

The general equations for the real, imaginary, and zero-sequence powers given in (3.94) to (3.99) are the basis for understanding the energy transfer in a three-

80 THE INSTANTANEOUS POWER THEORY

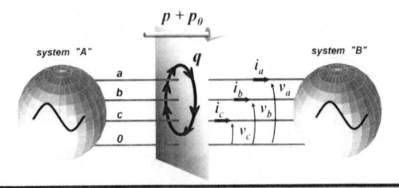

$p + p_0$: instantaneous *total* energy flow per time unit;
q : energy exchanged between the phases without transferring energy.

Figure 3-19. Physical meaning of the instantaneous powers defined in the α-β-0 reference frame.

- Zero-sequence components in the fundamental voltage and current and/or in the harmonics do not contribute to the real power p or to the imaginary power q.
- The total instantaneous energy flow per time unit, that is, the three-phase instantaneous active power, even in a distorted and unbalanced system, is always equal to the sum of the real power and the zero-sequence power ($p_{3\phi} = p + p_0$), and may contain average and oscillating parts.
- The imaginary power q, independent of the presence of harmonic or unbalances, represents the energy quantity that is being exchanged between the phases of the system. This means that the imaginary power does not contribute to energy transfer* between the source and the load at any time.

---

phase system. It is possible to understand how all the voltage and current components achieve energy transfer or induce energy exchange (imaginary power) in a three-phase circuit. Note that these voltages and currents may be at the same or different frequencies and at the same or different sequences, including the fundamental frequency.

3.3.5. Avoiding the Clarke Transformation in the p-q Theory

The p-q Theory in a three-phase, four-wire system defines the real power, the imaginary power, and the zero-sequence power as functions of voltages and currents in the $\alpha\beta0$-reference frames. Some expressions relating these powers in terms of phase mode, that is, the abc variables, were presented in the previous sections, and the two most useful ones are rewritten below:

---

*The term "energy transfer" is used here in a general manner, referring not only to the energy delivered to the load, but also to the energy oscillation between source and load as well.

$$p_{3\phi} = v_a i_a + v_b i_b + v_c i_c = v_\alpha i_\alpha + v_\beta i_\beta + v_0 i_0 = p + p_0 \qquad (3.100)$$

$$q = v_\beta i_\alpha - v_\alpha i_\beta = \frac{1}{\sqrt{3}}[(v_a - v_b)i_c + (v_b - v_c)i_a + (v_c - v_a)i_b] \qquad (3.101)$$

$$q = \frac{1}{\sqrt{3}}(v_{ab} i_c + v_{bc} i_a + v_{ca} i_b)$$

Moreover, it might be interesting to determine active (real) and reactive (imaginary) current components directly from the instantaneous *abc* voltages and currents. For convenience, the active (real) and reactive (imaginary) current decomposition defined in (3.22) is repeated here:

$$\begin{bmatrix} i_\alpha \\ i_\beta \end{bmatrix} = \underbrace{\frac{1}{v_\alpha^2 + v_\beta^2}\begin{bmatrix} v_\alpha & v_\beta \\ v_\beta & -v_\alpha \end{bmatrix}\begin{bmatrix} p \\ 0 \end{bmatrix}}_{\text{real currents}} + \underbrace{\frac{1}{v_\alpha^2 + v_\beta^2}\begin{bmatrix} v_\alpha & v_\beta \\ v_\beta & -v_\alpha \end{bmatrix}\begin{bmatrix} 0 \\ q \end{bmatrix}}_{\text{imaginary currents}} \qquad (3.102)$$

Applying the appropriate inverse Clarke transformation in (3.102) and taking only the first term in right side of the expression, determine the the *abc* real currents as

$$\begin{bmatrix} i_{ap} \\ i_{bp} \\ i_{cp} \end{bmatrix} = \sqrt{\frac{2}{3}} \begin{bmatrix} 1 & 0 \\ -\frac{1}{2} & \frac{\sqrt{3}}{2} \\ -\frac{1}{2} & -\frac{\sqrt{3}}{2} \end{bmatrix} \frac{v_\alpha i_\alpha + v_\beta i_\beta}{v_\alpha^2 + v_\beta^2} \begin{bmatrix} v_\alpha \\ v_\beta \end{bmatrix} \qquad (3.103)$$

The *abc* real currents in (3.103) can be calculated directly from the *abc* voltages and currents by using (3.5) and its corresponding matrix transformation for the currents. Thus, the $\alpha\beta$ variables can be set in terms of *abc* variables. After some manipulations, the real currents are found to be

$$\begin{bmatrix} i_{ap} \\ i_{bp} \\ i_{cp} \end{bmatrix} = \frac{(v_{ab} - v_{ca}) \cdot i_a + (v_{bc} - v_{ab}) \cdot i_b + (v_{ca} - v_{bc}) \cdot i_c}{v_{ab}^2 + v_{bc}^2 + v_{ca}^2} \begin{bmatrix} (v_{ab} - v_{ca})/3 \\ (v_{bc} - v_{ab})/3 \\ (v_{ca} - v_{bc})/3 \end{bmatrix} \qquad (3.104)$$

Note that line voltages, that is, $v_{ab} = v_a - v_b$, $v_{bc} = v_b - v_c$, $v_{ca} = v_c - v_a$, are used in (3.104). This expression confirms a fundamental concept introduced in the *p-q* Theory, which establishes that zero-sequence components do not contribute to the real power. Thus, the *abc* real currents i_{ap}, i_{bp}, and i_{cp} that correspond to the components in the original currents i_a, i_b, and i_c. Moreover, the real currents are not influenced by zero-sequence voltage or current components, neither in voltages nor in currents. The reason is that line voltages never contain zero-sequence components, because $v_{ab} + v_{bc} + v_{ca} = 0$. Further, even if zero-sequence currents are considered in (3.104), it follows that

$$(v_{ab} - v_{ca}) \cdot i_0 + (v_{bc} - v_{ab}) \cdot i_0 + (v_{ca} - v_{bc}) \cdot i_0 = 0 \quad (3.105)$$

which results in zero real current.

If the second term in (3.102) is considered, the *abc* imaginary currents can be determined by using the same procedure as the above. After some manipulations, they result in

$$\begin{bmatrix} i_{aq} \\ i_{bq} \\ i_{cq} \end{bmatrix} = \frac{v_{bc} \cdot i_a + v_{ca} \cdot i_b + v_{ab} \cdot i_c}{v_{ab}^2 + v_{bc}^2 + v_{ca}^2} \begin{bmatrix} v_{bc} \\ v_{ca} \\ v_{ab} \end{bmatrix} \quad (3.106)$$

As expected, (3.106) leads to the following conclusion: the *abc* imaginary currents i_{aq}, i_{bq}, and i_{cq} correspond to the components in the original currents i_a, i_b, i_c, which generate only imaginary power. They are not influenced by zero-sequence voltage or current components.

The inverse Clark transformation as given in (3.4) can be decomposed into the sum of two terms, as follows:

$$\begin{bmatrix} i_a \\ i_b \\ i_c \end{bmatrix} = \frac{1}{\sqrt{3}} \begin{bmatrix} i_0 \\ i_0 \\ i_0 \end{bmatrix} + \sqrt{\frac{2}{3}} \begin{bmatrix} 1 & 0 \\ -\frac{1}{2} & \frac{\sqrt{3}}{2} \\ -\frac{1}{2} & -\frac{\sqrt{3}}{2} \end{bmatrix} \begin{bmatrix} i_\alpha \\ i_\beta \end{bmatrix} \quad (3.107)$$

On the other hand, (3.102) separates i_α and i_β into two components, the real and the imaginary currents. One is dependent only on the real power, and the other is dependent only on the imaginary power. These current components can be calculated directly from the instantaneous *abc* voltages and currents by using (3.104) and (3.106). These equations together with (3.102) and (3.107) allow us to write

$$\begin{bmatrix} i_a \\ i_b \\ i_c \end{bmatrix} = \frac{1}{\sqrt{3}} \begin{bmatrix} i_0 \\ i_0 \\ i_0 \end{bmatrix} + \begin{bmatrix} i_{ap} \\ i_{bp} \\ i_{cp} \end{bmatrix} + \begin{bmatrix} i_{aq} \\ i_{bq} \\ i_{cq} \end{bmatrix} \quad (3.108)$$

3.3.6. Modified p-q Theory

This chapter describes the basic principles of shunt current compensation and series voltage compensation based on the *p-q* Theory, which are applicable to all controllers for active filters and active power-line conditioners that will be presented in

3.3. THE p-q THEORY IN THREE-PHASE, FOUR-WIRE SYSTEMS

the following chapters. The reader should notice that the authors use the $\alpha\beta 0$ transformation to deal properly with zero-sequence components that are separated from the $\alpha\beta$ components. In this way, zero-sequence voltage and current components are separated from the $\alpha\beta$ components, and treated as "single-phase variables." The positive- and negative-sequence components are naturally kept in the $\alpha\beta$ axes.

As will be shown in the next chapters, the approach adopted in the above sections is perfectly sufficient to deal with three-phase circuits for both three-wire and four-wire systems. Nevertheless, in 1994 and 1995, references [48] and [49] introduced a *modified p-q theory* in 1994 and 1995. In fact, they expanded the concept of the imaginary power q defined in (3.20).

A three-dimensional, instantaneous voltage vector $\mathbf{e}_{\alpha\beta 0}$ and a current vector $\mathbf{i}_{\alpha\beta 0}$ can be defined from the instantaneous voltages and currents transformed into the $\alpha\beta 0$ reference frames. These instantaneous vectors are defined as

$$\mathbf{e}_{\alpha\beta 0} = [v_\alpha, v_\beta, v_0]^T; \quad \mathbf{i}_{\alpha\beta 0} = [i_\alpha, i_\beta, i_0]^T \tag{3.109}$$

The original imaginary power q defined in (3.20) can be understood as the cross product of instantaneous vectors $\mathbf{e}$ and $\mathbf{i}$ defined in (3.10) and (3.11), respectively. On the other hand, the real power of the *p-q* Theory can be interpreted as the scalar product (dot product) of those vectors. Similarly, the modified *p-q* theory defines three instantaneous imaginary powers as derived from the cross product of the vectors defined in (3.109), whereas the three-phase instantaneous active power $p_{3\phi}$ defined in (3.71) represents the scalar product of these same vectors. Here, the three-phase instantaneous active power $p_{3\phi}$ is represented simply by the symbol p in the modified *p-q* theory. However, it should not be confused with the real power p defined in the original *p-q* Theory that excludes the zero-sequence power p_0 from the real power p. Thus, the modified *p-q* theory defines only a single instantaneous active power that is the sum of the real and zero-sequence powers in the original *p-q* Theory, that is,

$$p = \mathbf{e}_{\alpha\beta 0} \cdot \mathbf{i}_{\alpha\beta 0} = v_\alpha i_\alpha + v_\beta i_\beta + v_0 i_0 \tag{3.110}$$

On the other hand, the modified *p-q* theory defines an instantaneous imaginary-power vector composed of three elements, q_0, q_α, and q_β, as follows:

$$\mathbf{q} = \mathbf{e}_{\alpha\beta 0} \times \mathbf{i}_{\alpha\beta 0} = \begin{bmatrix} q_\alpha \\ q_\beta \\ q_0 \end{bmatrix} = \begin{bmatrix} \begin{vmatrix} v_\beta & v_0 \\ i_\beta & i_0 \end{vmatrix} \\ \begin{vmatrix} v_0 & v_\alpha \\ i_0 & i_\alpha \end{vmatrix} \\ \begin{vmatrix} v_\alpha & v_\beta \\ i_\alpha & i_\beta \end{vmatrix} \end{bmatrix} \tag{3.111}$$

The instantaneous powers defined above are combined in a matrix expression as follows:

$$\begin{bmatrix} p \\ q_\alpha \\ q_\beta \\ q_0 \end{bmatrix} = \begin{bmatrix} v_\alpha & v_\beta & v_0 \\ 0 & -v_0 & v_\beta \\ v_0 & 0 & -v_\alpha \\ -v_\beta & v_\alpha & 0 \end{bmatrix} \begin{bmatrix} i_\alpha \\ i_\beta \\ i_0 \end{bmatrix} \quad (3.112)$$

Note that the new imaginary power q_0 is the same as the original imaginary power q defined in (3.20). The other two imaginary powers q_α and q_β relate α and β components with a zero-sequence voltage and current, which are not considered in the original p-q Theory. The norm of the instantaneous imaginary-power vector expresses the "total" instantaneous imaginary power q as follows:

$$q = |\mathbf{q}| = \sqrt{q_\alpha^2 + q_\beta^2 + q_0^2} \quad (3.113)$$

The inverse transformation of (3.112) is performed as follows:

$$\begin{bmatrix} i_\alpha \\ i_\beta \\ i_0 \end{bmatrix} = \frac{1}{v_{\alpha\beta 0}^2} \begin{bmatrix} v_\alpha & 0 & v_0 & -v_\beta \\ v_\beta & -v_0 & 0 & v_\alpha \\ v_0 & v_\beta & -v_\alpha & 0 \end{bmatrix} \begin{bmatrix} p \\ q_\alpha \\ q_\beta \\ q_0 \end{bmatrix} \quad (3.114)$$

where

$$v_{\alpha\beta 0}^2 = v_\alpha^2 + v_\beta^2 + v_0^2 \quad (3.115)$$

From (3.114), active and reactive current components can be derived. For instance, the zero-sequence current i_0 is divided in its active part i_{0p} and reactive part i_{0q}, where $i_0 = i_{0p} + i_{0q}$. They are calculated as follows:

- Instantaneous zero-sequence active current i_{0p}

$$i_{0p} = \frac{v_0}{v_{\alpha\beta 0}^2} p \quad (3.116)$$

- Instantaneous zero-sequence reactive current i_{0q}

$$i_{0q} = \frac{v_\beta}{v_{\alpha\beta 0}^2} q_\alpha - \frac{v_\alpha}{v_{\alpha\beta 0}^2} q_\beta \quad (3.117)$$

Similarly, instantaneous active and reactive currents on the α and β axes can be redefined. Unlike the active and reactive current components defined in (3.23) to (3.26) in the original p-q Theory, they also depend on zero-sequence components [50,51]:

- Instantaneous active current on the α axis $i_{\alpha p}$

$$i_{\alpha p} = \frac{v_\alpha}{v_{\alpha\beta 0}^2} p \quad (3.118)$$

- Instantaneous reactive current on the α axis $i_{\alpha q}$

$$i_{\alpha q} = \frac{v_0}{v_{\alpha\beta 0}^2} q_\beta - \frac{v_\beta}{v_{\alpha\beta 0}^2} q_0 \qquad (3.119)$$

- Instantaneous active current on the β axis $i_{\beta p}$

$$i_{\beta p} = \frac{v_\beta}{v_{\alpha\beta 0}^2} p \qquad (3.120)$$

- Instantaneous reactive current on the β axis $i_{\beta q}$

$$i_{\beta q} = \frac{v_\alpha}{v_{\alpha\beta 0}^2} q_0 - \frac{v_0}{v_{\alpha\beta 0}^2} q_\alpha \qquad (3.121)$$

The modified p-q theory presented above provides the basic equations to develop a new control method for shunt current compensation. The method based on the p-q Theory, presented in Figure 3-11, can be modified by using (3.112) and (3.114). Fig. 3-20 shows the control method for shunt current compensation based on the modified p-q theory.

The control method based on the modified p-q theory has the same flexibility as that based on the p-q Theory. Powers or even portions of powers, like the average or oscillating component of the above-defined active power p, and the elements of the imaginary-power vector **q** can be selected and compensated independently from each other. However, it should be kept in mind that the zero-sequence voltage and current are now manipulated together with the positive-sequence and negative-sequence components. In some cases, this can produce undesirable compensation effects, which will be clarified in the next section.

If zero-sequence components—generically said to be homopolar mode—do not represent a problem when manipulated together with positive-sequence and negative-sequence components—generically said to be nonhomopolar modes—the modified p-q theory can be simplified. The use of the Clarke transformation can be avoided. Peng and his coauthors [52,53] established a set of power definitions directly in the abc-phase mode and correlated it with the modified p-q theory presented above. The basic idea consists in defining the three-phase instantaneous active power, as well as the instantaneous imaginary-power vector, directly from instantaneous voltage and current vectors formed by the instantaneous phase voltages v_a, v_b, and v_c, and the instantaneous line currents i_a, i_b, and i_c, instead of their corresponding $\alpha\beta 0$ variables. Both approaches in the modified theory differ from the approach preferred in this book in the sense that they consider the homopolar mode as containing the same properties as the nonhomopolar modes. It was shown that the Clarke transformation is useful to separate the nonhomopolar modes that form the $\alpha\beta$ variables, from the zero-sequence components.

In 1999, a deep discussion on these different definitions was presented in [51]. It shows that the definitions of the imaginary power, given in (3.20), as well as the

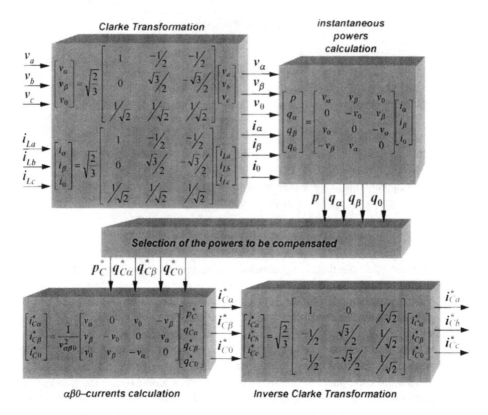

Figure 3-20. Control method for shunt current compensation based on the modified *p-q* theory.

imaginary-power vector, given in (3.111), or that one given in [52], are correct. However, they lead to some different interpretations. For instance, the *p-q* Theory suggests that zero-sequence currents should be completely eliminated and never treated as containing "active" and "reactive" portions that could be separated from each other. Contrarily, the modified theories allow the compensation of parts of the zero-sequence current, depending on the presence or absence of zero-sequence voltage. Moreover, the presence of zero-sequence components in voltage and current affect the $\alpha\beta$ variables—the nonhomopolar quantities—in the modified theories.

When $v_0 = 0$, the zero-sequence current is simply an "instantaneous current" that does not transfer any energy from the source to the load, since $p_0 = 0$. Contrarily, when $v_0 \neq 0$, the zero-sequence current becomes an "instantaneous active current," because p_0 is not zero, and energy is transferred from the source to the load in the same way as in a single-phase circuit [38]. The next chapter confirms the advantage of the *p-q* Theory when applied to controllers of three-phase, four-wire shunt active filters.

The imaginary-power vector given in (3.111) treats the zero-sequence current as an "instantaneous reactive current" when $v_0 = 0$. A compensator without an energy storage element can be designed to completely compensate for i_0, under the condition of $v_0 = 0$, by using either the p-q Theory or the modified p-q theory. However, the authors believe that working with zero-sequence components separated from the $\alpha\beta$ components makes it possible to better understand the physical meaning of the problem resulting from the presence of v_0 and i_0, and to take measures to eliminate them. For instance, as will be shown in the next chapters, the current i_0 can be compensated easily by a shunt active filter with or without the presence of v_0. In the case of series active filter, the zero-sequence voltage v_0 can be compensated in a dual way. References [48], [49], [50], [52] and [53] proposed expanded imaginary powers. However, they do not consider the elimination of v_0.

In 2004, Dai and coauthors [54] published a paper giving a more generic view of the definition of the "instantaneous reactive quantities for multiphase power systems." Instead of using a vector as result of the cross product, which is valid only for three-dimensional spaces, they define an imaginary-power tensor, which has the advantage of being applicable to generic multiphase systems. This is also an interesting approach for those who will work with power systems with more than three phases. However, this problem will not be addressed in this book.

3.4. INSTANTANEOUS abc THEORY

As explained in the introduction to this chapter, the instantaneous power theory can be separated into two groups: one defining the powers on the $\alpha\beta 0$-reference frame and the other defining them directly in the *abc* phases, that is, the use of instantaneous phase voltages and instantaneous line currents. A summary of the theories working directly in the *abc* axes are presented here. To maintain a clear contrast with the p-q Theory, these sets of power definitions and current decompositions in the *abc* axes will be called the *abc* Theory.

Until the end of the 1960s, the concept of reactive power was related to a time-independent variable. However, with an increase in the problems caused by power electronics devices, the term *"instantaneous reactive power"* appeared in the literature as an extension of the old concept [10,11]. The work presented by Erlicki and Emanuel-Eigeles, in 1968, introduced the idea of compensation of "instantaneous" reactive power [6]. However, the analysis was done in the frequency domain and the compensation in the time domain. This means that the concept of instantaneous reactive power was not well defined at that time. One of the first works to relate the possibility of reactive power generation by means of power electronic converters without direct association with the energy storage capacity of dc reactors of a thyristor converter was presented by Depenbrock, in 1962 [10]. Recently, all these issues have been well addressed by means of the p-q Theory.

The p-q Theory gave the well-defined concept of imaginary power a clear physical meaning. However, it requires the use of the $\alpha\beta 0$ transformation (Clarke transformation) in this theory. For some analysis, or even for some cases of harmonic elimination, this transformation may be seen as an extra calculation effort that should be avoided. In fact, it is possible to avoid the $\alpha\beta 0$ transformation in the p-q Theory, as shown in (3.104), (3.106), and (3.108).

A summary of some original research has been made, and will be presented below as an alternative approach to calculate the *active* and *nonactive* current portions of a generic load current. Instead of using the Clarke transformation and the real and imaginary power calculation, the following approach calculates the active and nonactive current components of a generic load current directly from the *abc*-phase voltages and line currents. This approach can be generally understood as being obtained from the application of a minimization method to the load current, as will be clarified in the following sections.

Although Fryze did not mention the concept of "*instantaneous*" reactive power in his original work, he used the idea of decomposition into active current component ("*Wirkstrom,*" in German) and reactive current component ("*Blindstrom,*" in German) [1]. An extension to active and reactive voltage components was also addressed. Based on these components, the active power, P_w, and the reactive power, P_q, were consistently defined, as summarized in (2.26) to (2.32).

The following *abc* Theory consists in determining instantaneously the active portion of a generic load current. In other words, a minimized, instantaneous, active current component is determined with the constraint that it should transfer the same amount of energy as the uncompensated (original) load current. The difference between the instantaneous original load current and the calculated instantaneous active current (minimized current) is the instantaneous nonactive current that is part of the original load current. To determine the instantaneous active current, a minimization method (the *Lagrange Multiplier Method*) formulation can be used.

A concept of *three-phase instantaneous reactive power* can be derived from the above-defined concept of active and nonactive currents as follows.

> The three-phase instantaneous reactive power comprises products of voltage and current components that do not contribute to the three-phase instantaneous active power.

Fryze used similar words to define its reactive power, P_q [1]. The difference is that here instantaneous values of voltages and currents are considered.

It should be remarked that here the traditional idea that reactive power is related to an oscillating energy flow, as introduced in Chapter 2 (Fig. 2-1), is abandoned. If an oscillating energy flow exists between two subsystems, irrespective of whether it is in a three-phase three-wire or four-wire system, it is treated here as a three-phase instantaneous *active* power portion that has a zero average value.

Since instantaneous values of voltages and current are used, the above concepts of instantaneous decomposition into active and nonactive currents, along with the instantaneous active and reactive powers, are valid during transient periods or in steady-state conditions. Further, no restrictions are imposed on their waveforms, and they can be used under nonsinusoidal or unbalanced conditions.

At this point, the reader could be asking the following questions. What is the difference between the *p-q* Theory and the *abc* Theory? What is the difference between the current decomposition made in (3.23) to (3.26) and the concept of active and nonactive current? What is the relation between the imaginary power and the above-defined concept of three-phase instantaneous reactive power?

The physical meaning related to the imaginary power in the *p-q* Theory cannot be directly extended to the above-defined three-phase instantaneous reactive power. The results are the same if no zero-sequence components are present, whereas they are different in the presence of zero-sequence components in current and/or voltage. During the presentation of the *abc* Theory, these differences from the *p-q* Theory will be clarified through hypothetical examples of current compensation.

3.4.1. Active and Nonactive Current Calculation by Means of a Minimization Method

The instantaneous nonactive current in a three-phase system is the component of the load current that does not produce three-phase instantaneous active power, although it increases the current amplitude and consequently also increases the losses in the network. The instantaneous nonactive current can be determined by applying a minimization method. For the formulation of the problem, a three-phase hypothetical load current i_k, $k = (a, b, c)$, is assumed to consist of an *active* portion i_{wk} and a *nonactive* portion i_{qk}, that is,

$$i_k = i_{wk} + i_{qk}; \qquad k = (a, b, c) \tag{3.122}$$

The method involves minimizing load currents under the constraint that the *nonactive* current components i_{qa}, i_{qb}, and i_{qc} do not generate three-phase instantaneous active power. Thus, the task consists of finding the minimum of

$$L(i_{qa}, i_{qb}, i_{qc}) = (i_a - i_{qa})^2 + (i_b - i_{qb})^2 + (i_c - i_{qc})^2$$

constrained by

$$g(i_{qa}, i_{qb}, i_{qc}) = v_a i_{qa} + v_b i_{qb} + v_c i_{qc} = 0 \tag{3.123}$$

The problem can be solved by applying the Lagrange Multiplier Method, which leads to the following system of equations:

$$\begin{bmatrix} 2 & 0 & 0 & v_a \\ 0 & 2 & 0 & v_b \\ 0 & 0 & 2 & v_c \\ v_a & v_b & v_c & 0 \end{bmatrix} \begin{bmatrix} i_{qa} \\ i_{qb} \\ i_{qc} \\ \lambda \end{bmatrix} = \begin{bmatrix} 2i_a \\ 2i_b \\ 2i_c \\ 0 \end{bmatrix} \tag{3.124}$$

Solving (3.124) for λ gives

$$\lambda = \frac{2(v_a i_a + v_b i_b + v_c i_c)}{v_a^2 + v_b^2 + v_c^2} = \frac{2p_{3\phi}}{v_a^2 + v_b^2 + v_c^2} \tag{3.125}$$

By replacing (3.125) in (3.124), the instantaneous nonactive currents are found to be:

$$\begin{bmatrix} i_{qa} \\ i_{qb} \\ i_{qc} \end{bmatrix} = \begin{bmatrix} i_a \\ i_b \\ i_c \end{bmatrix} - \frac{p_{3\phi}}{v_a^2 + v_b^2 + v_c^2} \begin{bmatrix} v_a \\ v_b \\ v_c \end{bmatrix} \qquad (3.126)$$

From (3.122) and (3.126), the instantaneous active currents are

$$\begin{bmatrix} i_{wa} \\ i_{wb} \\ i_{wc} \end{bmatrix} = \frac{p_{3\phi}}{v_a^2 + v_b^2 + v_c^2} \begin{bmatrix} v_a \\ v_b \\ v_c \end{bmatrix} \qquad (3.127)$$

Therefore, the restriction imposed in the minimization method forces the active currents calculated in (3.127) and the original load currents i_a, i_b, and i_c to produce the same three-phase instantaneous active power ($p_{3\phi}$) when multiplied by their respective phase voltage v_a, v_b, v_c, that is,

$$p_{3\phi} = v_a i_a + v_b i_b + v_c i_c = v_a i_{wa} + v_b i_{wb} + v_c i_{wc} \qquad (3.128)$$

Hence, they are equivalent from an energy transfer point of view. The difference is that the active currents i_{wa}, i_{wb}, and i_{wc} do not generate any three-phase instantaneous reactive power and have smaller rms values. The active currents determined in (3.127), contrary to the *abc* real currents given in (3.104), are influenced by zero-sequence components in the voltages and currents.

If i_{qa}, i_{qb}, and i_{qc} are compensated close to the load terminals, then the power generating system supplies only i_{wa}, i_{wb}, and i_{wc}, reducing losses in the network. The basic principle of shunt current compensation is illustrated in Fig. 3-10. This principle can be used for the compensation of nonactive currents as calculated in (3.126). In this case, the original load currents $i_a = i_{wa} + i_{qa}$, $i_b = i_{wb} + i_{qb}$ and $i_c = i_{wc} + i_{qc}$, are compensated by making $i_{Ca}^* = -i_{qa}$, $i_{Cb}^* = -i_{qb}$ and $i_{Cc}^* = -i_{qc}$, forcing the source currents to become $i_{Sa} = i_{wa}$, $i_{Sb} = i_{wb}$, $i_{Sc} = i_{wc}$. Since i_{qa}, i_{qb}, and i_{qc} do not produce active power, the compensator does not supply any energy to the system all the time. Hence, the compensator does not need any energy storage capability.

The *abc* Theory described above is based on the principle of load-current decomposition into active and nonactive current portions. The concept of three-phase instantaneous reactive power given above is derived from this current decomposition. In fact, there is no equation for directly calculating the three-phase instantaneous reactive power, in contrast to the well-defined imaginary power in the *p-q* Theory.

Much research on power theories can be found in the literature. Although the above-described *abc* Theory does not go in its particularities, it tries to summarize the results of the research. Some recent power formulations that resulted in formulations similar to those given in (3.126) and (3.127) were presented by Furuhashi et al. [41], Tenti et al. [42,43,44], van Wyk et al. [45,46,47], Depenbrock et al. [10,25,40] and Czarnecki [26,27].

If the three-phase system is balanced and sinusoidal, $p_{3\phi}$ is constant. In this case, the quadratic sum $(v_a^2 + v_b^2 + v_c^2)$ is also constant. Thus, the minimized instantaneous currents i_{wa}, i_{wb}, and i_{wc} are *proportional* to the phase voltages v_a, v_b, and v_c, respectively. Hence, the line current in a phase is sinusoidal and in phase with the corresponding phase (line-to-neutral) voltage. This gives rise to the idea that "*the best kind of load is a purely resistive one.*"

The main objectives of the compensation strategy based on the *abc* Theory are: (i) to obtain the compensated current proportional to the voltage and (ii) to obtain the compensated current with a minimum rms value, capable of delivering the same active power as that of the original current of the load.

However, in most cases under nonsinusoidal conditions, $p_{3\phi}$ and $(v_a^2 + v_b^2 + v_c^2)$ vary, so that the linearity between voltages and currents no longer exists. An example is given in Fig. 3-21. When the hypothetical phase voltages v_a, v_b and v_c, and the generic load currents i_a, i_b and i_c are used in (3.126) and (3.127), neither the waveforms of the active currents, nor those of the nonactive currents have any similarity to those of the voltages. Although the waveforms of the generic load currents and the minimized active currents i_{wa}, i_{wb}, and i_{wc} differ significantly, (3.128) is always valid and they produce the same three-phase instantaneous active power $(p_{3\phi})$, as shown in Fig. 3-21.

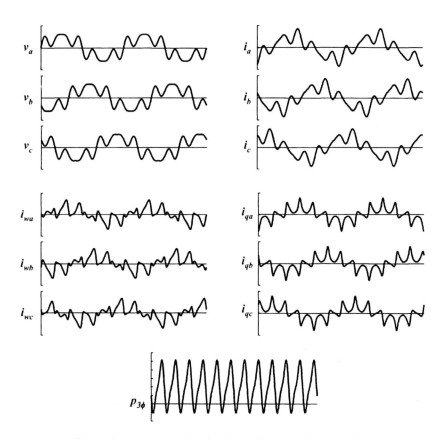

Figure 3-21. An example of minimized currents i_{wa}, i_{wb}, and i_{wc}.

92 THE INSTANTANEOUS POWER THEORY

Another point that should be reinforced is that the linearity between voltages and currents cannot guarantee *constant* three-phase instantaneous active power. An example is given in Fig. 3-22. This important feature must be kept in mind when designing controllers for active power-line conditioners. Although the linearity does not guarantee constant instantaneous active power, there are some special cases. In these cases, even under nonsinusoidal voltages, proportional currents can produce constant instantaneous active power, as well as zero three-phase instantaneous reactive power. One interesting case is given in Fig. 3-23. This figure suggests that, from an energy-flow point of view, rectangular supply voltages might be adequate to the voltage source for three-phase diode bridges. A problem might be encountered in that the rectangular voltages could not be transmitted for long distances, keeping the same waveshape along the transmission line.

In practice, most cases do not produce constant three-phase active power by making the currents proportional to the voltages. Anyway, the minimized active currents from (3.127) never generate three-phase instantaneous reactive power. This will be shown later by using the *p-q* Theory, because the *abc* Theory does not have a well-accepted expression for reactive-power calculation.

It was seen that the linearity between voltage and current waveforms can or cannot draw constant three-phase active power from a generic voltage source. To finalize this discussion, the case shown in Fig. 3-24 is an example of voltage and current waveforms that are different even though they draw constant three-phase active power from the source.

The three-phase phase voltages of the previous examples shown in Fig. 3-22 and Fig. 3-24 include a fundamental component and fifth- and seventh-harmonic components. Both harmonic components have equal amplitude and are negative- and positive-sequence components, respectively. Moreover, they have the same phase angle, that is, they are in phase with each other. This is a very particular case. These voltages produce a constant three-phase active power if multiplied by sinusoidal and balanced line currents, as can be seen in Fig. 3-24. Normally, nonsinusoidal voltages do not produce constant three-phase active power if multiplied by sinu-

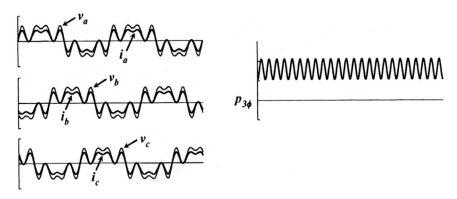

Figure 3-22. Currents proportional to the phase voltages.

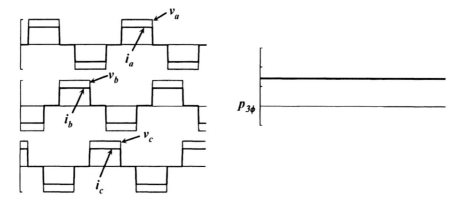

Figure 3-23. Special case of linearity between voltages and currents that produces constant three-phase active power.

soidal currents, or vice-versa. However, this is a very special case that can be better understood by using the general power equations in the *p-q* Theory in terms of symmetrical components. The currents in Fig. 3-24 consist only of a fundamental positive-sequence component, $\dot{I}_{+1} = I_{+1} \angle 0°$, whereas the voltages have three components: $\dot{V}_{+1} = V_{+1} \angle 0°$, $\dot{V}_{-5} = V \angle 0°$, and $\dot{V}_{+7} = V \angle 0°$. Thus, from (3.96) it is possible to see that the unique term that could produce constant imaginary power $\bar{q}$ is related to the phasors $\dot{I}_{+1}$ and $\dot{V}_{+1}$. However, they have the same phase angle and, therefore, $\bar{q}$ is zero. Contrarily, the constant real power $\bar{p}$ reaches the maximum due to the product of parallel phasors, as can be seen in (3.95).

The cross products of voltage and current phasors at different frequencies and/or from different sequence components appear in the oscillating real ($\tilde{p}$) and imaginary ($\tilde{q}$) powers. Surprisingly, as shown in Fig. 3-24 and proven in (3.98), the oscillating real power $\tilde{p}$ is zero, because $\dot{V}_{-5} = V \angle 0°$ and $\dot{V}_{+7} = V \angle 0°$ have the same amplitude and phase angle. Both products of $\dot{I}_{+1}$ with $\dot{V}_{-5}$, and with $\dot{V}_{+7}$ generate oscillating

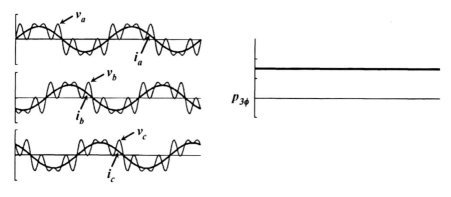

Figure 3-24. Special case under nonsinusoidal conditions.

powers at 6ω, where ω is the fundamental angular frequency. These power terms are summed up in (3.99) for determining the oscillating imaginary power $\tilde{q}$, whereas they are subtracted in (3.98) for calculation of $\tilde{p}$.

With certain restrictions, the calculation of the imaginary power in the p-q Theory is being used to analyze power behaviors derived from the decomposition of currents into active and nonactive parts. A significant difference would exist between the active currents i_{wa}, i_{wb}, and i_{wc} given in (3.127) and the real currents i_{ap}, i_{bp}, and i_{cp} determined from (3.104) if zero-sequence components were present. The same situation occurs with the nonactive currents i_{qa}, i_{qb}, and i_{qc} given in (3.126) and the imaginary currents i_{aq}, i_{bq}, and i_{cq} determined from (3.106). Hence, the imaginary power as calculated in the p-q Theory should be applied with limitations to the analysis of the concept of three-phase instantaneous reactive power as described above for the *abc* Theory.

3.4.2. Generalized Fryze Currents Minimization Method

If linearity between voltage and current is desired when performing power compensation, an extension of the minimization method shown above may be used to guarantee the linearity, even under distorted and/or unbalanced source voltages. This alternative method of compensation is known as *the generalized Fryze currents minimization method* [40,44].

The basic idea consists of determining the minimized currents from the *average value* of the three-phase instantaneous active power ($\bar{p}_{3\phi}$), instead of using the three-phase instantaneous active power ($p_{3\phi}$) as in (3.126) and (3.127). Moreover, in order to obtain an average value for the equivalent conductivity ($i = G_e v$), the instantaneous sum of squared phase voltages have also to be replaced by a sum of the squared rms values of voltages, as shown below. The new minimized currents—*the generalized Fryze currents*—are represented by the symbols $i_{\bar{w}a}$, $i_{\bar{w}b}$, and $i_{\bar{w}c}$, and are defined as:

$$i_{\bar{w}k} = G_e v_k; \quad k = (a, b, c) \qquad (3.129)$$

If the admittance G_e is represented by an average value instead of a varying instantaneous value, the linearity between voltage and current is ensured. No restriction should be given to the three-phase voltage, allowing it to be distorted and/or unbalanced. The average admittance G_e is obtained from an old concept of aggregate voltage,* as follows:

$$G_e = \frac{\bar{p}_{3\phi}}{V_\Sigma^2} \qquad (3.130)$$

---

*Buchholz [39] established the concept of aggregate current and voltage in 1919 to define apparent power for ac systems under nonsinusoidal conditions. Recently, Emanuel [22], Depenbrock [40], and Czarnecki [28] have also used this concept.

where

$$\bar{p}_{3\phi} = \frac{1}{T}\int_0^T p_{3\phi}(t)dt = \frac{1}{T}\int_0^T (v_a i_a + v_b i_b + v_c i_c)dt \quad (3.131)$$

and

$$V_\Sigma = \sqrt{V_a^2 + V_b^2 + V_c^2} = \sqrt{\frac{1}{T}\left(\int_0^T v_a^2(t)dt + \int_0^T v_b^2(t)dt + \int_0^T v_c^2(t)dt\right)} \quad (3.132)$$

Alternatively, in order to reduce computation efforts in a real implementation of a controller for an active power-line conditioner, the average admittance G_e can be obtained by means of a low-pass filter or a moving-average filter that determines

$$G_e = \frac{1}{T}\int_0^T g_e(t)dt = \frac{1}{T}\int_0^T \left(\frac{v_a i_a + v_b i_b + v_c i_c}{v_a^2 + v_b^2 + v_c^2}\right)dt \quad (3.133)$$

The currents in Fig. 3-22 and Fig. 3-23 could be considered to be the result of the above minimization method (generalized Fryze currents). Unfortunately, this extended method imposes some dynamics in a real implementation, because some time is needed to measure and calculate the rms values of the phase voltages that are used in the calculation of the aggregate voltage V_Σ. Some dynamics are also associated with the extraction of the average value $(\bar{p}_{3\phi})$ from the three-phase instantaneous active power. Alternatively, a low-pass filter can be used to determine the average admittance G_e, defined in (3.133). This is not the case if the method given by (3.127) is used, since it deals with instantaneous voltage and current values. Therefore, (3.127) can be considered as an "instantaneous" control algorithm for minimizing load currents, while (3.129) cannot.

In most cases, the generalized Fryze currents given in (3.129) and the minimization method given in (3.127) produce different compensated (minimized) currents under nonsinusoidal conditions. This point is evidenced in Fig. 3-25, where they are applied to compensate for the load currents under nonsinusoidal source voltages. Although the minimized currents produce different instantaneous powers, both currents produce exactly the same three-phase *average* active power $(\bar{p}_{3\phi})$ required by the load. Thus, the generic load currents i_a, i_b, and i_c, the active currents i_{wa}, i_{wb}, and i_{wc}, as well as the generalized Fryze currents $i_{\bar{w}a}$, $i_{\bar{w}b}$, and $i_{\bar{w}c}$ are well equivalent from the point of view of average energy transfer.

The generalized Fryze currents given in (3.129) and the active currents given in (3.127) do not produce any reactive power, whereas the generalized Fryze currents method offers the best result when reduction in losses is the main goal. Fig. 3-25 shows that this method produces the lowest rms currents values to transmit the same three-phase average active power. Further, it is possible to demonstrate that the aggregate values of the currents $i_{\bar{w}a}$, $i_{\bar{w}b}$, and $i_{\bar{w}c}$ are smaller than i_{wa}, i_{wb},

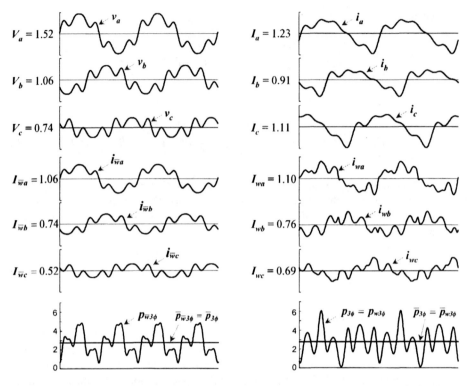

Figure 3-25. Comparison between the generalized Fryze currents and the minimization method.

and i_{wc}. For the sinusoidal case, (3.127) and (3.129) produce the same result. Thus,

$$I_{\bar{w}\Sigma} = \sqrt{I_{\bar{w}a}^2 + I_{\bar{w}b}^2 + I_{\bar{w}c}^2} \leq I_{w\Sigma} = \sqrt{I_{wa}^2 + I_{wb}^2 + I_{wc}^2} \qquad (3.134)$$

The example given above proves that both methods of minimization may not be used to compensate for the load currents if a *constant* three-phase instantaneous active power drawn from the source is required under nonsinusoidal conditions. Moreover, it has been proved that the linearity between voltage and current is not sufficient to guarantee an optimal (constant) power flow to the network.

Three-phase, four-wire systems may contain zero-sequence components. If these components are simultaneously present in voltage and current, they produce zero-sequence power. The zero-sequence power is a kind of active power and never contains any reactive portion according to the definition given above. Zero-sequence components should be avoided in three-phase systems, because they cannot produce three-phase constant power, as explained in the general equations of the p-q Theory in terms of symmetrical components. The minimization methods given in the *abc* Theory are inappropriate to deal with zero-sequence components. Some extra efforts have to be made with the above-described minimization methods in order to deal properly with zero-sequence components.

Figure 3-26 shows an example that evidences some drawbacks of the minimization methods defined in the *abc* Theory when applied in the presence of zero-sequence voltage components. This figure shows three-phase, arbitrarily chosen voltages with fundamental negative- and zero-sequence components, and distorted with second- and third-harmonic components. The load currents are also shown in Fig. 3-26. These load currents are similar to those in a three-phase thyristor rectifier. Thus, the load currents contain no zero-sequence component, although a three-phase four-wire compensator and network should be considered to agree with the results shown in Fig. 3-26. The phase angle of the currents corresponds to a firing angle of 45° in the thyristor rectifier. Therefore, the load currents produce a large amount of reactive power. For comparison, both minimization methods were applied to compensate for the reactive power of the load, since the fundamental component of compensated currents i_{wa}, i_{wb}, and i_{wc}, and $i_{\bar{w}a}$, $i_{\bar{w}b}$, and $i_{\bar{w}c}$ are in phase with the fundamental phase voltages. However, the sum $i_{wa} + i_{wb} + i_{wc} = i_{So}$, as well

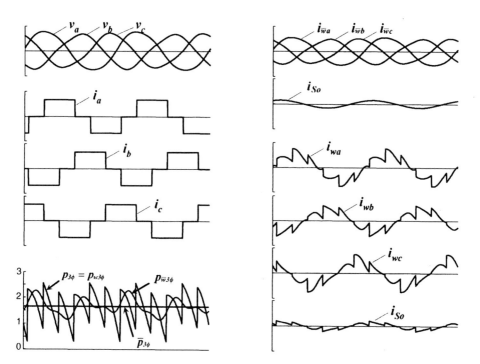

Figure 3-26. Application example of the generalized Fryze currents method and the active current minimization method in the presence of zero-sequence components only in the voltages.

as the sum $i_{\overline{w}a} + i_{\overline{w}b} + i_{\overline{w}c} = i_{So}$, are not equal to zero. In other words, the minimization methods force the compensator to draw undesirable zero-sequence currents (neutral currents) from the network, which are not present in the original load currents.

3.5. COMPARISONS BETWEEN THE p-q THEORY AND THE abc THEORY

The *p-q* Theory can identify voltage and current components on the $\alpha\beta$ axes, which are dependent only on the real power *p* or on the imaginary power *q*. In other words, it is possible to separate the real and the imaginary voltages, as well as the real and imaginary currents, which contribute only to the real or to the imaginary powers. Similarly, the *abc* Theory determines the active currents i_{wa}, i_{wb}, and i_{wc} that contribute only to the three-phase instantaneous active power, and the nonactive currents i_{qa}, i_{qb}, and i_{qc} that contribute only to the three-phase instantaneous reactive power. The current decomposition given in (3.22) is rewritten here as

$$\begin{bmatrix} i_\alpha \\ i_\beta \end{bmatrix} = \begin{bmatrix} i_{\alpha p} \\ i_{\beta p} \end{bmatrix} + \begin{bmatrix} i_{\alpha q} \\ i_{\beta q} \end{bmatrix} \quad (3.135)$$

A compensation algorithm to determine the instantaneous imaginary currents can be derived directly from (3.135) by just calculating the inverse transformation of $i_{\alpha q}$ and $i_{\beta q}$, that is,

$$\begin{bmatrix} i_{aq} \\ i_{bq} \\ i_{cq} \end{bmatrix} = \sqrt{\frac{2}{3}} \begin{bmatrix} 1 & 0 \\ -\frac{1}{2} & \frac{\sqrt{3}}{2} \\ -\frac{1}{2} & -\frac{\sqrt{3}}{2} \end{bmatrix} \begin{bmatrix} i_{\alpha q} \\ i_{\beta q} \end{bmatrix} \quad (3.136)$$

Therefore, in analogy to compensation of the nonactive current given by (3.126) in the *abc* Theory, a compensation algorithm based on the *p-q* Theory can be realized to compensate only for the imaginary power. If a shunt compensator, as shown in Fig. 3-10, draws the imaginary current $i_{Ca} = -i_{aq}$, $i_{Cb} = -i_{bq}$, and $i_{Cc} = -i_{cq}$, the network supplies only real current (i_{ap}, i_{bp}, and i_{cp}) of a generic load current (i_a, i_b, and i_c). This case disregards zero-sequence components. Figure 3-27 illustrates the whole algorithm that determines the instantaneous values of compensating currents that should be drawn by the shunt compensator. Although this control algorithm demands more calculation efforts than those for the nonactive current compensation based on the *abc* Theory, it involves only algebraic operations, that is, it contain no dynamic blocks. This means that the instantaneous imaginary currents, as well as the nonactive currents, can be determined instantaneously. If no zero-sequence components are present, the following relation for the compensation method based on the p-q Theory is valid:

$$\begin{bmatrix} i_{ap} \\ i_{bp} \\ i_{cp} \end{bmatrix} = \begin{bmatrix} i_a \\ i_b \\ i_c \end{bmatrix} - \begin{bmatrix} i_{aq} \\ i_{bq} \\ i_{cq} \end{bmatrix} \quad (3.137)$$

Otherwise, (3.108) should be considered.

In contrast to the instantaneous nonactive currents given in (3.126), the instantaneous imaginary currents obtained in Fig. 3-27 are not influenced by zero-sequence components. The abc-phase voltages, as well as the three-phase instantaneous active power $p_{3\phi}$ used in (3.126), can contain zero-sequence components, whereas the $\alpha\beta$ voltages and the imaginary power q in Fig. 3-27 are not affected by these components. This difference is evidenced in Fig. 3-28, where the compensated current (source current) derived from both methods are plotted together for the same case of phase voltages and load currents considered in Fig. 3-26. No neutral current is drawn by the shunt compensator if its controller is based on the p-q Theory, whereas an undesirable neutral current appears in the case of the abc Theory.

Although the load currents i_a, i_b, and i_c in Fig. 3-26 and in Fig. 3-28 are strongly distorted to compensate, they do not contain any zero-sequence current. That is, i_a+

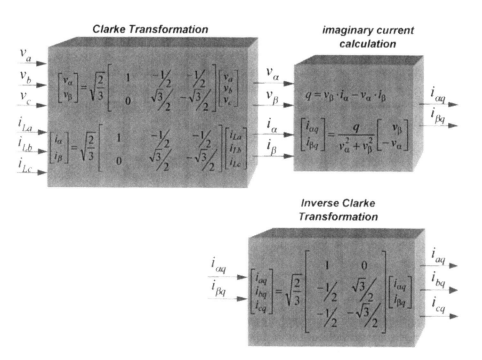

Figure 3-27. Control algorithm for imaginary current compensation based on the p-q Theory.

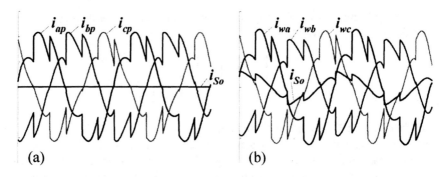

Figure 3-28. Compensated currents. (a) Real currents $i_{kp} = i_k - i_{kq}$, $k = a, b, c$ (p-q Theory). (b) Active currents $i_{wk} = i_k - i_{qk}$, $k = a, b, c$ (minimization method).

$i_b + i_c = 0$. Unfortunately, an undesirable neutral current would be generated by a shunt compensator controlled by the minimization method given in (3.126). This neutral current flows through the source, since $i_{So} \neq 0$, as shown in Fig. 3-28(b). Contrarily, the control algorithm based on the p-q Theory can compensate properly for the load currents in the presence of zero-sequence voltages, without generating zero-sequence currents to the source. That is, $i_{So} = i_{ap} + i_{bp} + i_{cp} = 0$, as can be seen in Fig. 3-28(a).

If no zero-sequence components are present in both phase voltages and line currents, the imaginary current determined from (3.136) is identical to the nonactive current from (3.126). Consequently, the active current $i_{wk} = i_k - i_{qk}$, $k = (a, b, c)$, and the real current $i_{kp} = i_k - i_{kq}$, $k=(a, b, c)$ are also the same in this case. However, in the presence of zero-sequence components, the compensation methods result in two different solutions, although neither i_{wk} nor i_{kp} produce imaginary power q, and neither i_{qk} nor i_{kq} produce active power, that is,

$$\begin{cases} q = \frac{1}{\sqrt{3}}(v_{ab}i_{wc} + v_{bc}i_{wa} + v_{ca}i_{wb}) = 0 \\ q = \frac{1}{\sqrt{3}}(v_{ab}i_{cp} + v_{bc}i_{ap} + v_{ca}i_{bp}) = 0 \end{cases} \quad (3.138)$$

$$\begin{cases} p_{3\phi} = v_a i_{qa} + v_b i_{qb} + v_c i_{qc} = 0 \\ p_{3\phi} = v_a i_{aq} + v_b i_{bq} + v_c i_{cq} = 0 \end{cases} \quad (3.139)$$

Equation (3.17) shows that $p_{3\phi} = p + p_0$. On the other hand (3.139) shows that the shunt compensator does not supply any energy to the load all the time. Therefore, the three-phase instantaneous active power ($p_{3\phi}$) of the source is identical to that of the load for both compensation methods used in Fig. 3-28. Since the minimized currents i_{wa}, i_{wb}, and i_{wc}, as well as the supply voltages, contain zero-sequence components, the source will supply a zero-sequence power through its own zero-sequence voltage components. Contrarily, the source does not supply a zero-sequence power if the shunt compensator is controlled by the algorithm given in Fig. 3-27. Therefore, the currents i_{wa}, i_{wb}, and i_{wc}, and i_{ap}, i_{bp}, and i_{cp} produce different real and zero-sequence powers. However, their sum, $p + p_0$, must be equal to $p_{3\phi}$ all the time. Note that the considered load currents i_a, i_b, and i_c do not contain zero-sequence components. Thus, the real power produced by i_a, i_b, and i_c, and i_{ap}, i_{bp}, and i_{cp} must be identical.

The example given above evidences that the p-q Theory is more efficient to deal with nonsinusoidal currents, particularly when zero-sequence components are present. One advantage of using the power definitions given in the p-q Theory is the possibility of compensating separately the powers p, q, and p_0. This is the reason why the p-q Theory can compensate load currents to provide constant instantaneous power to the source, even under distorted and/or unbalanced voltages. The controllers of the active power-line conditioners presented in the next chapters exploit widely this feature.

The next chapter deals with shunt active filters. It shows that the p-q Theory can deal properly with zero-sequence components. Moreover, it can compensate for load currents so as to produce sinusoidal currents to the source, even if the voltages are distorted and/or unbalanced. If desired, the p-q Theory can also be used to force the compensated currents to draw constant three-phase instantaneous active power from the source, even in the presence of distorted and/or unbalanced voltages.

3.5.1. Selection of Power Components to be Compensated

One significant advantage of using the p-q Theory in designing controllers for active power-line conditioners is the possibility of independently selecting the portions of real, imaginary, and zero-sequence powers to be compensated. Some times, it is convenient to separate these powers into their average and oscillating parts, that is,

$$
\begin{aligned}
\text{Real power:} \quad & p = \bar{p} + \tilde{p} \\
\text{Imaginary power:} \quad & q = \bar{q} + \tilde{q} \\
\text{Zero-sequence power:} \quad & p_0 = \bar{p}_0 + \tilde{p}_0
\end{aligned}
\quad (3.140)
$$

$$
\underbrace{\phantom{\bar{p}_0}}_{\text{Average powers}} \underbrace{\phantom{\tilde{p}_0}}_{\text{Oscillating powers}}
$$

The idea is to compensate all undesirable power components generated by nonlinear loads that can damage or make the power system overloaded or stressed by harmonic pollution. In this way, it would be desirable for a three-phase balanced power-generating system to supply only the average real power $\bar{p}$ of the load. Thus, all other power components required by the nonlinear load, that is, $\tilde{p}$, $\bar{q}$, $\tilde{q}$, $\bar{p}_0$, and $\tilde{p}_0$, should be compensated by a shunt compensator connected as close as possible to this load.

In Fig. 3-25 and Fig. 3-26, examples are shown to prove that the current minimization methods based on the *abc* Theory cannot guarantee constant instantaneous power drawn from the source, principally in the presence of zero-sequence voltage components. Even more difficult is to find a solution based on the current minimization methods to compensate zero-sequence currents in the presence of zero-sequence voltages, such that constant real (active) power is drawn from the source.

If there is no zero-sequence current to be compensated, as in the example given in Fig. 3-26, the zero-sequence power p_0 is always zero. In this case, a shunt compensator as shown in Fig. 3-10 should be controlled to compensate the oscillating real power $\tilde{p}$ and the whole imaginary power $q = \bar{q} + \tilde{q}$. This guarantees constant instantaneous power ($\bar{p}$) drawn from the source with reduced losses in the transmission system, since the imaginary power of the load is also being compensated. This is done by selecting the powers $\tilde{p}$ and q of the load to be compensated in the control algorithm shown in Fig. 3-11, thus making the following compensating current calculation:

$$\begin{bmatrix} i_{c\alpha} \\ i_{c\beta} \end{bmatrix} = \frac{1}{v_\alpha^2 + v_\beta^2} \begin{bmatrix} v_\alpha & v_\beta \\ v_\beta & -v_\alpha \end{bmatrix} \begin{bmatrix} -\tilde{p} \\ -q \end{bmatrix} \quad (3.141)$$

$$\begin{bmatrix} i_{Ca}^* \\ i_{Cb}^* \\ i_{Cc}^* \end{bmatrix} = \sqrt{\frac{2}{3}} \begin{bmatrix} 1 & 0 \\ -\frac{1}{2} & \frac{\sqrt{3}}{2} \\ -\frac{1}{2} & -\frac{\sqrt{3}}{2} \end{bmatrix} \begin{bmatrix} i_{c\alpha} \\ i_{c\beta} \end{bmatrix} \quad (3.142)$$

Now, the new compensated currents i_{Sa}, i_{Sb}, and i_{Sc} ($i_{Sk} = i_k - i_{Ck}$, $k = a, b, c$) that flow in the source are those shown in Fig. 3-29 if the same load currents and source voltages of the example given in Fig. 3-26 are considered.

In contrast to the currents compensated by the generalized Fryze current method, the compensation algorithm based on the *p-q* Theory forces the power source to provide currents that are not proportional to its phase voltages. Hence, a group of nonlinear loads compensated by a shunt compensator with a controller based on (3.141) and (3.142) are not "seen" as linear impedances or pure resistive loads by the power source. Nevertheless, this controller guarantees that the compensated currents draw *constant* instantaneous real power p from the source. Although the generalized Fryze current method provides the smallest rms values of currents to draw the same average active power, it cannot guarantee *constant* instantaneous real power p drawn from the source. Since the load currents do not have zero-sequence components, the three-phase instantaneous active power $p_{S3\phi}$ of the source is equal to the average real power $\bar{p}$ of the load.

3.6. SUMMARY

In this chapter, several important of power definitions have been covered, emphasizing their use in the control of active power-line conditioners. Some important conclusions are summarized below.

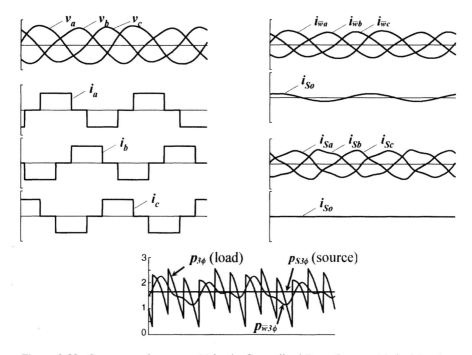

Figure 3-29. Compensated currents: (a) by the Generalized Fryze Currents Method ($i_{\bar{w}a}$, $i_{\bar{w}b}$, and $i_{\bar{w}c}$, and source power $p_{\bar{w}3\phi}$), and (b) by the p-q Theory (i_{Sa}, i_{Sb}, i_{Sc}, and source power $p_{S3\phi}$).

1. The instantaneous real and imaginary powers defined in the time domain form a consistent basis for efficient algorithms to be applied to the control of active power-line conditioners.
2. Clear physical meanings are assigned to the real power, the imaginary power, and the zero-sequence power in the p-q Theory.
3. The compensation algorithms established through minimization methods are relatively simple to implement. However, they are inapplicable to three-phase, four-wire systems, and cannot guarantee constant active power to the source.
4. The generalized Fryze current method results in line currents proportional to the phase voltages and gives the smallest rms value for the compensated currents. However, it is not an "instantaneous" algorithm (it is not an algebraic set of equations).
5. The compensation algorithm based on the p-q Theory is very flexible. The undesirable powers to be compensated can be conveniently selected. The instantaneous imaginary power is calculated without time delay ("instantaneously"). The compensation algorithm using the $\alpha\beta 0$ transformation can allow three-phase loads to provide constant instantaneous active power to the source, even if the supply voltages are unbalanced and/or contain harmonics.

However, it is more complex than the other methods mentioned in (3) and (4).

REFERENCES

[1] S. Fryze, "Wirk-, Blind- und Scheinleistung in elektrischen Stromkreisen mit nicht-sinusförmigem Verlauf von Strom und Spannung," *ETZ-Arch. Elektrotech.*, vol. 53, 1932, pp. 596–599, 625–627, 700–702.

[2] H. Akagi, Y. Kanazawa, and A. Nabae, "Principles and Compensation Effectiveness of a Instantaneous Reactive Power Compensator Devices," in *Meeting of the Power Semiconductor Converters Researchers—IEE-Japan*, SPC-82-16, 1982 (in Japanese).

[3] H. Akagi, Y. Kanazawa, and A. Nabae, "Generalized Theory of Instantaneous Reactive Power and Its Applications," *Transactions of the IEE-Japan, Part B*, vol. 103, no.7, 1983, pp. 483–490 (in Japanese).

[4] H. Akagi, Y. Kanazawa, and A. Nabae, "Generalized Theory of the Instantaneous Reactive Power in Three-Phase Circuits," in *IPEC'83—International Power Electronics Conference*, Tokyo, Japan, 1983, pp. 1375–1386.

[5] H. Akagi, Y. Kanazawa, and A. Nabae, "Instantaneous Reactive Power Compensator Comprising Switching Devices Without Energy Storage Components," *IEEE Trans. Ind. Appl.*, vol. IA-20, no. 3, 1984, pp. 625–630.

[6] M. S. Erlicki and A. Emanuel-Eigeles, "New Aspects of Power Factor Improvements Part I—Theoretical Basis," *IEEE Trans. on Industry and General Applications*, vol. IGA-4, 1968, July/August, pp. 441–446.

[7] H. Sasaki and T. Machida, "A New Method to Eliminate AC Harmonic by Magnetic Compensation—Consideration on Basic Design," *IEEE Trans. on Power Apparatus and Syst.*, vol. 90, no. 5, 1970, pp. 2009–2019.

[8] T. Fukao, H. Iida, and S. Miyairi, "Improvements of the Power Factor of Distorted Waveforms by Thyristor Based Switching Filter," *Transactions of the IEE-Japan, Part B*, vol. 92, no.6, 1972, pp. 342–349 (in Japanese).

[9] L. Gyugyi and B. R. Pelly, *Static Power Frequency Changers: Theory, Performance and Application*, Wiley, 1976.

[10] M. Depenbrock, *Untersuchungen über die Spannungs- und Leistungsverhältnisse bei Umrichtern ohne Energiespeicher*, Ph.D. Thesis, TH Hannover, Germany, 1962.

[11] F. Harashima, H Inaba, and K. Tsuboi, "A Closed-loop Control System for the Reduction of Reactive Power Required by Electronic Converters," IEEE Trans. IECI, vol. 23, no. 2, May 1976, pp. 162–166.

[12] L. Gyugyi and E. C. Strycula, "Active ac Power Filters," in *Proceedings IEEE Industrial Applications Annual Meeting*, vol. 19-C, 1976, pp. 529–535.

[13] I. Takahashi and A. Nabae, "Universal Reactive Power Compensator," in *IEEE—Industry Application Society Annual Meeting Conference Record*, 1980, pp. 858–863.

[14] I. Takahashi, K. Fujiwara, and A. Nabae, "Distorted Current Compensation System Using Thyristor Based Line Commutated Converters," *Transactions of the IEE-Japan, Part B*, vol. 101, no. 3, 1981, pp. 121–128 (in Japanese).

[15] E. Clarke, *Circuit Analysis of A-C Power Systems, Vol. I—Symmetrical and Related Components*, Wiley, 1943.

[16] J. L. Willems, "Instantaneous Sinusoidal and Harmonic Active and Deactive Currents in Three-phase Power Systems," *ETEP—Eur. Trans. Elect. Power Eng.*, vol. 4, no. 5, Sep./Oct. 1994, pp. 335–346.

[17] W. V. Lyon, "Reactive Power and Unbalanced Circuits," *Electrical World*, vol. 75, no. 25, 1920, pp. 1417–1420.

[18] F. Buchholz, "Die Drehstrom-Scheinleistung bei ungleichmäßiger Belastung der drei Zweige," *Licht und Kraft, Zeitschrift für elekt. Energie-Nutzung*, no. 2, Jan. 1922, pp. 9–11.

[19] C. I. Budeanu, "Puissances reactives et fictives," *Instytut Romain de l'Energie*, pub. no. 2, Bucharest, 1927.

[20] C. I. Budeanu, "The Different Options and Conceptions Regarding Active Power in Non-sinusoidal Systems", *Instytut Romain de l'Energie*, pub. no. 4, Bucharest, 1927.

[21] R. M. Kerchner, G. F. Corcoran, *Alternating-Current Circuits*, Wiley, 1938.

[22] A. E. Emanuel, "Apparent and Reactive Powers in Three-Phase Systems: In Search of a Physical Meaning and a Better Resolution," *ETEP—Eur. Trans. Elect. Power Eng.*, vol. 3, no. 1, Jan./Feb. 1993, pp. 7–14.

[23] L. S. Czarnecki, "What Is Wrong with the Budeanu Concept of Reactive and Distortion Power and Why It Should Be Abandoned," *IEEE Trans. Instr. Meas.*, vol. IM-36, no. 3, 1987, pp. 834–837.

[24] P. Filipski and R. Arseneau, "Definition and Measurement of Apparent Power under Distorted Waveform Conditions," *IEEE Tutorial Course on Non-sinusoidal Situations*, 90EH0327-7, 1990, pp. 37–42.

[25] M. Depenbrock, D. A. Marshall, and J. D. van Wyk, "Formulating Requirements for a Universally Applicable Power Theory as Control Algorithm in Power Compensators," *ETEP—Eur. Trans. Elect. Power Eng.*, vol. 4, no. 6, Nov./Dec. 1994, pp. 445–455.

[26] L. S. Czarnecki, "Comparison of Power Definitions for Circuits with Nonsinusoidal Waveforms," *IEEE Tutorial Course on Non-sinusoidal Situations*, 90EH0327-7, 1990, pp. 43–50.

[27] L. S. Czarnecki, "Misinterpretations of Some Power Properties of Electric Circuits," *IEEE Trans. Power Delivery*, vol. 9, no. 4, Oct. 1994, pp. 1760–1769.

[28] L. S. Czarnecki, "Power Related Phenomena in Three-Phase Unbalanced Systems," *IEEE Trans. Power Delivery*, vol. 10, no. 3, July 1995, pp. 1168–1176.

[29] *IEEE Standard Dictionary of Electrical and Electronics Terms*, 3rd ed., IEEE/Wiley-Interscience, 1984.

[30] H. Schering, "Die Definition der Schein- und Blindleistung sowie des Leistungsfaktors bei Mehrphasenstrom," *Elektrotechnische Zeitschrift*, vol. 27., July 1924, pp. 710–712.

[31] V. N. Nedelcu, "Die einheitlische Leistungstheorie der unsymmetrischen und mehrwelligen Mehrphasensysteme," *ETZ-Arch. Elektrotech.* vol. 84, no. 5, 1963, pp. 153–157.

[32] Z. Nowomiejski, "Generalized Theory of Electric Power," *Arch. für Elektrotech.*, vol. 63, 1981, pp. 177–182.

[33] A. E. Emanuel, "Energetical Factors in Power Systems with Nonlinear Loads," *Arch. für Elektrotech.*, vol. 59, 1977, pp. 183–189.

[34] R. Arseneau and P. Filipski, "Application of a Three-Phase Nonsinusoidal Calibration System for Testing Energy and Demand Meters under Simulated Field Conditions," *IEEE Trans. Power Delivery*, vol. 3, July 1988, pp. 874–879.

[35] E. W. Kimbark, *Direct Current Transmission*, vol. 1, Wiley, 1971.

[36] C. L. Fortescue, "Method of Symmetrical Co-ordinates Applied to the Solution of Polyphase Networks," *A.I.E.E. Trans.*, vol. 37, June 1918, pp. 1027–1140.

[37] E. H. Watanabe, R. M. Stephan, and M. Aredes, "New Concepts of Instantaneous Active and Reactive Powers in Electrical Systems with Generic Loads," *IEEE Trans. Power Delivery*, vol. 8, no. 2, Apr. 1993, pp. 697–703.

[38] M. Aredes and E. H. Watanabe, "New Control Algorithms for Series and Shunt Three-Phase Four-Wire Active Power Filters," *IEEE Trans. Power Delivery*, vol. 10, no. 3, July 1995, pp. 1649–1656.

[39] F. Buchholz, "Die Darstellung der Begriffe Scheinleistung und Scheinarbeit bei Mehrphasenstrom," *Elektro-J. 9*, 1929, pp. 15–21.

[40] M. Depenbrock, "The FBD-Method, a Generally Applicable Tool for Analysing Power Relations," *IEEE Trans. Power Systems*, vol. 8, no. 2, May 1993, pp. 381–387.

[41] T. Furuhashi, S. Okuma, and Y. Uchikawa, "A Study on the Theory of Instantaneous Reactive Power," *IEEE Trans. Ind. Elect.*, vol. 37, no. 1, Feb. 1990, pp. 86–90.

[42] L. Malesani, L. Rossetto, and P. Tenti, "Active Filter for Reactive Power and Harmonics Compensation," in *IEEE-PESC'86—Power Electronics Special Conference*, 86CH2310-1, pp. 321–330.

[43] L. Rossetto and P. Tenti, "Evaluation of Instantaneous Power Terms in Multi-Phase Systems: Techniques and Application to Power-Conditioning Equipments," *ETEP—Eur. Trans. Elect. Power Eng.*, vol. 4, no. 6, Nov./Dec. 1994, pp. 469–475.

[44] P. Tenti, "The Relation Between Global Performance Indexes: Power Factor and FBD Factor," *ETEP—European Transactions on Electrical Power Engineering*, vol. 4, no. 6, Nov./Dec. 1994, pp. 505–507.

[45] J. H. R. Enslin and J. D. van Wyk, "Measurement and Compensation of Fictitious Power under Nonsinusoidal Voltage and Current Conditions," *IEEE Trans. Instr. Meas.*, vol. IM-37, no. 3, 1988, pp. 403–408.

[46] M. A. Slonim and J. D. van Wyk, "Power Components in Systems with Sinusoidal and Nonsinusoidal Voltage and/or Current Conditions," *IEE Proc.*, vol. 135, Pt. B, no. 2, Mar. 1988, pp. 76–84.

[47] D. A. Marshall and J. D. van Wyk, "An Evaluation of the Real-Time Compensation of Fictitious Power in Electric Energy Networks," *IEEE Trans. Power Delivery*, vol. 6, no. 4, Oct. 1991, pp. 1774–1780.

[48] S. Togasawa, T. Murase, H. Nakano, and A. Nabae, "Reactive Power Compensation Based on a Novel Cross-Vector Theory," *Trans. on Ind. App. of the IEE—Japan (IEEJ)*, vol. 114, no. 3, Mar. 1994, pp. 340–341 (in Japanese).

[49] A. Nabae, L. Cao, and T. Tanaka, "An instantaneous distortion current compensator without any coordinate transformation," in *Proceedings of the International Power Electronics Conference (IPEC-Yokohama)*, 1995, pp. 1651–1655.

[50] A. Nabae and T. Tanaka, "New definition of instantaneous active-reactive current and power based on instantaneous space vectors on polar coordinates in three-phase circuits," *IEEE Trans. Power Delivery*, vol. 11, no. 3, pp. 1238–1243, July 1996.

[51] H. Akagi, S. Ogasawara, and H. Kim, "The Theory of Instantaneous Power in Three-Phase Four-Wire Systems: A Comprehensive Approach," in *IEEE IAS Annual Meeting*, 1999, pp. 431–439.

[52] F. Z. Peng and J. S. Lai, "Generalized Instantaneous Reactive Power Theory for Three-

Phase Power System," *IEEE Trans. on Instr. and Meas.,* vol. 45, no. 1, Feb. 1996, pp. 293–297.

[53] F. Z. Peng, G. W. Ott Jr., and D. J. Adams, "Harmonic and Reactive Power Compensation Based on the Generalized Instantaneous Reactive Power Theory for Three-Phase Four-Wire Systems," *IEEE Trans. on Power Electronics,* vol. 13, no. 6, Nov. 1998, pp. 1174–1181.

[54] X. Dai, G. Liu, and R. Gretsh, "Generalized Theory of Instantaneous Reactive Quantity for Multiphase Power System," *IEEE Trans. on Power Delivery,* vol. 19, no. 3, July 2004, pp. 965–972.

CHAPTER 4

SHUNT ACTIVE FILTERS

The concept of shunt active filtering was first introduced by Gyugyi and Strycula in 1976 [1]. Nowadays, a shunt active filter is not a dream but a reality, and many shunt active filters are in commercial operation all over the world. Their controllers determine *in real time* the compensating current reference, and force a power converter to synthesize it accurately. In this way, the active filtering can be selective and adaptive. In other words, a shunt active filter can compensate only for the harmonic current of a selected nonlinear load, and can continuously track changes in its harmonic content. This chapter will present several approaches to shunt active filters, including applications to three-phase grounded or ungrounded systems. All shunt active filters described in this chapter have controllers based on the instantaneous active and reactive power theory (the *p-q* Theory) presented in Chapter 3.

Harmonic currents are generated mainly due to the presence of:

1. Nonlinear loads
2. Harmonic voltages in the power system

Figure 4-1 summarizes the basic concept of shunt active filtering. A nonlinear load draws a fundamental current component I_{LF} and a harmonic current I_{Lh} from the power system. The harmonic current I_{Sh} is induced by the source harmonic voltage V_{Sh}. A shunt active filter can compensate both harmonic currents I_{Sh} and I_{Lh}. However, the principal function of a shunt active filter is compensation of the load harmonic current I_{Lh}. This means that the active filter confines the load harmonic current at the load terminals, hindering its penetration into the power system. For simplicity, the power system is represented only by an equivalent impedance X_L in Fig. 4-1. If the load harmonic current I_{Lh} flows through the power system, it pro-

Instantaneous Power Theory and Applications to Power Conditioning. By Akagi, Watanabe, & Aredes
Copyright © 2007 the Institute of Electrical and Electronics Engineers, Inc.

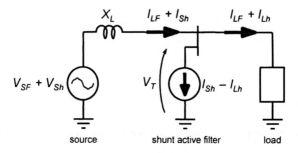

Figure 4-1. Principle of shunt current compensation.

duces an additional harmonic voltage drop equal to $X_L \cdot I_{Lh}$ that further degenerates the load terminal voltage V_T.

The principle of shunt current compensation shown in Fig. 4-1 is very effective in compensating harmonic currents of loads. However, a shunt active filter that realizes this principle of shunt current compensation should also draw an additional harmonic current I_{Sh}, in order to keep the load terminal voltage sinusoidal and equal to $V_T = V_{SF} - X_L \cdot I_{LF}$. The harmonic voltage drop appearing across the equivalent impedance becomes equal to the source harmonic voltage if $V_{Sh} = X_L \cdot I_{Sh}$. In this case, the harmonic voltage components cancel each other, so that the terminal voltage V_T is kept sinusoidal.

If the system impedance X_L is low, the harmonic current I_{Sh} that should be drawn by the shunt active filter can be very high [2]. This can strongly increase the power rating of the shunt active filter, making it impractical. Therefore, if the power system has a high short-circuit capacity, which is the same as saying that it has a low equivalent impedance X_L, or if it has an already significant level of voltage distortion, the active filtering of current I_{Sh} should be left for other filter configurations. For instance, an interesting solution is to install a series active filter at the load terminals for direct compensation of the harmonic voltage V_{Sh}, instead of the use of a shunt active filter to drain the harmonic current I_{Sh} from the power system. Note that the principle of series voltage compensation is the complement shunt current compensation. In other words, if the series active filter generates a compensating voltage equal to V_{Sh}, it forces the harmonic current I_{Sh} to become zero. On the other hand, as mentioned above, if the shunt active filter draws a compensating current equal to $-I_{Lh}$, it confines the load harmonic current at the load terminals, hindering its penetration into the power system. Series active filters will be introduced in the next chapter.

The shunt active filter can be properly controlled to present a selective compensation characteristic. In other words, it is possible to select what current is to be compensated. That is, it can compensate the source current I_{Sh} and/or the load current I_{Lh}, or even an arbitrarily chosen set of harmonic components of them. Most applications of shunt active filters are intended to compensate for the load current harmonics produced by a specific load [4–17].

Another interesting compensation function that a shunt active filter can realize is to provide harmonic damping in power lines, in order to avoid harmonic propagation resulting from harmonic resonances between the series inductances and shunt capacitors [22,23]. Note that this is not a case of direct compensation of voltage harmonics like V_{Sh} in Fig. 4-1, but rather a mitigation of voltage harmonics because of harmonic damping. The following sections will present several configurations of shunt active filters to provide load current compensation as well as harmonic damping.

4.1. GENERAL DESCRIPTION OF SHUNT ACTIVE FILTERS

Shunt active filters generally consist of two distinct main blocks:

1. The PWM converter (power processing)
2. The active filter controller (signal processing)

The PWM converter is responsible for power processing in synthesizing the compensating current that should be drawn from the power system. The active filter controller is responsible for signal processing in determining *in real time* the instantaneous compensating current references, which are continuously passed to the PWM converter. Figure 4-2 shows the basic configuration of a shunt active filter for harmonic current compensation of a specific load. It consists of a voltage-fed converter with a PWM current controller and an active filter controller that realizes an almost instantaneous control algorithm. The shunt active filter controller works in a

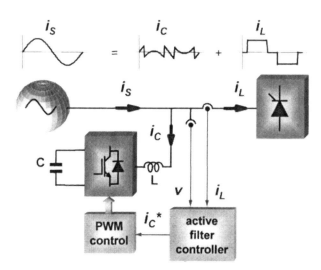

Figure 4-2. Basic configuration of a shunt active filter.

closed-loop manner, continuously sensing the load current i_L, and calculating the instantaneous values of the compensating current reference i_C^* for the PWM converter. In an ideal case, the PWM converter may be considered as a linear power amplifier, where the compensating current i_C tracks correctly its reference i_C^*.

The PWM converter should have a high switching frequency (f_{PWM}) in order to reproduce accurately the compensating currents. Normally, $f_{PWM} > 10 f_{hmax}$, where f_{hmax} represents the frequency of the highest order of harmonic current that is to be compensated. In Fig. 4-2, the dc capacitor and the IGBT (insulated gate bipolar transistor) with antiparallel diode are used to indicate a shunt active filter that is built up from a voltage-source converter (VCS). In fact, voltage-source converters or current-source converters can be used in shunt active filters [14,17,18]. Nowadays, almost all shunt active filters in commercial operation use voltage-source converters [18]. Voltage-source converter (VSC) and voltage-fed converter are synonymous. However, this converter is commonly referred to as the VSC. Thus, this book will do so. Similarly, CSC will be used to refer to current-source converters. Regardless if it is based on a VSC or a CSC, the PWM control must have a minor *current* feedback loop to force the power converter to behave as a controlled, nonsinusoidal current source.

4.1.1. PWM Converters for Shunt Active Filters

Figure 4-3 shows three-phase power converters for implementing shunt active filters. Figure 4-3(a) is a voltage-source converter (VSC) and Fig. 4-3(b) is a current-source converter (CSC). The associated PWM current controllers of each converter have different designs. However, both PWM controllers have the same functionality: to force the converter to behave as a controlled current source. It should be noted that no power supply, only an energy storage element (capacitor for the VSC and inductor for the CSC) is connected at the dc side of the converters. The reason is that the principal function of a shunt active filter is to behave as a compensator. In other words, the *average energy* exchanged between the active filter and the power system should be zero. In addition, the active filter controller should be designed to

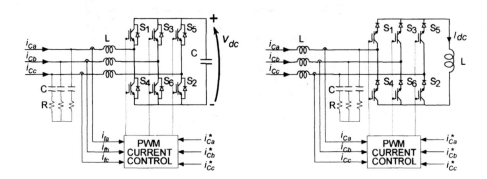

Figure 4-3. PWM converters for shunt active filters.

keep constant the average dc voltage of the VSC, or the average dc current of the CSC, and it should force the power system to supply the losses in the power converter.

PWM converters generate undesirable current harmonics around the switching frequency and its multiples. If the switching frequency of the PWM converter is sufficiently high, these undesirable current harmonics can be easily filtered out by using small, passive high-pass filters represented by R and C in Fig. 4-3. Ideally, the switching-frequency current harmonics are fully cut out, and the compensating currents i_{Ck} correctly track its references i^*_{Ck} ($k = a, b, c$).

Sometimes, it is convenient to implement three-phase active filters using three single-phase converters instead of a single three-phase converter shown in Fig. 4-3. However, in order to compensate for the three-phase instantaneous reactive power (imaginary power in the p-q Theory) without the need for energy storage elements, as explained in Chapter 3, a single dc capacitance in the case of using a VSC, or a single dc inductance in the case of using a CSC, should be employed with the three single-phase converters connected in parallel to it.

A critical comparison of the performance of VSCs and CSCs when used as the power converters of shunt active filters is beyond scope of this book. One may prefer the CSC due to its robustness [1,8] or the VSC due to its high efficiency, low initial cost, and smaller physical size [18,20]. Moreover, the IGBT module that is now available on the market is more suitable for the voltage-source PWM converter because a free-wheeling diode is connected in antiparallel with each IGBT. This means that the IGBT does not need to provide the capability of reverse voltage blocking in itself, thus bringing more flexibility to device design in a compromise between conducting and switching losses and short-circuit capability than the reverse-blocking IGBT. On the other hand, the current-source PWM converter requires either series connection of a traditional IGBT and a reverse-blocking diode as shown in Fig. 4-3(b), or the reverse-blocking IGBT that leads to more complicated device design and fabrication, and slightly worse device characteristics than the traditional IGBT without reverse-blocking capability. In fact, almost all active filters that have been put into practical applications have adopted the voltage-source PWM converter equipped with the dc capacitor as the power circuit [19].

The authors of [21] describe shunt active filters using a voltage-source PWM converter and a current-source PWM converter, and focus on comparing them from various points of view.

4.1.2. Active Filter Controllers

The control algorithm implemented in the controller of the shunt active filter determines the compensation characteristics of the shunt active filter. There are many ways to design a control algorithm for active filtering. Certainly, the p-q Theory forms a very efficient basis for designing active filter controllers. Some control strategies for active filters based on this theory will be discussed here.

The general principles of shunt current compensation described in Chapter 3 are the guidelines for explaining the following design methods of active filter controllers. The controller design is particularly difficult if the shunt active filter is applied in power systems in which the supply voltage itself has been already distorted and/or unbalanced. The general expressions of the *p-q* Theory show that it is impossible to compensate the load current and force the compensated source current to satisfy *simultaneously* the following three "optimal" compensation characteristics if the power system contains voltage harmonics and/or imbalances at the fundamental frequency:

1. Draw a constant instantaneous active power from the source
2. Draw a sinusoidal current from the source
3. Draw the minimum rms value of the source current that transports the same energy to the load with minimum losses along the transmission line. This means that the source has current waveforms proportional to the corresponding voltages

Under three-phase sinusoidal balanced voltages, it is possible to satisfy simultaneously the three optimal compensation characteristics given above. Example #3 in Chapter 3 shows a compensation case in which the compensated source current becomes sinusoidal, and has the minimum rms value to supply $\bar{p}$ to the load. However, under nonsinusoidal and/or unbalanced system voltages, the shunt active filter can compensate load currents to guarantee *only one* optimal compensation characteristic [24]. Therefore, a choice must be made before designing the controller of a shunt active filter. This is the reason to derive three different control strategies:

1. Constant instantaneous power control strategy
2. Sinusoidal current control strategy
3. Generalized Fryze current control strategy

The following sections will present implementation details of these control strategies.

Under sinusoidal, balanced system voltages, the three control strategies can produce the same results. However, under nonsinusoidal and/or unbalanced system voltages, each control strategy guarantees its respective compensation characteristic. Hence, the resultant compensated source currents are different. The following example clarifies this point, showing simplified results from ideal cases.

Example #1

Suppose that the three-phase voltage of a hypothetical infinite bus is unbalanced and distorted as shown in Fig. 4-4 (upper left). The level of imbalance and harmonic content were exaggerated to make easier the following explanation. The three different control strategies mentioned before were applied to compensate for the same hypothetical load current, which is not shown in Fig. 4-4. Ideal shunt active filters were considered, and the power losses were neglected. Therefore, the three

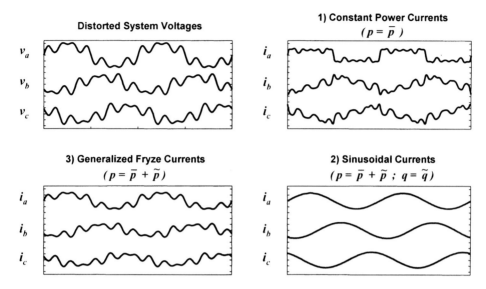

Figure 4-4. Results from three different current compensation methods under nonsinusoidal system voltages. 1) Constant instantaneous power control strategy. 2) Sinusoidal current control strategy. 3) Generalized Fryze currents control strategy.

compensation methods produce source currents (compensated currents) that are equivalent from an average energy transfer point of view. In other words, all compensated currents produce the same average power ($\bar{p}$).

The first control strategy guarantees that only this portion of power is drawn from the source. According to the p-q Theory, to draw constant instantaneous active power from the source means that the shunt active filter must compensate for the oscillating real power ($\tilde{p}$). Additionally, the rms value of the compensated current is minimized by the compensation of the total imaginary power $q = \bar{q} + \tilde{q}$ of the load. There is no zero-sequence power because a three-phase, three-wire system is being considered. If the system voltage contains harmonics and/or imbalance at the fundamental frequency, the compensated current cannot be sinusoidal to guarantee constant real power, $\bar{p}$, that is drawn from the source. Figure 4-4 (upper right) shows that the compensated source currents are fully asymmetrical and distorted, although they produce the same three-phase constant power $\bar{p}$ as that produced by the original load current. This case might be interesting if no real-power oscillation between the source and the load is desired.

The second control strategy is applied if the shunt active filter must compensate the load current to guarantee balanced, sinusoidal current drawn from the power system. Additionally, the active filter is compensating reactive power so that the compensated current is in phase with the fundamental positive-sequence component of the voltage. This case is shown in the bottom right of Fig. 4-4. However, this current does not produce constant real power as long as the system voltage is nonsinusoidal and unbalanced. The reason was well explained by (3.94) to (3.99) in Chapter 3. Therefore, the portions of oscillating real and imaginary power are drawn from the power system, accompanied by increased ohmic losses in the transmission system, because the source current does not have the minimum rms value that transfers the same energy, represented by $\bar{p}$, as the original load current.

It was seen in Chapter 3 that the generalized Fryze current method minimizes the compensated current, and gives the minimum rms value of current that can transfer the same amount of energy as the uncompensated current. Hence, this minimum rms current produces minimum ohmic losses in the transmission line. This feature is associated with the above-listed compensation characteristic #3, although it cannot guarantee any sinusoidal compensated current or constant instantaneous active power drawn from the power system. This control strategy makes the compensated line current proportional to the corresponding phase voltage, that is, they have the same waveform and behave like a "pure resistive" load, as shown in the bottom left of Fig. 4-4. These currents contain harmonics, which may not be a problem, but the instantaneous real power contains oscillating components. Oscillating components in three-phase active power may be accompanied by mechanical vibration in electric machines.

From the above analysis, we can see that harmonic compensation can have different functionalities. The solution may be different if the objective is to:

1. Eliminate real power oscillations
2. Improve power factor
3. Eliminate current harmonics
4. Provide harmonic damping

For load-current compensation, one principal task of the control method is to determine "instantaneously" the current harmonics from the distorted load current, whereas voltage harmonics should be detected for harmonic damping.

A special controller should be designed for a three-phase, four-wire active filter. In this case, the active filter deals not only with the real and imaginary power, but also with the zero-sequence power of the load.

All compensation aspects mentioned above will be seen in some applications of shunt active filter in the next sections, comprising three-phase, three-wire shunt active filters as well as three-phase, four-wire shunt active filters.

4.2. THREE-PHASE, THREE-WIRE SHUNT ACTIVE FILTERS

A particular characteristic of three-phase, three-wire systems is the absence of the neutral conductor and, consequently, the absence of zero-sequence current components. Thus, the zero-sequence power is always zero in these systems.

Figure 4-5 shows the most important parts of a three-phase, three-wire shunt active filter for current compensation. The control block that calculates the instanta-

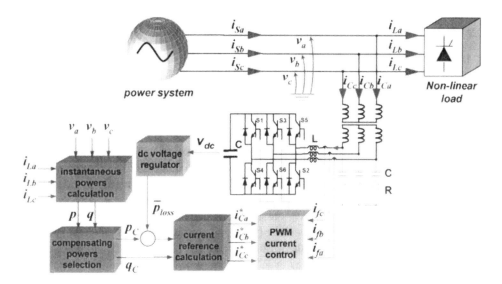

Figure 4-5. The three-phase, three-wire shunt active filter.

neous power takes as inputs the phase-voltages at the point of common coupling (PCC) and the line currents of the nonlinear load that should be compensated. This means that the shunt active filter has a selective compensation characteristic. In other words, it behaves as an open circuit for harmonic currents generated by other adjacent nonlinear loads.

The shunt active filter for load-current compensation, as shown in Fig. 4-5, is one of the most common active filters. As mentioned before, the shunt active filter can also provide harmonic damping throughout the power line to avoid "harmonic propagation" resulting from harmonic resonance between the series inductance of the power system and shunt capacitors for power-factor correction. Later, this shunt active filter for harmonic damping will be described.

The active filter controller consists of four functional control blocks:

1. Instantaneous-power calculation
2. Power-compensating selection
3. dc-voltage regulator
4. Current reference calculation

The first block calculates the instantaneous powers of the nonlinear load. According to the *p-q* Theory, only the real and imaginary powers exist, because the zero-sequence power is always zero. The second block determines the behavior of the shunt active filter. In other words, it selects the parts of the real and imaginary powers of the nonlinear load that should be compensated by the shunt active filter. Additionally, the dc voltage regulator determines an extra amount of *real power,* represented by $\bar{p}_{loss}$ in Fig. 4-5, that causes an additional flow of energy to (from) the dc capacitor in order to keep its voltage around a fixed reference value. This real power $\bar{p}_{loss}$ is added to the compensating real power p_C, which, together with the compensating imaginary power q_C, are passed to the current-reference calculation block. It determines the instantaneous compensating current references from the compensating powers and voltages. The control block structure differs slightly from the three control strategies mentioned in Section 4.1.2. All control blocks in Fig. 4-5 will be detailed later, according to each compensation strategy.

The power circuit of the shunt active filter consists of a three-phase voltage-source converter made up of IGBTs and antiparallel diodes. The PWM current control forces the VSC to behave as a controlled *current* source. In order to avoid high di/dt, the coupling of a VSC to the power system must be made through a series inductor, commonly known as a commutation inductor or a coupling inductor. In some cases, the leakage inductance of a normal power transformer is enough to provide di/dt limitation, so that the series inductor can be eliminated. In this case, the small passive filter, represented by R and C in Fig. 4-5, for filtering the current ripples around the switching frequency should be installed at the primary side of the transformer.

4.2.1. Active Filters for Constant Power Compensation

The constant power compensation control strategy for a shunt active filter was the first strategy developed based on the *p-q* Theory, and was introduced by Akagi et al. in 1983 [3,4]. The principles of this compensation method are described in Chapter 3. In terms of real and imaginary power, in order to draw a constant instantaneous power from the source, the shunt active filter should be installed as close as possible to the nonlinear load, and should compensate the oscillating real power $\tilde{p}$ of this load. Note that a three-phase system without neutral wire is being considered and, therefore, the zero-sequence power is zero. Hence, the shunt active filter should supply the oscillating portion of the instantaneous active current of the load, that is,

- Oscillating portion of the instantaneous active current on the α axis $i_{\alpha\tilde{p}}$:

$$i_{\alpha\tilde{p}} = \frac{v_\alpha}{v_\alpha^2 + v_\beta^2}(-\tilde{p}) \qquad (4.1)$$

- Oscillating portion of the instantaneous active current on the β axis $i_{\beta\tilde{p}}$:

$$i_{\beta\tilde{p}} = \frac{v_\beta}{v_\alpha^2 + v_\beta^2}(-\tilde{p}) \qquad (4.2)$$

The reason for adding a minus sign to the real power in the above equations is to match them with the current directions adopted in Fig. 4-5. The load and the active filter currents have positive values when flowing into the load and into the active filter, that is, the "load current convention" is adopted. If the shunt active filter draws a current that produces exactly $-\tilde{p}$ of the load, the power system would supply only the constant portion of the real power ($\bar{p}$) of the load. In order to compensate ($-\tilde{p}$), which implies an oscillating flow of energy, the dc capacitor of the PWM converter (see Fig. 4-5) must be made large enough to behave as an energy storage element, so as not to experience large voltage variations. Remember that if the dc voltage gets lower than the amplitude of the ac voltage, this kind of PWM converter (a boost type converter) loses its controllability.

If desired, the shunt active filter can optimize further the compensated currents by also filtering the portion of load current that produce imaginary power. In this case, it should also compensate the instantaneous reactive currents $i_{\alpha q}$ and $i_{\beta q}$:

- Instantaneous reactive current on the α axis $i_{\alpha q}$:

$$i_{\alpha q} = \frac{v_\beta}{v_\alpha^2 + v_\beta^2}(-q) \quad (4.3)$$

- Instantaneous reactive current on the β axis $i_{\beta q}$:

$$i_\beta = \frac{-v_\alpha}{v_\alpha^2 + v_\beta^2}(-q) \quad (4.4)$$

Note that the total imaginary power ($-q = -\bar{q} - \tilde{q}$) is being compensated. The reason for the minus sign is the same as explained for the real oscillating power compensation. Contrarily to compensation of ($-\tilde{p}$), compensation of the total imaginary power ($-q$) does not require any energy storage elements, as explained in Chapter 3.

If the shunt active filter compensates the oscillating real and imaginary power of the load, it guarantees that only a constant real power $\bar{p}$ (average real power of the load) is drawn from the power system. Therefore, the *constant instantaneous power control strategy* provides optimal compensation from a power flow point of view, even under nonsinusoidal or unbalanced system voltages. Figure 4-6 illustrates the idea in terms of "$\alpha\beta$ wires" and the *p-q* Theory.

As mentioned before, a dc voltage regulator should be added to the control strategyin a real implementation, as shown in Fig. 4-5. In fact, a small amount of average real power ($\bar{p}_{loss}$), not represented in Fig. 4-6, must be drawn continuously from the power system to supply switching and ohmic losses in the PWM converter. Otherwise, this energy would be supplied by the dc capacitor, which would discharge continuously. The power converter of the shunt active filter is a boost-type converter. This means that the dc voltage must be kept higher than the peak value of the ac-bus voltage, in order to guarantee the controllability of the PWM current control.

Figure 4-6 suggests that the real power of the nonlinear load should be continuously measured, and somehow "instantaneously" separated into its average ($\bar{p}$) and oscillating ($\tilde{p}$) parts. This would be the function of the block named "selection of the powers to be compensated" in Fig. 3-11. In a real implementation, the separation of $\bar{p}$ and $\tilde{p}$ from p is realized through a low-pass filter. The low-pass filter and

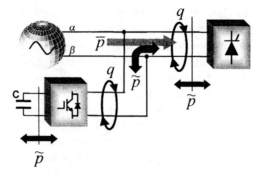

Figure 4-6. Optimal power flow provided by a shunt current compensation.

its cutoff frequency must be selected carefully as to the inherent dynamics that lead to compensation errors during transients. Unfortunately, the unavoidable time delay introduced by the low-pass filter may degenerate the entire performance of the shunt active filter during transients. Some simulation results will be shown later to clarify this point. In practice, a fifth-order Butterworth low-pass filter with a cutoff frequency between 20 and 100 Hz has been used successfully to separate $\bar{p}$ from p. In a digital implementation, a moving-average filter is also a very simple and effective solution. The oscillating real power is determined by the difference, that is, $\tilde{p} = p - \bar{p}$. A lower cutoff frequency in the low-pass filter may be required, depending on the spectral components included in $\tilde{p}$ that is to be compensated.

Figure 4-7 summarizes the complete algorithm of a controller for a three-phase, three-wire shunt active filter that compensates the oscillating real power and the imaginary power of the load (constant instantaneous power control strategy).

Next, the influence of the low-pass filter dynamic and the PI-Controller in the dc voltage regulator will be addressed during transients. Suppose that a nonlinear load draws the line currents shown in Fig. 4-8. In this example, the power system frequency is 50 Hz, and the nonlinear load is a six-pulse thyristor bridge (three-phase full bridge). The firing angle is set to be 30°, and the blocking/unblocking bridge control has a period of 120 ms. An appropriate thyristors' pair in the same leg is fired to block the bridge, and to provide a bypass to the dc current during the blocking periods, impeding the current to flow through the ac side.

Now, a question arises, "what is the period to be considered to calculate the average real power ($\bar{p}$) of the load?" This affects the set point of the cutoff frequency of the low-pass filter in Fig. 4-7. If a low cutoff frequency is chosen, a long period is considered. First, it was set to be 1 Hz. In this case, certainly a period longer than 120 ms is considered to determine the average real power $\bar{p}$. In the steady state, the real power of the load, p, and the oscillating real power that the shunt active filter compensates is determined by: $p_C = p - \bar{p}$, as shown in Fig. 4-9. Actually, due to the current direction adopted in Fig. 4-5, the shunt active filter should compensate for $-\tilde{p}$. Note that the active filter controller in Fig. 4-7 uses $-\tilde{p}$ to calculate the compensating current references.

4.2. THREE-PHASE, THREE-WIRE SHUNT ACTIVE FILTERS

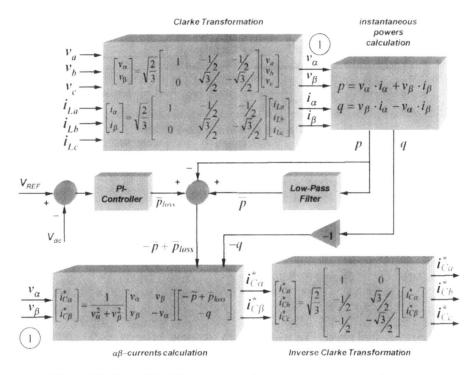

Figure 4-7. Control block for the constant instantaneous power control strategy.

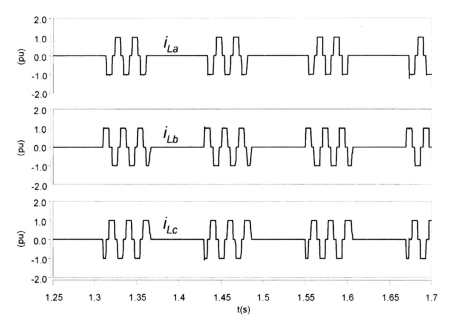

Figure 4-8. Line currents of a nonlinear load.

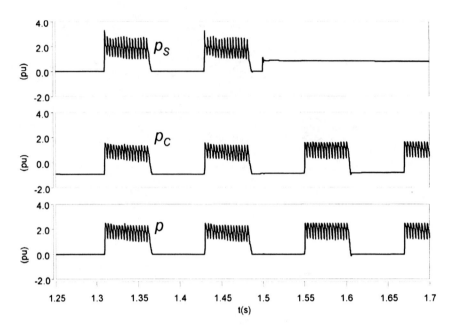

Figure 4-9. Real power in the source (upper), compensating power reference (middle), and the load power (bottom).

The shunt active filter starts at $t = 1.5$ s. After this instant, the active filter compensates $-\tilde{p}$ and the power system supplies only the constant real power $\bar{p}$ of the load, as can be seen in the curve of p_S in Fig. 4-9. The controller of the shunt active filter of Fig. 4-7 also compensates the imaginary power of the load, that is, it compensates $-q_C$, which corresponds to the opposite curve of q_C shown in Fig. 4-10, due to the same reason above for compensating $-\tilde{p}$. This happens after $t = 1.5$ s, as can be seen in the curve of q_S (imaginary power of the source) in Fig. 4-10.

The compensating currents of the shunt active filter and the currents drawn from the source are given in Fig. 4-11 and Fig. 4-12, respectively. After $t = 1.5$ s, the source current becomes sinusoidal as a result of compensation for $\tilde{p}$ and $\tilde{q}$, and it becomes in phase with the voltage due to compensation for $\bar{q}$. The sinusoidal waveform of the compensated current results from the balanced, sinusoidal supply voltage that was considered. In this case, only the fundamental positive-sequence component of the load current (i_{+1}) can produce the constant real power that is the goal of this control strategy. It should be remarked again that this is the main characteristic of the compensation strategy implemented in Fig. 4-7. If the supply voltage were unbalanced and/or distorted, the compensated current would not be sinusoidal, and would not draw constant real power. Examples of this feature were given in Chapter 3.

The very low cutoff frequency (1 Hz) of the low-pass filter makes the shunt active filter consider the blocking and unblocking work cycles of the thyristor bridge as a low-frequency, oscillating real power of the load that should be compensated.

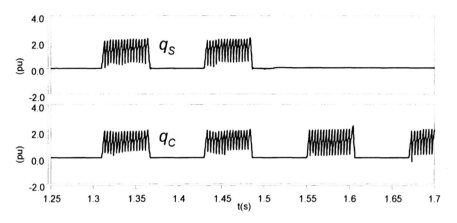

Figure 4-10. Imaginary powers.

The blocking/unblocking cycle has a period of 120 ms, where during 60 ms it is blocked and 60 ms unblocked. This means that a very large energy storage element (very large dc capacitor), or even a dc source (for instance, a battery) should be connected to the active filter. The large energy storage element should absorb energy during the "long period," equal to 60 ms, without causing overvoltage in the dc capacitor. Then, it should release the same amount of energy during the subsequent semicycle.

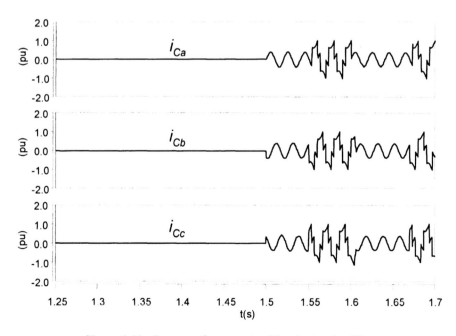

Figure 4-11. Compensating currents of the shunt active filter.

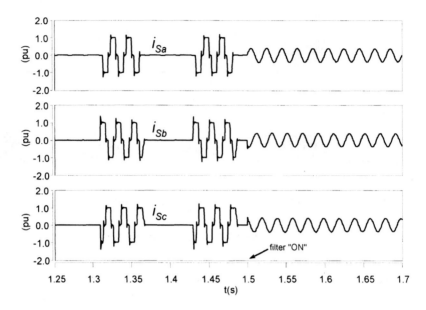

Figure 4-12. Compensated currents of the source.

For simplicity, the dc capacitor was replaced by an ideal dc voltage source, eliminating the dc voltage regulator (see Fig. 4-7) that generates the signal $\bar{p}_{loss}$ in the above simulationin. This is not a practical solution, principally due to economical reasons. The next measurements, Fig. 4-13 to Fig. 4-17, show dc voltage fluctuations when the low-frequency oscillating power of the load is compensated and a dc capacitor is used as an energy storage element in the shunt active filter. The following discussion is based on experimental results from one of the first prototypes of a shunt active filter based on the *p-q* Theory [43].

The experimental prototype of the shunt active filter described in [43] had a power rating equal to 7 kVA (200 V). The power circuit consisted of a quadruple voltage-source PWM converter. Four three-phase PWM converter modules are combined through a magnetic interface, also called a coupling transformer, at the ac side, and they are connected in parallel at the dc side. The PWM current control (see Fig. 4-5) of each phase of each converter compares the actual current with its reference, and the output of the comparator is sampled and held at a frequency of 33 kHz (1/30 μs). The sequence of sampling for each converter is delayed by 30/4 μs from each other. This technique of phase shifting produces an equivalent switching frequency at the primary side of the coupling transformer roughly four times as high as that of each converter.

The experimental compensation system that was mounted to test the active filter prototype is similar to that shown in Fig. 4-5. Only the power circuit of the shunt active filter differs substantially, as described above. The nonlinear load consisted of a three-phase thyristor bridge converter of 20 kVA. To discuss the compensating characteristics in transient states, the firing angle of the thyristor bridge converter is controlled to generate the following dc current:

$$i_{dc} = I_d + I_{do} \sin(2\pi f_0 t) \tag{4.5}$$

In the experiment, I_d was set to 50 A and I_{do} to 30 A. The ac system frequency is 50 Hz and the frequency of the pulsating dc current portion was set to $f_0 = 10$ Hz. This pulsating dc current affects the ac current, and introduces subharmonics, as well as superharmonics, besides all the characteristic harmonics of a six-pulse thyristor bridge operating at a constant firing angle. For instance, 40-Hz subharmonic and 60-Hz superharmonic current components appear, which are very difficult to filter by means of passive filtering. The following discussion is focused on active filtering of such harmonic current components.

In Fig. 4-13, v_a is the a-phase voltage, i_{dc} is the dc current of the thyristor bridge (nonlinear load), i_{La} is the a-phase ac line current, p_L is the real power, and q_L is the imaginary power of the thyristor bridge. Instead of the originally proposed symbol "IVA" [4], this book adopts the symbol "vai" for the unit of the imaginary power. The imaginary power of the load has a positive average value ($\bar{q} > 0$), because the thyristor bridge produces lagging currents, representing an "inductive" load. Figure 4-13 concludes that it is possible to conclude that the 10-Hz firing angle oscillation introduced in the control of the thyristor bridge causes almost only imaginary power ($\tilde{q}$) oscillation. Although relatively small, only the oscillating real power ($\tilde{p}$) of the load, which actually represents energy flow oscillation, is responsible for dc voltage fluctuation on the energy storage element of the shunt active filter.

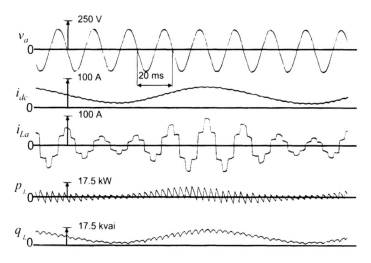

Figure 4-13. Voltage and current waveforms for a thyristorbridge with a 10 Hz oscillating firing angle.

In the controller of the active filter prototype, a fifth-order Butterworth low-pass filter was used to extract $\bar{p}$ from p_L, and a second one, not represented in Fig. 4-7, was used to extract $\bar{q}$ from q_L. Thus, the shunt active filter prototype is compensating only the oscillating powers of the load, that is, it compensates only $(-\tilde{p})$ and $(-\tilde{q})$, determined from p_L and q_L of the load. First, the cutoff frequency of both Butterworth filters was set at 150 Hz. In this case, the active filter compensates only high-frequency current harmonics that produce power oscillations at frequencies greater than 150 Hz. Fig. 4-14 shows the results with the cutoff frequency of both Butterworth filters at 150 Hz. Hence, the 10-Hz power oscillation and other low-frequency components caused by the firing angle pulsation are not being compensated, as can be seen in the curve of i_{Sa} of the compensated current in the source. The high content of fifth and seventh harmonics, characteristic of six-pulse thyristor bridges, produces $\tilde{p}$ and $\tilde{q}$ at frequencies higher than 150 Hz, which are being compensated. Thus, the compensated current i_{Sa} becomes almost sinusoidal, although it still has an amplitude modulation at 10 Hz. It is not in phase with the supply voltage, because $\bar{q}$ is not being compensated. Since the active filter does not compensate low-frequency real power, the voltage v_{dc} across the large dc capacitor does not fluctuate too much.

In order to evidence low-frequency dc voltage fluctuations, the cut-off frequency of the Butterworth filter for extraction of $\tilde{p}_L$ is reduced to 0.9 Hz. The other, for extraction of $\tilde{q}_L$, was kept at 150 Hz. The results are shown in Fig. 4-15. It is possible to see a 10-Hz dc voltage fluctuation much more pronounced than that in Fig. 4-14. Since now almost all oscillating *real* power components, that is, all frequencies in $\tilde{p}_L$ that are greater than 0.9 Hz are being compensated, the compensated current i_{Sa} does not produce any power-flow oscillation in the power source. However, a great

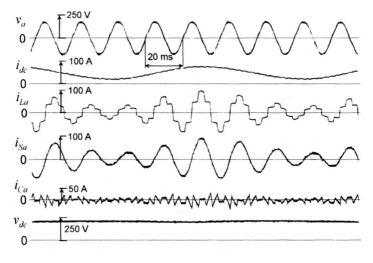

Figure 4-14. Compensation results when the cutoff frequencies of both Butterworth filters for $\tilde{p}_L$ and $\tilde{q}_L$ calculation are set to 150 Hz.

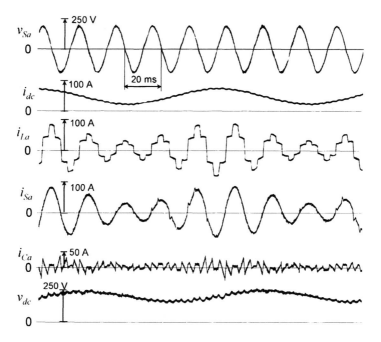

Figure 4-15. Compensation results when the cutoff frequency of the Butterworth filter for extraction of $\tilde{p}_L$ is set to 0.9 Hz, and the other for extraction of $\tilde{q}_L$ is set to 150 Hz.

portion of 10-Hz amplitude modulation component is still present in i_{Sa} due to the remaining, uncompensated, 10-Hz-oscillating, imaginary power. The current spikes present in part of the source current i_{Sa} reveal temporary incompleteness in current control [45]. During short time intervals, the shunt active filter cannot compensate fully the high di/dt present in the load current. Low dc-voltage levels together with peak values of the ac voltage contribute to this temporary incompleteness in current control.

Another interesting case is given in Fig. 4-16, where the cutoff frequency only for extraction of $\tilde{q}_L$ is reduced to 0.9 Hz and the other for extraction of $\tilde{p}_L$ is set to 150 Hz. Now, all oscillating imaginary power components that are related to oscillation in firing angle at 10 Hz are being compensated. This portion of power compensation comprises almost all undesirable effects at 10 Hz. Hence, the amplitude of the compensated current of the source becomes almost constant, although the low-frequency real power is not being compensated. As expected, the energy storage element (dc capacitor) of the shunt active filter is not required to absorb and inject low-frequency power, so that the dc voltage fluctuation is very low.

Finally, the cutoff frequencies of both Butterworth filters are made equal to 0.9 Hz, and the results are given in Fig. 4-17. The amplitude of the compensated current i_{Sa} becomes constant, although i_{Sa} is not in phase with the source a-phase voltage, because the average imaginary power ($\bar{q}_L$) is not being compensated. However, frequency components in $\tilde{q}_L$ that are greater than 0.9 Hz are being compensated.

128 SHUNT ACTIVE FILTERS

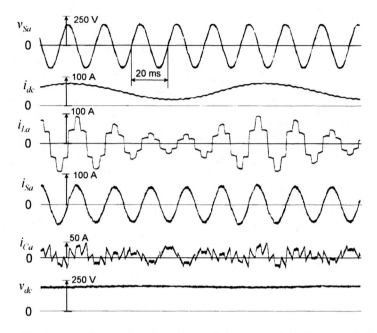

Figure 4-16. Compensation results when the cutoff frequency of the Butterworth filter for extraction of $\tilde{p}_L$ is set to 150 Hz and the other for extraction of $\tilde{q}_L$ is set to 0.9 Hz.

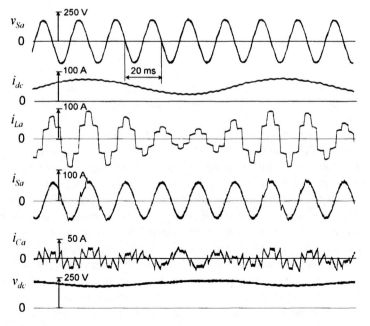

Figure 4-17. Compensation results when the cutoff frequencies of both Butterworth filters for $\tilde{p}_L$ and $\tilde{q}_L$ calculation are set to 0.9 Hz.

Again, spikes occur in i_{Sa} due to temporary incompleteness of current control during low dc voltage v_{dc}, as shown in Fig. 4-17. During these periods, high di/dt in the load current is not fully compensated, so that current spikes appear in i_{Sa}.

In a more realistic case, the capacity of the energy storage element of a shunt active filter should be reduced mainly for economical reasons. In this case, oscillating real power components at low frequencies cannot be compensated. The smaller the dc capacitor, the higher should be the cutoff frequency of the Butterworth filter in the active filter controller. Additionally, the gains of the PI controller in the dc-voltage regulator that generates the control signal $\bar{p}_{loss}$ in Fig. 4-7 should be increased to allow faster response, so that it avoids low-frequency voltage fluctuations.

Another simulation case derived from the same system and load currents as shown in Fig. 4-8 to Fig. 4-12 was carried out. Now, dc capacitance was significantly reduced, while the gains in dc voltage regulator were adjusted to produce faster response. To give an idea how small this reduced capacitance is, a factor called as the unit capacitance constant, UCC, defined as

$$UCC = \frac{CV_C^2}{2S} \qquad (4.6)$$

is used in analogy to the inertia constant of rotating machines. For the following simulation case, the calculated UCC is only 1.9 ms. This means that if the dc capacitor is charged at the rated voltage, it has an amount of energy to supply *theoretically* the rated power of the shunt active filter only for 1.9 ms, until it is completely discharged. Of course, the actual available UCC is small in a real implementation, because a voltage-fed converter that is normally used as a power circuit of a shunt active filter requires dc voltage greater than the ac peak voltage to operate without loss of controllability. In contrast, power generators and large synchronous machines have much more kinetic energy stored in their rotors. For these generators, the inertia constant, defined as $H = W_{kinetic}/S_{rated}$, is on the order of hundreds milliseconds for turbo-generators, or even few seconds for hydro-generators.

The signal $\bar{p}_{loss}$ determined by the dc voltage regulator, as shown in Fig. 4-7, forces the shunt active filter to draw/inject real power to act against low-frequency dc-voltage fluctuations. This signal is added to the compensating power $-\tilde{p}$ of the load to compose the compensating current references. Figure 4-18 shows results from the same nonlinear load shown in Fig. 4-8 and for the same cutoff frequency (1 Hz) for extraction of $\tilde{p}$. However, a reduced dc capacitor is now applied, and a faster response in the dc voltage regulator is required. Although the low cutoff frequency in the extraction of $\tilde{p}$ tries to compensate low-frequency components that are present in the real power of the load, the signal $\bar{p}_{loss}$ counteracts the low-frequency dc-voltage fluctuations. Therefore, $\bar{p}_{loss}$ inhibits the compensation of low-frequency components in $-\tilde{p}$ when both are added. In fact, in this case, $\bar{p}_{loss}$ is not only compensating for the losses, but also suppresses the dc-voltage variations caused by low-frequency components in $-\tilde{p}$. Hence, the low-frequency components produced by the blocking/unblocking of the thyristor bridge with a cycle of 120 ms period cannot be compensated. Note that the average value of the compensating real

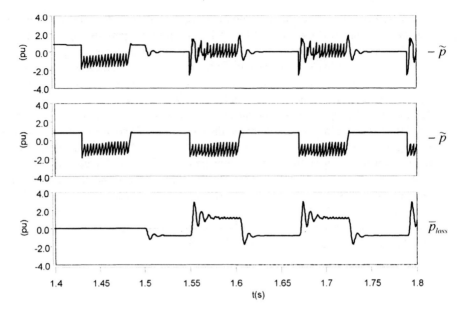

Figure 4-18. Compensating power signals for a reduced dc capacitor and a fast response in the dc-voltage regulator.

power given by $(-\tilde{p} + \bar{p}_{loss})$ is zero after the start of the active filter at $t = 1.5$ s. This happens even if short windows of time, like 10 ms, are considered to calculate the average real power.

Only high-order current harmonics, like the fifth, seventh, and higher, are being compensated, as seen in Fig. 4-19. Although the UCC of the dc capacitor is very small, the fast response of the PI controller in the dc voltage regulator reacts quickly against dc-voltage fluctuations, while it avoids overvoltages or undervoltages for long periods. In the steady state, the PI controller keeps the average value of the dc voltage v_{dc} equal to its reference value, 3.8 pu.

In the steady state, when the signal $\bar{p}_{loss}$ stabilizes, the compensated line currents are sinusoidal, or become zero. Although the gains in the PI controller that determine $\bar{p}_{loss}$ are not optimized for the present case of an intermittent load, it proves to be very effective. However, it inhibits the active filter to compensate low-frequency components in $\tilde{p}$. In contrast to the previous case, the compensated current does not draw constant real power from the source, as can be seen in Fig. 4-20.

The influence of voltage imbalance and/or distortion will be addressed in the following discussion. In Chapter 3, it was shown that the compensated currents cannot be sinusoidal if the supply voltage is unbalanced, and a constant real power drawn from the power system is required. The first case considers an imbalance of 10% of the fundamental, negative-sequence component. As a result, the phase voltages are represented by a positive-sequence phasor $\dot{V}_{+1} = 1 \angle 0°$ and a negative-sequence

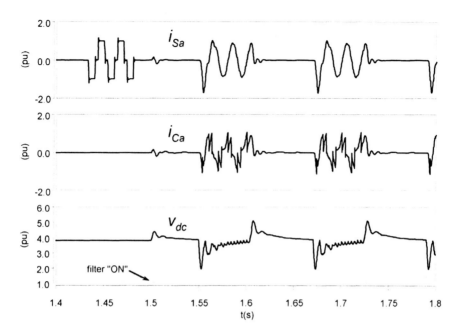

Figure 4-19. Source current, active filter current, and dc voltage.

phasor $\dot{V}_{-1} = 0.1 \angle 90°$. At no-load conditions, the phase voltages v_a, v_b, and v_c that appear at the load terminals are shown in Fig. 4-21.

The equivalent reactance of the power system is equal to $X_S = 0.1$ pu, while the commutation reactance of the thyristor bridge is made equal to $X_L = 0.05$ pu. This is a special case. A relation of $X_S \ll X_L$ exists in many cases. The constant-power com-

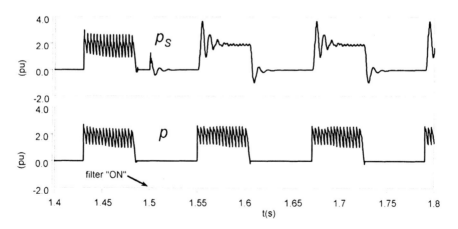

Figure 4-20. Real power drawn from the power system (p_s) and real power of the nonlinear load (p).

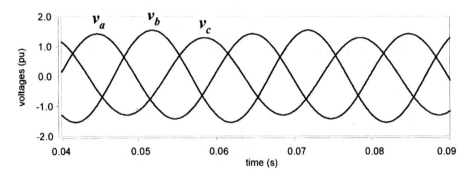

Figure 4-21. Unbalanced supply voltages.

pensation control strategy given in Fig. 4-7 is implemented in the controller. It should compensate the load current, as shown in Fig. 4-22. The unblocking of the thyristor bridge happens after $t = 150$ ms, and the active filter starts to compensate the currents at $t = 200$ ms. The firing angle of the thyristors is 30°. After the unblocking, notches in the phase voltages occur during the commutations of the thyristors, as can be seen in Fig. 4-22. For simplicity, a stiff dc-current source is considered.

Figure 4-23 shows the compensated currents. From the start of the thyristor bridge at 160 ms to the start of the shunt active filter at 200 ms, the currents present high di/dt that causes notches in the phase voltages at the load terminals.

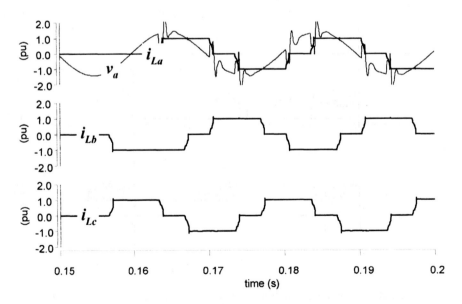

Figure 4-22. Phase voltage and load current waveforms.

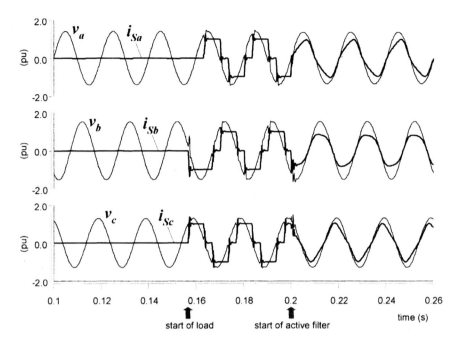

Figure 4-23. Compensated line currents.

These notches affect the performance of the active filter controller, since it uses these voltages to determine the compensating current references. In the worst case, it may cause instability. Therefore, filtering of such notches is necessary in the active filter controller. In fact, there is also a possibility of resonance, which may lead to instability. The PWM converter should not excite the system at the resonant frequency determined by the series-equivalent impedance of the power system and the small capacitor of the shunt passive filter represented by an RC branch (high-pass filter), as shown in Fig. 4-5. Remember that this passive filter is installed for filtering the current ripples in the converter current at the switching frequency. In the present case, filtering the measured phase voltages v_a, v_b, and v_c using first order low-pass filters with cutoff frequency equal to 600 Hz have solved the problems of instability and resonance. Their transfer function is given by $G(s) = (2\pi.600)/(s + 2\pi.600)$. In fact, the phase voltages shown in Fig. 4-23 are filtered by these low-pass filters.

After the start of the shunt active filter at $t = 200$ ms, the compensated currents do not contain high di/dt anymore. Hence, the phase voltages become almost sinusoidal. However, the compensated currents are not sinusoidal, although they draw constant real power from the unbalanced supply voltages, as can be seen in the waveform of p_S in Fig. 4-24. The voltage imbalance at the fundamental frequency causes a low-frequency (100 Hz) oscillating real power in the load, which is being compensated by the active filter. To avoid high dc voltage fluctuation, the dc capacitor was increased to give a UCC equal to 5.7 ms as defined in (4.6).

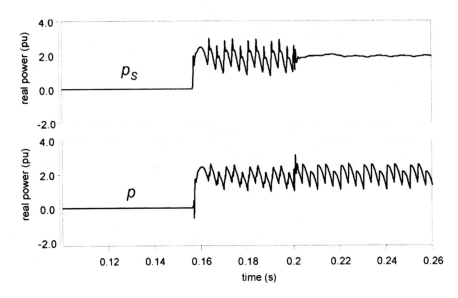

Figure 4-24. Real power drawn from the source (p_S) and load real power (p).

Figure 4-25 shows the imaginary powers q of the load and q_S of the source. A residual positive value of $\bar{q}_S$ remains in the compensated imaginary power q_S. Unfortunately, the low-pass filters used in the controller for filtering the phase voltages cause errors in the calculation of q_S. These filters introduce a phase delay in the fundamental component of the actual phase voltages, decreasing the phase-angle displacement between the filtered voltage and the lagging (inductive) load current. This makes the active filter controller calculate a smaller value of $\bar{q}$ than the actual value. The imaginary power as shown in Fig. 4-25 is calculated with the actual voltages at the load terminals. The above reason for incorrect compensation of $\bar{q}$ is confirmed if the source imaginary power q_S is calculated with the filtered phase voltages and the compensated source currents. In this case, q_S would give exactly zero after the start of the active filter at $t = 200$ ms.

4.2.2. Active Filters for Sinusoidal Current Control

The sinusoidal source current control strategy is a compensation method that makes the active filter compensate the current of a nonlinear load to force the compensated source current to become sinusoidal and balanced. In the presence of voltage imbalances or voltage distortions, it is impossible to satisfy both conditions simultaneously: to have *sinusoidal and balanced* compensated currents and to draw only a *constant real power* from the source. This point is extensively discussed in Chapter 3. Thus, a decision has to be made: to guarantee constant real power drawn from the power system, or to guarantee sinusoidal and balanced compensated current to the source. In other words, one should decide to implement the constant instantaneous power control strategy, or the sinusoidal current control strategy.

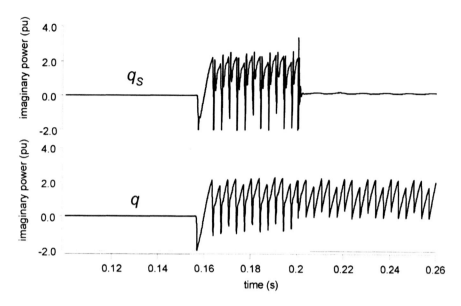

Figure 4-25. Source imaginary power q_S and load imaginary power q.

In order to make the compensated current become sinusoidal and balanced, the shunt active filter should compensate all harmonic components as well as the fundamental components that differ from the fundamental positive-sequence current i_{+1}. Only this component is left to be supplied by the source. In order to determine the fundamental positive-sequence component of the load current, a positive-sequence detector is needed in the active filter controller. The control block diagram for the sinusoidal current control strategy [24] is shown in Fig. 4-26. If compared with the control block diagram of the constant instantaneous power control strategy (Fig. 4-7), the unique difference is the insertion of a positive-sequence detector block, which extracts "instantaneously" the fundamental positive-sequence voltages v'_a, v'_b, and v'_c that correspond to the phasor $\dot{V}_{+1}$ of the system voltage v_a, v_b, and v_c. This positive-sequence detector is described in the next section. It replaces the low-pass filter $G(s) = (2\pi.600)/(s + 2\pi.600)$ that is used for filtering the measured system voltage in the previous control strategy. Therefore, the rest of the active filter controller shown in Fig. 4-26 determines the compensating current references if the system voltage contains only a fundamental positive-sequence component, instead of using filtered system voltages in the previous control strategy.

The real and imaginary powers calculated in Fig. 4-26 do not match the actual powers of the load, because the imbalance at the fundamental frequency (fundamental negative sequence) and the harmonics, which eventually may be present in the system voltage, are not being considered. However, the calculated real (p) and imaginary (q) powers are still useful for determining all imbalances and harmonics present in the load current. The general equations of powers in terms of symmetri-

SHUNT ACTIVE FILTERS

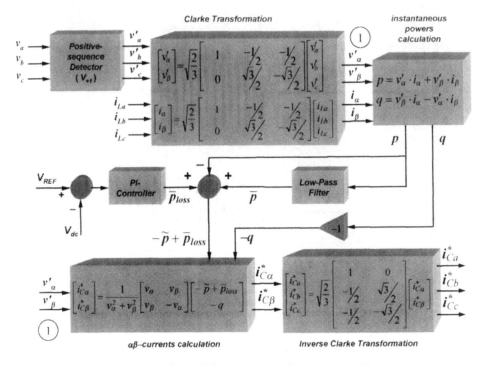

Figure 4-26. The control block diagram for the sinusoidal current control strategy.

cal components given in Chapter 3, equations (3.94) to (3.99), are helpful for understanding this point.

First, note that the instantaneous phase voltages v'_a, v'_b, and v'_c that correspond to the phasor $\dot{V}_{+1}$ of the fundamental positive-sequence voltage are transformed into the $\alpha\beta$ coordinates by means of the Clark transformation block. Then, they are used in both the instantaneous powers calculation block and in the $\alpha\beta$ currents calculation block. Together with the power being compensated, that is, $-\tilde{p}$ and $-q$, the $\alpha\beta$ currents calculation block determines exactly all current components in the load current that produce $-\tilde{p}$ and $q = \bar{q} + \tilde{q}$ with $\dot{V}_{+1}$. In terms of symmetrical components, since only $\dot{V}_{+1}$ is being considered, only $\dot{I}_{+1}$ produces $\bar{p}$ and $\bar{q}$ [see the general equations (3.94) to (3.99), in Chapter 3]. Therefore, if the shunt active filter compensates the power portions $\tilde{p}$ and $\tilde{q}$ of the calculated powers p and q, certainly it is compensating all components in the load current that are different from $\dot{I}_{+1}$ of the load. Note that this includes also the fundamental negative-sequence component $\dot{I}_{-1}$.

The positive-sequence detector that will be described in the next section, and used later in some simulations and experiments, can determine "instantaneously" and accurately the amplitude, the frequency, as well as the phase angle of voltage component $\dot{V}_{+1}$. The amplitude and phase angle of $\dot{V}_{+1}$ may be necessary in other controllers, although they are not important in the control algorithm implemented in Fig. 4-26. In order to extract the fundamental negative-sequence current and all current harmonics from the load current, it is necessary only that v'_a, v'_b, and v'_c have the same frequency as the actual system voltage. The amplitude could be arbitrarily chosen but must be equal in all three phases. The phase angle could also be arbitrarily chosen, but the displacement angles between phases must be maintained equal to $2\pi/3$. However, in order to properly compensate the portion of the fundamental positive-sequence current that is orthogonal to the fundamental positive-sequence voltage, the *phase angle* and the *frequency* of the fundamental positive-sequence voltage $\dot{V}_{+1}$ must be accurately determined. Otherwise, the active filter controller cannot exactly determine the *fundamental* reactive power of the load ($\bar{q}$) that, in turn, cannot produce ac currents ($\dot{I}_{+1}$) *orthogonal* to the ac voltages ($\dot{V}_{+1}$) to produce only $\bar{q}$. Fortunately, the positive-sequence voltage detector used in Fig. 4-26, as will be explained later in this section, can determine accurately the voltage component $\dot{V}_{+1}$.

The compensating powers $\tilde{p}$ and q in the active filter controller include all fundamental negative-sequence power, the fundamental reactive power, as well as the harmonic power. In other words, the active filter controller handles the load as "connected to a sinusoidal balanced voltage source." Thus, if $\tilde{p}$, $\bar{q}$, and $\tilde{q}$ are compensated by the shunt active filter, the source currents must be sinusoidal and contain only the active portion of the fundamental positive-sequence current that is in phase with $\dot{V}_{+1}$.

It is important to remark that the dc voltage regulator that determines the signal $\bar{p}_{loss}$ in Fig. 4-26 has an additional task besides those mentioned in the last section. Now it must correct errors in power compensation. This is because the active filter controller is now unable to calculate correctly the actual value of $\bar{p}$ of the load. This mistake in power compensation can be better understood in the following simple example.

Suppose that the system voltages and load currents comprise only fundamental positive- and negative-sequence components, that is, only $\dot{V}_{+1}, \dot{V}_{-1}, \dot{I}_{+1},$ and $\dot{I}_{-1}$ are present. In this case, the controller of Fig. 4-26 makes the active filter supply the whole negative-sequence current to the load, because it produces $\tilde{p}$ and $\tilde{q}$ with the voltage signals v'_a and v'_b that contain only $\dot{V}_{+1}$. However, the actual voltages at the load terminal also contain a fundamental negative-sequence component. If it is not orthogonal to $\dot{I}_{-1}$, the active filter will supply or absorb a nonzero *average negative-sequence* power contained in $\bar{p}$, as determined in (3.95). This causes voltage variation in the dc capacitors, where the feedback control loop of the dc voltage regulator that has a slower response than that of the $\tilde{p}$ calculation that will sense it. The signal $\bar{p}_{loss}$ will make the active filter absorb or supply *positive-sequence* power, in an attempt to neutralize the voltage variation. This occurs because the active filter current references are calculated only from $\dot{V}_{+1}$ (v'_a, v'_b, and v'_c output from the positive-sequence detector). In other words, if the shunt active filter supplies (absorbs) an average negative-sequence power to (from) the load, has to absorb (supply) the same amount of average positive-sequence power from the power system to keep constant the dc voltage.

The slower feedback loop provided by $\bar{p}_{loss}$ is also useful for correcting voltage variations due to compensation errors that occur during transient responses of the shunt active filter. This point is discussed in the previous section for the constant instantaneous power control strategy, and will be seen later, when the dynamic performance of a shunt active filter realizing the sinusoidal current control strategy is analyzed. Therefore, the signal $\bar{p}_{loss}$ is very important for providing energy balance inside the active filter.

4.2.2.A. Positive-Sequence Voltage Detector

The phase voltages v_a, v_b, and v_c at the load terminal consists mainly of the positive-sequence component ($\dot{V}_{+1}$), but can be unbalanced (containing negative- and zero-sequence components at fundamental frequency), and can also contain harmonics from any sequence component. The detection of the *fundamental positive-sequence component* of v_a, v_b, and v_c is necessary in the sinusoidal current control strategy shown in Fig. 4-26. This control strategy makes the shunt active filter to compensate load currents, so that only the active portion of the fundamental positive-sequence component, which produces average real power $\bar{p}$ only (that is, only that portion of $\dot{I}_{+1}$ being in phase with $\dot{V}_{+1}$) is supplied by the source.

4.2.2.A.1. Main Circuit of the Voltage Detector. Figure 4-27 shows the complete functional control block diagram of the fundamental positive-sequence volt-

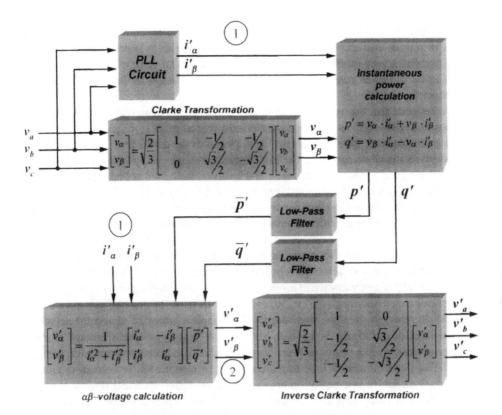

Figure 4-27. Fundamental positive-sequence voltage detector.

age detector. An important part of the positive-sequence detector is the phase-locked-loop (PLL) circuit. The PLL circuit is detailed in the next section.

The fundamental positive-sequence voltage detector is based on the dual p-q theory and the concepts of voltage compensation given in Fig. 3-18. The voltages v_a, v_b, and v_c are transformed into the $\alpha\beta$ axes to determine v_α and v_β. They are used together with auxiliary currents i'_α and i'_β, produced in the PLL circuit, to calculate the auxiliary powers p' and q', as shown in Fig. 4-27. It is assumed that the auxiliary currents i'_α and i'_β with any magnitude are derived only from an auxiliary positive-sequence current I'_{+1} at the fundamental frequency, detected by the PLL circuit. In terms of symmetrical components, the general equation from Chapter 3 gives

$$\begin{cases} i_\alpha = \sum_{n=1}^{\infty} \sqrt{3}I_{+n} \sin(\omega_n t + \delta_{+n}) + \sum_{n=1}^{\infty} \sqrt{3}I_{-n} \sin(\omega_n t + \delta_{-n}) \\ i_\beta = \sum_{n=1}^{\infty} -\sqrt{3}I_{+n} \cos(\omega_n t + \delta_{+n}) + \sum_{n=1}^{\infty} \sqrt{3}I_{-n} \cos(\omega_n t + \delta_{-n}) \\ i_0 = \sum_{n=1}^{\infty} \sqrt{6}I_{0n} \sin(\omega_n t + \delta_{0n}) \end{cases} \quad (4.7)$$

Thus, as for the fundamental ($n = 1$) positive-sequence component, (4.7) results in

$$\begin{cases} i_\alpha = \sqrt{3}I_{+1} \sin(\omega_1 t + \delta_{+1}) \\ i_\beta = -\sqrt{3}I_{+1} \cos(\omega_1 t + \delta_{+1}) \end{cases} \quad (4.8)$$

For extracting the fundamental positive-sequence voltage with the dual method shown in Fig. 4-27, the amplitude of the auxiliary currents i'_α and i'_β are not important, and can be chosen arbitrarily. For simplicity, they are set to unity. In other words, the absolute values of the auxiliary powers p' and q', as the auxiliary currents i'_α and i'_β, have no physical meaning. Note that they are used together in the voltage calculation, represented by the $\alpha\beta$ voltage calculation block in Fig. 4-27.

The phase angle δ_{+1} of i'_α and i'_β can also be chosen arbitrarily, because different values of δ_{+1} vary simultaneously the auxiliary powers p' and q', so that the results from the $\alpha\beta$ voltage calculation block will be always the same. Again, for simplicity, they are set to zero. Only the fundamental frequency (ω_1) must be determined accurately by the PLL. Thus, (4.8) is simplified to form the needed auxiliary currents:

$$\begin{cases} i'_\alpha = \sin(\omega_1 t) \\ i'_\beta = -\cos(\omega_1 t) \end{cases} \quad (4.9)$$

An important conclusion that can be drawn from the general equations (3.94) to (3.99) is useful for determining the fundamental positive-sequence component $\dot{V}_{+1}$ of the system voltages. From those equations, it is possible to see that *only* the fundamental positive-sequence voltage component $\dot{V}_{+1}$ contributes to the *average* values of the auxiliary powers p' and q', represented by $\bar{p}'$ and $\bar{q}'$ in Fig. 4-27. This is assured because (4.9) represents auxiliary currents in the $\alpha\beta$ axes composed only from $\dot{I}_{+1}$. The influence of the fundamental negative-sequence $\dot{V}_{-1}$ and other voltage harmonics will appear only in the oscillating components $\tilde{p}'$ and $\tilde{q}'$ of the auxiliary powers, which are being excluded from the inverse voltage calculation. Two fifth order Butterworth low-pass filters with cutoff frequency at 50 Hz are used for obtaining the average powers $\bar{p}'$ and $\bar{q}'$.

The $\alpha\beta$ voltage calculation block of Fig. 4-27 calculates the instantaneous voltages v'_α and v'_β, which correspond to time functions of the fundamental positive-sequence phasor $\dot{V}_{+1}$ of the system voltage:

$$\begin{bmatrix} v'_\alpha \\ v'_\beta \end{bmatrix} = \frac{1}{i'^2_\alpha + i'^2_\beta} \begin{bmatrix} i'_\alpha & -i'_\beta \\ i'_\beta & i'_\alpha \end{bmatrix} \begin{bmatrix} \bar{p}' \\ \bar{q}' \end{bmatrix} \qquad (4.10)$$

For several applications, the transformed voltages from (4.10) are useful, and may be directly used in an active filter controller. Especially in the case of using (4.10) in a real implementation of a controller, it should be considered that, in the steady state,

$$i'^2_\alpha + i'^2_\beta = \sin^2(\omega_1 t) + \cos^2(\omega_1 t) = 1 \qquad (4.11)$$

and the division in (4.10) may be avoided.

If necessary, the instantaneous abc phase voltages v'_a, v'_b, and v'_c can be calculated by applying the inverse Clarke transformation, disregarding the zero-sequence component v_0:

$$\begin{bmatrix} v'_a \\ v'_b \\ v'_c \end{bmatrix} = \sqrt{\frac{2}{3}} \begin{bmatrix} 1 & 0 \\ -\frac{1}{2} & \frac{\sqrt{3}}{2} \\ -\frac{1}{2} & -\frac{\sqrt{3}}{2} \end{bmatrix} \begin{bmatrix} v'_\alpha \\ v'_\beta \end{bmatrix} \qquad (4.12)$$

Experimental results have shown that the fundamental positive-sequence voltage detector presented above has a good dynamic and a satisfactory accuracy even under nonsinusoidal conditions. The PLL circuit must provide accurately the auxiliary currents i'_α and i'_β, corresponding to sinusoidal functions at the fundamental frequency. The next section presents a PLL circuit that can work properly, even under significant levels of imbalance and/or distortion in the system voltages. If the PLL supplies accurately i'_α and i'_β, the algorithm in Fig. 4-27 can calculate exactly the *amplitude,* the *frequency,* and the *phase angle* of $\dot{V}_{+1}$, which are given in the form of continuous-time functions v'_a, v'_b, and v'_c, or v'_α and v'_β. The controllers of series and shunt active filters, as well as other advanced compensators, such as custom power devices and FACTS devices [26–30], can exploit the advantages offered by the above positive-sequence detector.

4.2.2.A.2. Phase-Locked-Loop (PLL) Circuit.

The PLL circuit tracks continuously the fundamental frequency of the measured system voltages. The appropriate design of the PLL should allow proper operation under distorted and unbalanced voltage waveforms. An interesting design of a PLL circuit that is almost insensitive to imbalances and distortions in the voltage waveforms is described in [25]. This synchronizing circuit (PLL circuit) determines automatically the system *frequency* and the *phase angle* of the *fundamental positive-sequence component* of a three-phase generic input signal. In the present case, it is the three-phase measured phase voltages v_a, v_b, and v_c. Figure 4-28 illustrates its functional block diagram. This circuit has proved to be very effective, even under highly distorted system voltages. It has been used successfully, by the authors, in several power electronic devices, like active filters [31] and FACTS devices [32]. The algorithm is based on a fictitious instantaneous active power expression:

$$p'_{3\phi} = v_a i'_a + v_b i'_b + v_c i'_c = v_{ab} i'_a + v_{cb} i'_c \qquad (4.13)$$

Note that $i'_a + i'_b + i'_c = 0$ is considered in (4.13). In fact, $p'_{3\phi}$ is not related to any instantaneous active power of the power system, although it could be considered as a variable in the PLL circuit with a dimension of power. This is why it is called *fictitious* power. As no current is measured from the power circuit, one may find it difficult to understand how this circuit works. The fictitious current feedback signals $i'_a(\omega t) = \sin(\omega t)$ and $i'_c(\omega t) = \sin(\omega t + 2\pi/3)$ of Fig. 4-28 are built up by the PLL circuit just calculating the time integral of the output ω of the PI controller. The PLL can reach a stable point of operation only if the input $p'_{3\phi}$ of the PI controller has, in the steady state, a zero average value, that is, $\bar{p}'_{3\phi} = 0$. Moreover, the control circuit should minimize oscillations in $p'_{3\phi}$ at low frequencies. The oscillating portion $\tilde{p}'_{3\phi}$, where $p'_{3\phi} = \bar{p}'_{3\phi} + \tilde{p}'_{3\phi}$, at low frequencies is not well attenuated by the PI controller and may bring instability to the PLL control circuit. Recalling that the average three-phase power $P'_{3\phi} = \bar{p}'_{3\phi}$, in terms of phasors, is given by

$$P'_{3\phi} = \bar{p}'_{3\phi} = 3V_{+1}I'_{+1} \cos \phi \qquad (4.14)$$

the above constraints are found only if ω equals the system frequency, and the current $i'_a(\omega t)$ becomes orthogonal to the fundamental positive-sequence component of the measured three-phase voltages v_a, v_b, and v_c.

However, if the point where $i'_a(\omega t)$ lags v_a by 90° is reached, this is still an unstable point of operation. At this point, an eventual disturbance that slightly increases the system frequency will increase the frequency of v_{ab} and v_{cb} in Fig. 4-28, and will make the voltage phasor $\dot{V}_{+1}$ rotate faster than the current phasor corresponding to the generated feedback signal $i'_a(\omega t)$. Hence, the displacement angle between v_a and $i'_a(\omega t)$, given by $\cos \phi$ in (4.14), becomes greater than 90°. This leads to a negative average input $\bar{p}'_{3\phi} < 0$ and, consequently, decreases the signal ω, making the phase angle between v_a and $i'_a(\omega t)$ even greater. This positive feedback characterizes an unstable point of operation. Thus, the PLL has only one stable point of operation, that is, $i'_a(\omega t)$ leading by 90° the phase voltage v_a. Now, if the same disturbance is

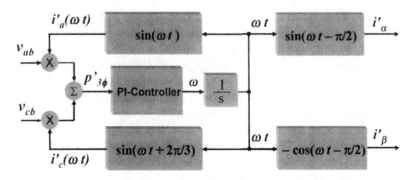

Figure 4-28. Functional block diagram of the PLL circuit [25].

verified, the displacement angle between the voltage and current phasors will be reduced so that the average power in (4.14) will be positive. This will increase the output signal ω and force the current phasor to rotate faster, maintaining the orthogonality (leading currents) between $\dot{V}_{+1}$ and the generated $\dot{I}'_{+1}$ [$i'_a(\omega t)$]. This fundamental characteristic of this PLL circuit can be exploited in many ways. For instance, to compose properly phase-shifted sinusoidal functions and to determine directly the firing angle of thyristor bridges or gate pulses for multipulse converters. Here, it is used to form auxiliary signals i'_α and i'_β that are needed in the positive-sequence detector of Fig. 4-27.

As explained above, $i'_a(\omega t) = \sin(\omega t)$ in Fig. 4-28 leads by 90° the fundamental positive-sequence component $\dot{V}_{+1}$ of the measured system voltages. Thus, the generated auxiliary current $i'_\alpha = \sin(\omega t - \pi/2)$ is in phase with $\dot{V}_{+1}$. The following simulation case shows this fundamental characteristic of the PLL circuit.

Suppose that the system voltage is strongly unbalanced by a 30% fundamental negative-sequence component, and is strongly distorted by 30% of the second harmonic. The second harmonic is made purely from the negative-sequence component. The phase angle of the fundamental positive-sequence component is set to zero, and the other components are orthogonal to it. Thus, time functions corresponding to this hypothetical system voltages are:

$$\begin{cases} v_a = \sin(2\pi 60 t) + 0.3 \sin\left(2\pi 60 t + \dfrac{\pi}{2}\right) + 0.3 \sin\left(4\pi 60 t + \dfrac{\pi}{2}\right) \\ v_b = \sin\left(2\pi 60 t + \dfrac{2\pi}{3}\right) + 0.3 \sin\left(2\pi 60 t + \dfrac{\pi}{2} + \dfrac{2\pi}{3}\right) + 0.3 \sin\left(4\pi 60 t + \dfrac{\pi}{2} + \dfrac{2\pi}{3}\right) \\ v_c = \sin\left(2\pi 60 t + \dfrac{2\pi}{3}\right) + 0.3 \sin\left(2\pi 60 t + \dfrac{\pi}{2} - \dfrac{2\pi}{3}\right) + 0.3 \sin\left(4\pi 60 t + \dfrac{\pi}{2} - \dfrac{2\pi}{3}\right) \end{cases} \quad (4.15)$$

Positive sequence, $\dot{V}_{+1} = 1 \angle 0°$

Negative sequence, $\dot{V}_{-1} = 0.3 \angle 90°$

Negative sequence, $\dot{V}_{-2} = 0.3 \angle 90°$

Figure 4-29 shows the curves built from (4.15). This figure shows also the dynamic behavior of signal i'_α, as determined in Fig. 4-28. The reference is given by the curve $v_{+1} = \sin(2\pi 60t)$ that corresponds to the fundamental positive-sequence component present in the a-phase voltage. Fig. 4-29 shows that the PLL circuit, even if working under highly distorted and unbalanced system voltages, reaches the steady state in less than 200 ms. As mentioned before, the unique stable point of operation of the PLL circuit is when the auxiliary current $i'_a(\omega t)$ leads 90° the fundamental positive-sequence component $i_a(\omega t)$ of phase voltage v_a. Hence, this point corresponds to i'_α in phase with v_{+1}.

As explained above, a fundamental characteristic of the PLL (Fig. 4-28) is the ability to tracking the frequency and phase angle of the fundamental positive-sequence component, instead of the phase angle of the a-phase voltage. For instance, if the second-order harmonic did not exist, the phase voltages would become sinusoidal. However, the phase angle of the a-phase voltage would remain different from that of the fundamental positive-sequence component. In this case, the phase angle of v_a would be +16.7° (leading), whereas the phase angle of $\dot{V}_{+1}$ is zero, that is,

$$\dot{V}_a = \dot{V}_{+1} + \dot{V}_{-1} = 1 + j0.3 = 1.044 \angle 16.7° \tag{4.16}$$

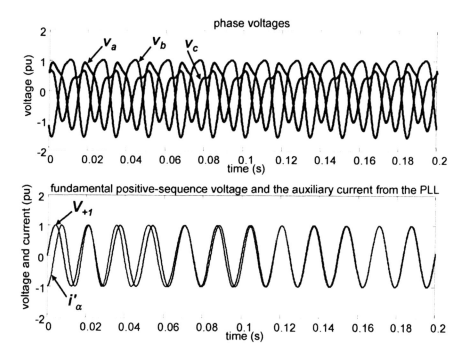

Figure 4-29. Distorted system voltages and the dynamic behavior of signal i'_α (see Fig. 4-28).

The ability to track the positive-sequence component is verified also with faulty system voltages. For the same case of system voltages shown above, a single phase-to-ground fault is simulated during $0.5 < t < 0.6$ s. Figure 4-30 shows the results. The signal i'_α continues tracking accurately the fundamental positive-sequence component of the faulty system voltages. Note that, disregarding the second harmonic, the three-phase voltage is given by:

$$\begin{cases} v_a = \sin(2\pi 60 t) + 0.3 \sin\left(2\pi 60 t + \dfrac{\pi}{2}\right) \\ v_b = 0 \\ v_c = \sin\left(2\pi 60 t + \dfrac{2\pi}{3}\right) + 0.3 \sin\left(2\pi 60 t + \dfrac{\pi}{2} - \dfrac{2\pi}{3}\right) \end{cases} \quad (4.17)$$

that in terms of symmetrical components, only considering the fundamental frequency results in

$$\begin{bmatrix} \dot{V}_{01} \\ \dot{V}_{+1} \\ \dot{V}_{-1} \end{bmatrix} = \frac{1}{3} \begin{bmatrix} 1 & 1 & 1 \\ 1 & \alpha & \alpha^2 \\ 1 & \alpha^2 & \alpha \end{bmatrix} \begin{bmatrix} 1 \angle 0° + 0.3 \angle 90° \\ 0 \\ 1 \angle 120° + 0.3 \angle -30° \end{bmatrix} = \begin{bmatrix} 0.42 \angle 53.21° \\ 0.58 \angle 4.93° \\ 0.19 \angle -28.02° \end{bmatrix} \quad (4.18)$$

This is the reason why the signal i'_α leads slightly the reference signal $v_{+1} = \sin(2\pi 60 t)$ that actually does not correspond to the fundamental positive-sequence

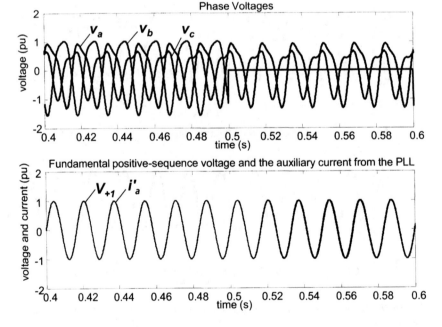

Figure 4-30. Dynamic behavior of the PLL circuit under a phase-to-ground fault.

component during the phase-to-ground fault simulation. During $0.5 < t < 0.6$ s, $\dot{V}_{+1}$ = $0.58 \angle 4.93°$, as given in (4.18).

4.2.2.B. Simulation Results

For the same case of load current compensation under unbalanced system voltage shown in Fig. 4-21 to Fig. 4-25, another simulation case was performed using the sinusoidal current control strategy in the shunt active filter controller.

The shunt active filter starts at $t = 0.2$ s, and the resulting currents of the shunt active filter are shown in Fig. 4-31. They are used to compensate the load currents shown in Fig. 4-22. The compensated currents drawn from the power system (source) are determined by the sums $i_{Sa} = i_{La} + i_{Ca}$, $i_{Sb} = i_{Lb} + i_{Cb}$, and $i_{Sc} = i_{Lc} + i_{Cc}$, as shown in Fig. 4-32. After the start of compensation, the source currents become in phase with the voltages, since the fundamental reactive power $\bar{q}$, included in $q = \bar{q} + \tilde{q}$, is being compensated. As expected, the compensated currents are sinusoidal. However, they do not drain constant real power from the source, as shown in Fig. 4-33.

4.2.3. Active Filters for Current Minimization

Despite of the usefulness and flexibility of the *p-q* Theory to design controllers for active power line conditioners, other approaches may be found also suitable, de-

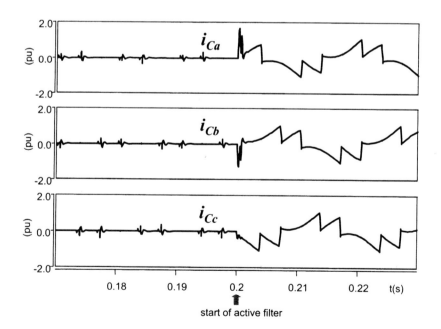

Figure 4-31. Compensating currents of the shunt active filter.

146 SHUNT ACTIVE FILTERS

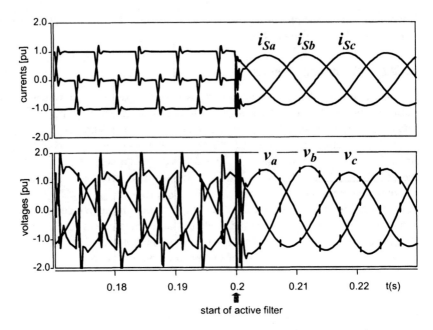

Figure 4-32. Compensated currents of the source and the system voltages.

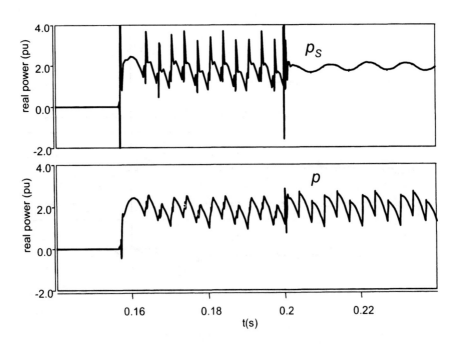

Figure 4-33. Real power p_S drawn from the source and load real power p.

pending on the objectives to be accomplished. For instance, the decomposition of the load current into active and nonactive current portions through the current-minimization methods introduced in Chapter 3 can be used to design controllers for shunt active filters.

Controllers for shunt active filters that guarantee compensated currents proportional to the supply voltages can be implemented by using the concepts of the current minimization method introduced in Chapter 3. In this case, if the supply voltage is sinusoidal and balanced, the compensated source current will be also sinusoidal and balanced. It was seen that the current compensated by means of the generalized Fryze currents method presents a minimum rms value to draw the same three-phase *average* active power from the source as the original load current. This reduces ohmic losses in the transmission system. In terms of the p-q Theory, it is the same as to saying that the shunt active filter is compensating the whole imaginary power ($q = \bar{q} + \tilde{q}$) of the load. However, the minimization methods and the methods based on the p-q Theory differ a lot in the presence of zero-sequence components, as described at the end of Chapter 3.

It should be remarked that the main purpose of the following control strategy, implemented under the concepts of *generalized Fryze currents,* is to guarantee linearity between the supply voltage and the compensated current. An example following this control strategy is illustrated in Fig. 4-4, in which it is compared with other control strategies, for the case of nonsinusoidal system voltages.

The background for designing *the generalized Fryze current control strategy* for shunt current compensation is detailed in Chapter 3. Here, only additional controls for solving problems in a real implementation are detailed. In the control strategies based on the p-q Theory, the signal $\bar{p}_{loss}$ shown in Fig. 4-7 and in Fig. 4-26 is fundamental for keeping the dc voltage of the active filter around its reference value. Here, similar control for providing dc voltage regulation has to be implemented. An advantage of the generalized Fryze current control is the reduced calculation effort, since it works directly with the *abc*-phase voltages and line currents. The elimination of the Clarke transformation makes this control strategy simple. Figure 4-34 shows the complete control circuit for a real implementation of the generalized Fryze current control strategy. The instantaneous equivalent conductance G_e is calculated from the three-phase instantaneous active power in (2.38), and the squared instantaneous aggregate voltage from (2.47). The *average* conductance $\bar{G}_e$ is obtained, passing G_e through a low-pass filter. Note that this way of calculation is slightly different from that proposed in (3.130), and is preferred here due to its simplicity.

The instantaneous active portions $i_{\overline{w}a}$, $i_{\overline{w}b}$, and $i_{\overline{w}c}$ of the load current are directly obtained by multiplying $\bar{G}_e$ by the phase voltages v_a, v_b, and v_c, respectively, as given in (3.129). The control strategy in Fig. 4-34 follows the convention of current directions as given in Fig. 4-2. This is the same as saying that the shunt active filter should draw the inverse of the nonactive current of the load, that is, $i_{Ca}^* = -i_{qa} = (i_{\overline{w}a} - i_{La})$. An extra *active* portion of current is added, in order to draw a small amount of active power to compensate for switching and conducting losses in the shunt active filter, which tend to discharge the dc capacitor. This is realized by the addition of the signal $\bar{G}_{loss}$ to the average conductance $\bar{G}_e$.

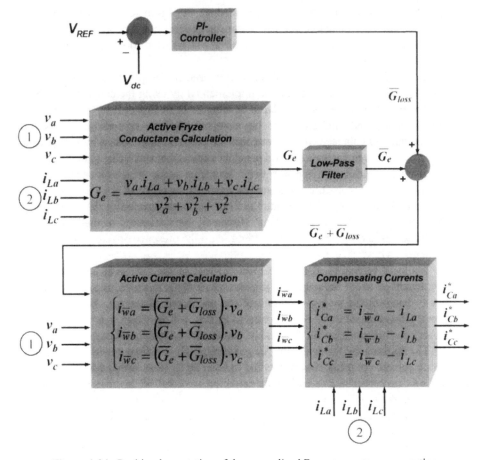

Figure 4-34. Real implementation of the generalized Fryze currents compensation.

In Chapter 3, two control strategies are derived from the *abc* Theory: the generalized Fryze current control (Fig. 4-34) and the instantaneous nonactive current compensation, given in (3.126). Note that this latter control strategy can be implemented simply by eliminating the low-pass filter in Fig. 4-34 and by using the instantaneous equivalent conductance G_e in the compensating current calculation.

For comparison, a simulation case is presented, using the same supply voltages and load currents as those considered in the control strategies based on the *p-q* Theory. The results from the constant instantaneous power control strategy (Fig. 4-23) and from the sinusoidal current control strategy (Fig. 4-32) are related to the same supply voltages of Fig. 4-21 and the same load currents of Fig. 4-22. Figure 4-35 presents the compensated currents applying the generalized Fryze current control strategy. After the start of the shunt active filter at $t = 0.2$ s, the source currents be-

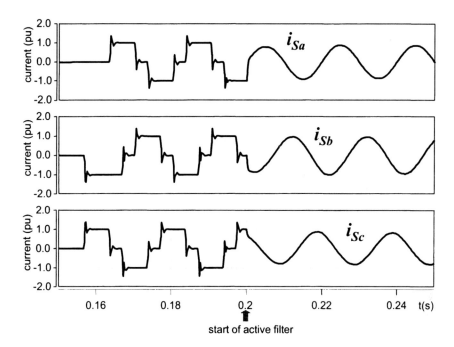

Figure 4-35. Compensated currents of the source.

come sinusoidal, although they remain unbalanced as do the supply voltages, since the compensated currents are proportional to the corresponding phase voltages.

Ideally, the compensated current should be purely sinusoidal, since the supply voltage consists of a fundamental positive-sequence component $\dot{V}_{+1}$ and a fundamental negative-sequence component $\dot{V}_{-1}$. It does not contain harmonics. Thus, the compensated current should contain, ideally, only $\dot{I}_{+1}$ and $\dot{I}_{-1}$. However, in a real implementation, a dc voltage regulator must be present to keep the dc voltage around its reference value, which is represented by the signal $\overline{G}_{loss}$ in Fig. 4-34. There is a compromise between gains in the PI controller for determining $\overline{G}_{loss}$ and the size of the energy storage element of the active filter. Generally, the higher the gains, the smaller can be the energy storage capacity. However, there is a limitation for these gains, because the PI controller must act as a low-pass filter and cannot allow the signal $\overline{G}_{loss}$ to react to the lowest frequency of active power oscillation of the load that is desired to be compensated. The instantaneous active powers p of the load, p_C of the shunt active filter, and p_S of the source are shown in Fig. 4-36. From the p-q Theory, one can see that the simultaneous presence of $\dot{V}_{+1}$, $\dot{V}_{-1}$, $\dot{I}_{+1}$ and $\dot{I}_{-1}$ produces an average active power and an oscillating portion at twice the system frequency. The active filter supplies the difference in active power between the source and the load. Hence, there is an oscillating component at twice the system frequency in the power p_C of the active filter, thus leading to dc voltage fluctuation at same

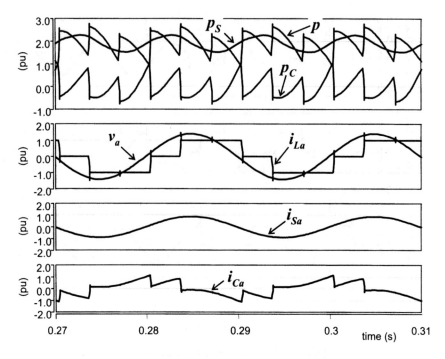

Figure 4-36. Three-phase instantaneous active power p of the load, p_C of the shunt active filter, and p_S of the source; the load current i_{La}; the source current i_{Sa}; and the compensating current i_{Ca}.

frequency. The adjusted gains in the PI controller, for the present simulation case, make it slightly sensitive for the 100 Hz voltage fluctuation. Therefore, $\overline{G}_{loss}$ responds, although in an attenuated form, to this frequency of voltage fluctuation. The 100 Hz oscillation in $\overline{G}_{loss}$ distorts slightly the compensated current, as can be seen in Fig. 4-35.

The *abc* Theory has no consistent expression to calculate the three-phase instantaneous reactive power. Thus, the imaginary power calculation from the *p-q* Theory is used to show that the generalized Fryze current control strategy is able to compensate for the imaginary power of the load, as can be seen in Fig. 4-37.

4.2.4. Active Filters for Harmonic Damping

Electric power utilities have the responsibility for harmonic damping throughout power distribution systems, while individual customers and end users are responsible for harmonic compensation of their own nonlinear loads. The previously detailed control strategies were designed to compensate for currents from nonlinear loads. They have a singular compensation property: they perform selective harmonic current compensation. They compensate for the extracted harmonic current from a selected nonlinear load. In contrast to passive filters, the shunt active filter does

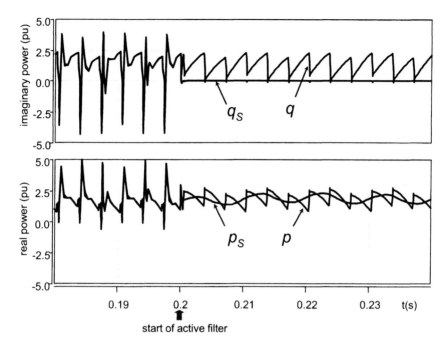

Figure 4-37. Compensated real (p_S) and imaginary (q_S) powers supplied by the source, and the real (p) and imaginary (q) powers of the load.

not compensate any harmonic current from other nonlinear loads that may be connected in the vicinity of the active filter. In other words, this minimizes the risk of overloading the shunt active filter due to the presence of other unidentified nonlinear loads.

Next, another approach based on harmonic voltage detection is presented to damp out harmonic propagation caused by resonance effects. For instance, a serious problem of harmonic propagation may arise in long distribution feeders with large capacitor banks installed for power-factor correction. In this case, performing harmonic damping would be effective in removing harmonic pollution as well as providing harmonic compensation [46]. This type of control strategy is more adequate to solve problems of electric power utilities.

4.2.4.A. Shunt Active Filter Based on Voltage Detection

The purpose of the active filter based on voltage detection is to damp out harmonic propagation throughout feeders in distribution systems. The active filter detects voltage harmonics v_h at the point of installation, and then injects a compensating current i_C^* as follows:

$$i_C^* = K_V \cdot v_h \qquad (4.18)$$

where K_V is a control gain in the active filter controller. The above equation implies that the active filter behaves like a resistor of $1/K_V$ Ω in the external circuit for all harmonic frequencies. In contrast, the active filter makes no contribution to the external circuit for the fundamental frequency. For the fundamental frequency, it behaves as an infinite impedance. Figure 4-38 illustrates the compensation principle of the shunt active filter based on voltage detection. Note that the shunt active filters presented in the previous sections are based on load-current detection. Hence, they constitute a "feedforward" control between the detected harmonic current and the compensating current. Contrarily, an active filter based on voltage detection forms a "feedback" control loop between the detected harmonic voltage and the compensating current. Therefore, time and phase delays in the active filter controller might contribute to instability or might deteriorate harmonic-damping performance of the active filter based on voltage detection.

Damping of harmonic propagation in distribution systems is the responsibility of electric power utilities. It is very important to select the proper location to install a shunt active filter in order to achieve the most cost-effective harmonic damping [47,48]. In general, installing the active filter on the end bus of the feeder, and properly setting the control gain K_V (Fig. 4-38) makes it possible to damp out harmonic propagation. This procedure can be understood as an active filter that is acting as a "harmonic terminator," just like a 50-Ω terminator installed on the end terminal of a signal transmission line. The value of K_V should be carefully determined. Otherwise, the installation of an active filter acting as a "harmonic resistance" on long-distance power distribution feeders may result in a strange phenomenon: voltage harmonics are mitigated at the point of installation, whereas they are magnified on other busses where no filter is connected [51]. This point is discussed later, in an application example.

4.2.4.B. Active Filter Controller Based on Voltage Detection

The controller of the shunt active filter of Fig. 4-38 works as a complement of the control circuit for detecting the fundamental positive-sequence voltage (v_F) pre-

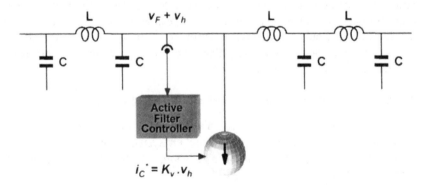

Figure 4-38. Compensation principle of the shunt active filter based on voltage detection.

sented in Fig. 4-27. The controller based on voltage detection must determine "instantaneously" and continuously the harmonic content (v_h) in the measured voltage ($v_F + v_h$) on the bus where the shunt active filter is connected. If the bus voltage is balanced, that is, does not contain negative-sequence and/or zero-sequence components at the fundamental frequency, the *difference* between the measured voltage and the output of that positive-sequence detector of Fig. 4-27 corresponds to the harmonic voltage v_h, that is, $v_{hk} = v_k - v'_k$; $k = a, b, c$. The harmonic voltage v_h is determined to form the compensating current reference as $i^*_C = K_V \cdot v_h$.

The harmonic voltage v_h can be calculated not only as mentioned above, but also by replacing the low-pass filters in Fig. 4-27 by high-pass filters to form a new controller, as shown in Fig. 4-39. The low-pass filters separate the average values of the auxiliary real and imaginary powers calculated with the auxiliary currents i'_α and i'_β, whereas the high-pass filters separate the oscillating portions of those auxiliary powers. The signals i'_α and i'_β are generated in the PLL circuit as built up from an auxiliary positive-sequence current component at the fundamental frequency ($\dot{I}_{+1}$). Therefore, the average real power $\bar{p}'$ together with the average imaginary power $\bar{q}'$ comprise the fundamental positive-sequence voltage component ($\dot{V}_{+1}$) of the mea-

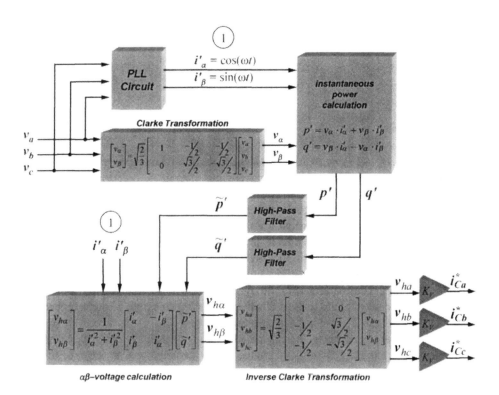

Figure 4-39. Active filter controller based on voltage detection.

sured bus voltage. The general equations presented in Chapter 3, (3.94) to (3.99), can help to better understand this property of the *p-q* Theory. Therefore, by replacing the low-pass filters by high-pass filters, the resulting voltage v_{ha}, v_{hb}, and v_{hc} (Fig. 4-39) comprises all other voltage components that differ from the fundamental positive-sequence voltage. If the bus voltage is unbalanced due to the fundamental negative-sequence voltage $\dot{V}_{-1}$, the imbalance will be present in v_{ha}, v_{hb}, and v_{hc}, although it is not a real "harmonic" component. However, it produces with $\dot{I}_{+1}$ oscillating powers at double the system frequency, which is included in the above compensation method. In this case, the shunt active filter would draw an undesirable negative-sequence current at the fundamental frequency.

If necessary, an additional negative-sequence voltage detector can be implemented just by replacing the signals i'_α and i'_β by sinusoidal functions that represent an auxiliary, negative-sequence, current component at the fundamental frequency. A negative-sequence detector will be presented in the following discussion.

Fig. 4-27 and Fig. 4-28 suggest the use of

$$\begin{cases} i'_\alpha = \sin\left(\omega t - \dfrac{\pi}{2}\right) = -\cos(\omega t) \\ i'_\beta = -\cos\left(\omega t - \dfrac{\pi}{2}\right) = -\sin(\omega t) \end{cases} \quad (4.19)$$

instead of (4.9), whereas the following expression is used in Fig. 4-39:

$$\begin{cases} i'_\alpha = \sin\left(\omega t + \dfrac{\pi}{2}\right) = \cos(\omega t) \\ i'_\beta = -\cos\left(\omega t + \dfrac{\pi}{2}\right) = \sin(\omega t) \end{cases} \quad (4.20)$$

In fact, (4.9), (4.19), or (4.20) can be used arbitrarily in the control circuits of Fig. 4-27 or Fig. 4-39. The resultant voltage-signal outputs v_{ha}, v_{hb}, and v_{hc} will be the same. The amplitude of the sinusoidal functions in (4.19) and (4.20) are set to unity, although the results would be the same if these amplitudes were multiplied by a factor k. The reason is that the same arbitrarily chosen time functions with a predefined amplitude are applied to both "direct" and "inverse" calculations. That is, (4.19) or (4.20) with the amplitudes multiplied by the arbitrary factor k is applied to the instantaneous power calculation block, and to the $\alpha\beta$ voltage calculation block. The factor k varies the amplitudes of the auxiliary currents and the auxiliary powers, although it does not alter the final results that are the fundamental positive-sequence voltage in Fig. 4-27 and the harmonic voltage in Fig. 4-39.

In both control circuits, the calculated auxiliary real and imaginary powers have no physical meaning. They only guarantee that the average real power $\bar{p}'$ together with the average imaginary power $\bar{q}'$ comprises only the fundamental positive-sequence voltage component of the measured bus voltage. Consequently, the oscillating real power $\tilde{q}'$ together with the oscillating imaginary power $\tilde{p}'$ comprises all

4.2. THREE-PHASE, THREE-WIRE SHUNT ACTIVE FILTERS

harmonics and imbalances present in the measured bus voltage. In conclusion, the signals i'_α and i'_β must contain only the positive-sequence component at the fundamental frequency. The phase angle of this fundamental positive-sequence component is not important. Both control circuits operate properly with an arbitrarily-chosen phase angle. The amplitude and the displacement angle that are made equal to $-\pi/2$ in (4.19) and equal to $+\pi/2$ in (4.20) are not important, and can be chosen arbitrarily, since they do not affect the final result. Note that (4.19) is the Clarke transformation of

$$\begin{cases} i'_a = \sqrt{\dfrac{2}{3}} \sin\left(\omega t - \dfrac{\pi}{2}\right) \\ i'_b = \sqrt{\dfrac{2}{3}} \sin\left(\omega t - \dfrac{\pi}{2} - \dfrac{2\pi}{3}\right) \\ i'_c = \sqrt{\dfrac{2}{3}} \sin\left(\omega t - \dfrac{\pi}{2} + \dfrac{2\pi}{3}\right) \end{cases} \quad (4.21)$$

whereas (4.20) is the Clarke transformation of

$$\begin{cases} i'_a = \sqrt{\dfrac{2}{3}} \cos(\omega t) \\ i'_b = \sqrt{\dfrac{2}{3}} \cos\left(\omega t - \dfrac{2\pi}{3}\right) \\ i'_c = \sqrt{\dfrac{2}{3}} \cos\left(\omega t + \dfrac{2\pi}{3}\right) \end{cases} \quad (4.22)$$

and (4.21) and (4.22) comprise only a fundamental positive-sequence component.

From the above explanation, it is easy to formulate some relations between the p-q Theory and the transformation of three-phase currents and voltages in the synchronous reference frames. The p-q Theory uses the Clarke transformation, also known as three-to-two phases transformation, whereas the synchronous reference frame transformation uses the d-q reference frame rotating at the synchronous angular velocity, also known as the Park transformation. In the theory of electric machines, it is common to use the following Park transformation that gives d-q components directly from the instantaneous abc voltages:

$$\begin{bmatrix} v_d \\ v_q \end{bmatrix} = \sqrt{\dfrac{2}{3}} \begin{bmatrix} \cos(\theta) & \cos\left(\theta - \dfrac{2\pi}{3}\right) & \cos\left(\theta + \dfrac{2\pi}{3}\right) \\ -\sin(\theta) & -\sin\left(\theta - \dfrac{2\pi}{3}\right) & -\sin\left(\theta + \dfrac{2\pi}{3}\right) \end{bmatrix} \begin{bmatrix} v_a \\ v_b \\ v_c \end{bmatrix} \quad (4.23)$$

A similar equation holds for the current transformation. In motor drives or generally in controllers of power electronics devices, it is common to use a combination of the Clarke and Park transformations. First, the Clarke transformation is applied to achieve the $\alpha\beta$ components in the stationary reference frames:

$$\begin{bmatrix} v_\alpha \\ v_\beta \end{bmatrix} = \sqrt{\frac{2}{3}} \begin{bmatrix} 1 & -\frac{1}{2} & -\frac{1}{2} \\ 0 & \frac{\sqrt{3}}{2} & -\frac{\sqrt{3}}{2} \end{bmatrix} \begin{bmatrix} v_a \\ v_b \\ v_c \end{bmatrix} \quad (4.24)$$

In this approach, the zero-sequence components in voltages and currents are neglected. Then, a modified Park transformation is applied to calculate the dq components in the synchronous reference frame from the $\alpha\beta$ components:

$$\begin{bmatrix} v_d \\ v_q \end{bmatrix} = \begin{bmatrix} \cos(\theta) & \sin(\theta) \\ -\sin(\theta) & \cos(\theta) \end{bmatrix} \begin{bmatrix} v_\alpha \\ v_\beta \end{bmatrix} \quad (4.25)$$

The expressions in (4.24) and (4.25) are graphically represented in Fig. 4-40.

The angle θ in the Park transformation of (4.25) is the synchronous angular position. It can be made equal to the time integral of the fundamental angular frequency ω, determined by a PLL circuit (phase-locked loop circuit), that is, $\theta = \omega t$. In this case, (4.20) and (4.25), give the following relation is given:

$$\begin{bmatrix} v_d \\ v_q \end{bmatrix} = \begin{bmatrix} i'_\alpha & i'_\beta \\ -i'_\beta & i'_\alpha \end{bmatrix} \begin{bmatrix} v_\alpha \\ v_\beta \end{bmatrix} \quad (4.26)$$

Now, it is possible to make a direct correlation between the control circuits of Fig. 4-39, based on the *p-q* Theory, with an equivalent control circuit based on the synchronous reference frames, since the following relation is valid:

$$\underbrace{\begin{bmatrix} v_d \\ v_q \end{bmatrix} = \begin{bmatrix} \cos(\theta) & \sin(\theta) \\ -\sin(\theta) & \cos(\theta) \end{bmatrix} \begin{bmatrix} v_\alpha \\ v_\beta \end{bmatrix}}_{\text{Park transformations}} = \underbrace{\begin{bmatrix} p' \\ q' \end{bmatrix} = \begin{bmatrix} i'_\alpha & i'_\beta \\ -i'_\beta & i'_\alpha \end{bmatrix} \begin{bmatrix} v_\alpha \\ v_\beta \end{bmatrix}}_{p\text{-}q \text{ Theory}} \quad (4.27)$$

Thus, the auxiliary real power p' in a controller based on the *p-q* Theory corresponds to the voltage component v_d on the d axis in a controller based on the synchronous reference frames, whereas the auxiliary imaginary power q' corresponds to the voltage component v_q on the q axis. The inverse Park transformation from (4.25) and the equation in the $\alpha\beta$-voltage-calculation block in Fig. 4-39 are also equivalent, that is,

$$\begin{bmatrix} v_\alpha \\ v_\beta \end{bmatrix} = \begin{bmatrix} \cos(\theta) & -\sin(\theta) \\ \sin(\theta) & \cos(\theta) \end{bmatrix} \begin{bmatrix} v_d \\ v_q \end{bmatrix} = \frac{1}{i'^2_\alpha + i'^2_\beta} \begin{bmatrix} i'_\alpha & -i'_\beta \\ i'_\beta & i'_\alpha \end{bmatrix} \begin{bmatrix} p' \\ q' \end{bmatrix} \quad (4.28)$$

In conclusion, one can say that control circuits based on the synchronous reference frame are analogous to the case of the control circuit based on the *p-q* Theory,

4.2. THREE-PHASE, THREE-WIRE SHUNT ACTIVE FILTERS

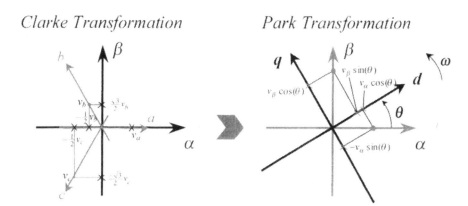

Figure 4-40. Graphical representation of the Clarke (left) and Park (right) transformations.

where only the fundamental, positive-sequence, current (or voltage) component with unitary amplitude is considered.

Analogous equations can be derived if it is desired to build a control circuit based on a fundamental, *negative-sequence* voltage detector. In this case, (4.20) should be replaced by

$$\begin{cases} i'_\alpha = \sin(\omega t) \\ i'_\beta = \cos(\omega t) \end{cases} \quad (4.29)$$

which corresponds to the Clarke transformation of an auxiliary, fundamental, negative-sequence current.

In some controllers, it is necessary to generate current references at the fundamental frequency that are exactly in phase or in quadrature with the system voltage. That is the function, for instance, of the signal $\bar{p}_{loss}$ in Fig. 4-26, which creates current references in phase with the bus voltage to draw energy to regulate the dc capacitor of the shunt active filter without producing reactive power. When using PLL circuits and the synchronous reference frames, special care should be taken to properly set up the angular position $\theta = \omega t$ of the Park transformation in order to guarantee that the direct-axis current component i_d produces only real power (active power) with the system voltage. In this case, the quadrature axis i_q produces only imaginary power (reactive power). This is the principal reason to set up $\theta = \omega t$ in phase with the fundamental, positive-sequence voltage.

4.2.4.C. An Application Case of an Active Filter for Harmonic Damping

Harmonic propagation as a result of harmonic resonance between line inductors and capacitors for power factor correction has made voltage harmonics a serious concern in power distribution systems and industrial power plants [53,54]. Actual measurements showing that harmonic propagation occurs frequently in a 6.6-kV power distribution system under light-load conditions at night has been reported in [55,56]. In addition, it has been pointed out by actual measurements that the fifth-harmonic voltage at the end bus is magnified by 3.5 times as large as that at the beginning bus in a 6.6-kV, 17-km-long power distribution feeder having capacitors with a total capacity of 245 kVA.

Installation of an active or passive filter on a long power distribution feeder may result in a strange phenomenon. That is, the voltage harmonics are mitigated at the point of installation, whereas they are magnified on other busses where no filter is connected. This phenomenon can be well understood by recalling the fundamentals of transmission line theory. For instance, a well-known resonance effect appears at the end terminal of a transmission line with a physical length corresponding to one-quarter of the electric wavelength at a given frequency. The higher the frequency, the shorter the wavelength. Moreover, the higher the shunt capacitor for power-factor correction, the greater the equivalent electric wavelength of the compensated line. Hence, long distribution feeders with bulk shunt reactive-power compensation can experience resonance phenomena at relatively low frequencies.

Installing the active filter on the end bus of the feeder makes it possible to damp out harmonic propagation. The purpose of the following analysis is not only to investigate resonance effects in long-distance distribution feeders, but also to discuss the installation effect of a voltage-detection-based shunt active filter characterized by acting as a harmonic terminator.

4.2.4.C.1. The Power Distribution Line for the Test Case. For simplicity, it is assumed that line inductors and capacitors are uniformly dispersed in a radial trunk feeder having no branch feeders. This assumption allows the use of distributed-constant circuit theory. The reason why the harmonic pollution is magnified in some busses in the trunk feeder can be clarified by both theoretical analysis based on the circuit theory and computer simulation. As a result, this analysis leads to the following two essential conclusions:

1. Installation of the active filter on the end bus does not cause any harmonic magnification at a specified harmonic frequency when the feeder length is shorter than a quarter of wavelength at that frequency.
2. As long as the active filter has an adequate gain related to the characteristic impedance of the feeder, no harmonic magnification occurs, irrespective of the feeder length.

These conclusions are verified by experimental results obtained from a three-phase, 200-V, 60 Hz, 20-kVA laboratory system [50].

Fig. 4-41 shows a three-phase power distribution line simulator used for the following laboratory experiments. The line simulator rated at 200 V, 60 Hz, and 20 kW is characterized by simplifying a real radial overhead distribution line rated at 6.6 kV and 3 MW in Japan. Hence, the line simulator is adequate to justify the effectiveness of the active filter for the purpose of harmonic termination.

A lumped *RL* circuit can represent the real overhead line between a bus and an adjacent bus, because it is reasonable to neglect the effect of the stray capacitors of the

4.2. THREE-PHASE, THREE-WIRE SHUNT ACTIVE FILTERS

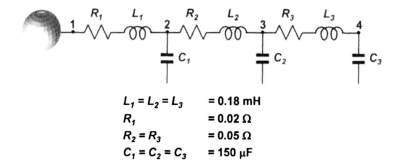

Figure 4-41. Three-phase power distribution line simulator rated at 200 V, 60 HZ, and 20 kW.

line for the fifth- and seventh-harmonic voltage and current, the parameters of which depend on the length and geometry of the line. In Fig. 4-41, L_1 corresponds to a leakage inductance of a distribution transformer, and L_2 and L_3 to line inductances. Eleven capacitors for power factor correction, which are dispersed by high-power consumers on the real overhead line, are represented by three capacitors, C_1, C_2, and C_3 in the line simulator. The total capacity of the capacitors is 450 μF (7 kVA).

Harmonic propagation results from series and/or parallel resonance between the inductive reactances and the capacitive reactances. The most serious harmonic propagation in Fig. 4-41 occurs around the seventh-harmonic frequency (420 Hz) under no-load conditions.

4.2.4.C.2. The Active Filter for Damping of Harmonic Propagation. Figure 4-42 shows a power circuit of the active filter used for the experiment. The active filter is installed on a bus in the 200-V, 20-kW line simulator via a three-phase

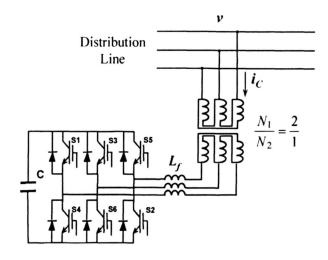

Figure 4-42. Power circuit of the active filter.

transformer with a turn ratio of 2:1. An electrolytic capacitor of 3300 μF is connected to the dc side, and the dc voltage is 260 V, whereas three inductors of $L_f =$ 1.0 mH (1.8% on a 3ϕ, 100-V, 60-Hz, 500-VA base) are connected to the ac side.

Figure 4-43 shows the control circuit of the active filter. Three-phase voltages, which are detected at the point of installation, are transformed to v_d and v_q on the dq-synchronous reference frames. Two first-order, high-pass filters (HPFs) with the same cutoff frequency as 5 Hz extract ac components $\tilde{v}_d$ and $\tilde{v}_q$ from v_d and v_q. The ac components are applied to the inverse dq-transformation circuit, so that the control circuit yields three-phase harmonic voltages at the point of installation. Multiplying each harmonic voltage by a gain K_V produces current references as follows:

$$i^*_{Ck} = K_V \cdot v_{hk}; \quad (k = a, b, c) \quad (4.30)$$

The above equation implies that the active filter behaves like a resistor of $1/K_V$ [Ω] to the external circuit for harmonic frequencies. The active filter does not contribute to the external circuit for the fundamental frequency. The gain K_V is set to 1 S, as discussed in the following section. Three-phase actual currents i_{Ca}, i_{Cb}, and i_{Cc} are controlled in such a way as to follow their current references.

Figure 4-44 shows a current control circuit of the active filter. This circuit compares the reference current with its actual current, and then amplifies the error signal between the two currents by a proportional gain K_I. Each phase voltage is detected at the point of active filter installation, and this voltage v is added to each magnified error signal, thus constituting a feedforward compensation to improve current controllability. As a result, three current controllers yield three-phase voltage references. Then, each voltage reference v^*_i is compared with a 10-kHz repetitive triangular waveform to generate the gate signals for the insulated gate bipolar transistors (IGBTs). The loop gain K_I in the current feedback controller is set to 150 V/A.

4.2.4.C.3. Experimental Results

Experimental Conditions. In reality, harmonic-producing loads are dispersed on a power distribution line, and the distribution line itself is a dynamic system that varies with the passage of time, day, season, and/or year, thus making it difficult to

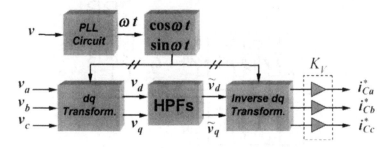

Figure 4-43. Control circuit of the active filter.

4.2. THREE-PHASE, THREE-WIRE SHUNT ACTIVE FILTERS

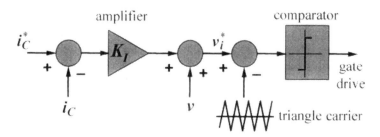

Figure 4-44. Current control circuit per phase.

perform experiments under realistic conditions in the laboratory. Therefore, the following ideal conditions, rather than realistic conditions, are taken. The following ideal conditions make it easier to evaluate the installation effect of the active filter with a gain $K_V = 1.0$ S, applied to (4.30):

- Although the fifth-harmonic voltage and current are the most dominant in a real 6.6-kV power distribution line in Japan, only the seventh-harmonic voltage and current are taken into account because harmonic propagation occurs around this harmonic frequency in Fig. 4-41.
- Experiments are performed under the no-load condition because this condition causes more severe harmonic propagation than realistic, light-load conditions.
- Either a seventh-harmonic voltage source of 1.7 V (1.5%) or a seventh-harmonic current source of 1.8 A (3.0%) is connected to the line simulator. Both harmonic sources are independent of each other, because the principle of superposition is applicable.

Seventh Harmonic Voltage Source Upstream of Bus 1. Figure 4-45 shows an experimental system in which a seventh-harmonic voltage source of 1.7 V (1.5%) is

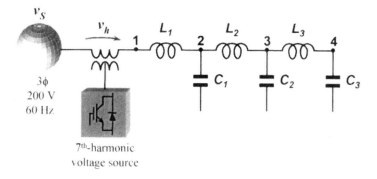

Figure 4-45. Power distribution line simulator with seventh-harmonic voltage source of V_h = 1.7 V (1.5%) upstream of bus 1.

162 SHUNT ACTIVE FILTERS

connected upstream of bus 1 in Fig. 4-41. Note that this harmonic voltage source can be considered as a background harmonic voltage existing upstream of a primary distribution transformer in an actual power system. The seventh-harmonic voltage source is implemented with a three-phase, voltage-fed pulse-width-modulation (PWM) converter that is connected in series to the 200-V power system via three single-phase transformers. Here, no harmonic-producing load exists in the power distribution line simulator.

Figures 4-46 to 4-48 show experimental waveforms under the circuit configuration depicted in Fig. 4-45. Table 4.1 summarizes actual measurements of the seventh-harmonic currents and voltages contained in the waveforms of Figures 4-46 to 4-48. Table 4.2 shows the ratio of the seventh-harmonic voltage at each bus with respect to that at bus 1, which implies a voltage-magnifying factor at each bus.

Figure 4-46 shows experimental waveforms when no active filter is connected. Harmonic voltage propagation magnifies the seventh-harmonic voltage at bus 4 by 5.7 times as large as that at bus 1. Figure 4-47 shows experimental waveforms when the active filter is installed on bus 2. The active filter attenuates the harmonic voltage propagation at bus 2. It, however, has no capability of harmonic damping throughout the power distribution line, because 4.0 V (3.5%) at bus 4 is much larger than 1.7 V (1.5%) at bus 1.

Figure 4-48 shows experimental waveforms when the active filter is installed on bus 4. Table 4.1 and Table 4.2 verify that the installation of the active filter on bus

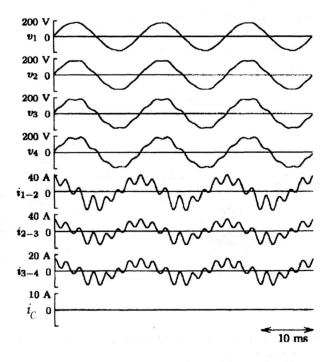

Figure 4-46. Experimental waveforms without active filter.

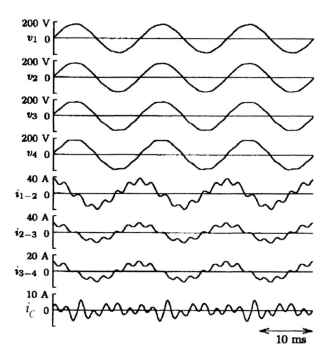

Figure 4-47. Experimental waveforms with active filter connected to bus 2.

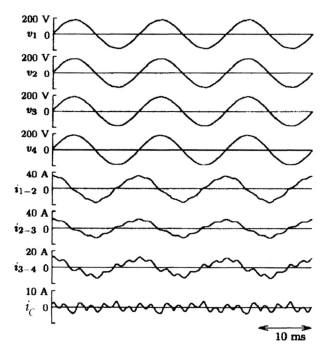

Figure 4-48. Experimental waveforms with active filter connected to bus 4.

Table 4.1. Actual measurements of seventh-harmonic currents and voltages when the active filter is disconnected or connected

A	Disconnected	Connected	
		Bus 2	Bus 4
I_{1-2}	8.2	4.1	1.2
I_{2-3}	6.7	2.8	1.3
I_{3-4}	3.8	1.6	1.4
I_C	—	1.8	1.3

V	Disconnected	Connected	
		Bus 2	Bus 4
$V_1 (=V_h)$	1.7	1.7	1.7
V_2	3.7	1.7	1.6
V_3	7.4	3.3	1.3
V_4	9.7	4.0	1.1

4, that is, on the end bus of the power distribution line, leads to achieving harmonic damping throughout the line. Paying attention to the active filter current in Table 4.1, one can conclude that its installation on bus 4 makes the required current rating of the active filter smaller than does installation on bus 2. In other words, the required volt-ampere rating of the active filter installed on bus 2 is 624 VA, whereas that of the active filter installed on bus 4 is 450 VA. When the active filter is installed on bus 4, it draws the following amount of seventh harmonic power from bus 4:

$$3 \times 1.3^2 \times \frac{1}{1.0} = 5.1 \text{ W} \quad (4.31)$$

which is only 1.1% of the active filter rating of 450 VA.

Seventh-Harmonic Current Source Downstream of Bus 2. Figure 4-49 shows an experimental system in which a seventh-harmonic current source of 1.8 A (3.0%) is con-

Table 4.2. Magnifying factors of seventh-harmonic voltages when the active filter is disconnected or connected

A	Disconnected	Connected	
		Bus 2	Bus 4
V_2/V_1	2.2	1.0	0.9
V_3/V_1	4.4	1.9	0.8
V_4/V_1	5.7	2.4	0.6

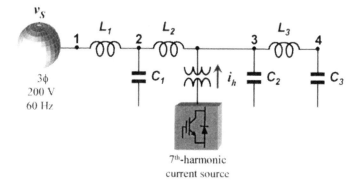

Figure 4-49. Power distribution line simulator, where a seventh-harmonic current source of $I_h = 1.8$ A (3.0%) exists on bus 3.

nected in parallel to bus 3. For the sake of simplicity, many harmonic-producing loads dispersed on the real power distribution line are represented by the single harmonic current source depicted in Fig. 4-49. The current source is implemented with a three-phase, voltage-fed PWM converter with a minor current loop, as if it were a shunt active filter. Note that no harmonic voltage source is connected upstream of bus 1.

Figures 4-50 to 4-52 show experimental waveforms under the circuit configura-

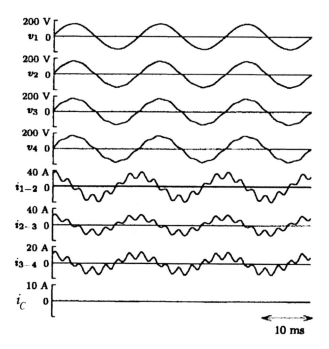

Figure 4-50. Experimental waveforms without active filter connected.

166 SHUNT ACTIVE FILTERS

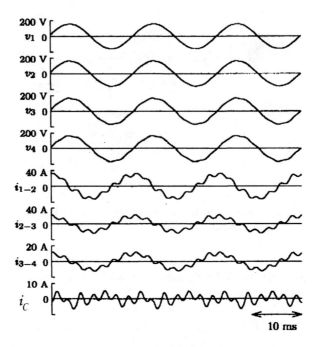

Figure 4-51. Experimental waveforms with active filter connected on bus 2.

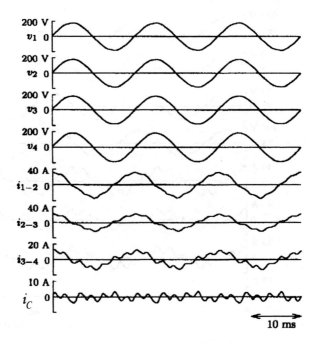

Figure 4-52. Experimental waveforms with active filter connected to bus 4.

Table 4.3. Actual measurements of seventh-harmonic currents and voltages when the active filter is disconnected or connected

		Connected	
A	Disconnected	Bus 2	Bus 4
I_{1-2}	5.8	3.5	1.3
I_{2-3}	4.5	3.4	1.0
I_{3-4}	2.8	1.6	1.6
I_h	1.8	1.8	1.8
I_C	—	1.9	1.5

		Connected	
V	Disconnected	Bus 2	Bus 4
V_2	2.8	1.6	1.0
V_3	5.0	3.1	1.3
V_4	6.3	3.8	1.3

tion depicted in Fig. 4-49. Table 4.3 summarizes actual measurements of the seventh-harmonic currents and voltages from the waveforms of Figures 4-50 to 4-52. Table 4.4 shows the ratio of the seventh-harmonic current flowing between a bus and an adjacent bus with respect to $I_h = 1.8$ A, which implies a current-magnifying factor. Note that no seventh-harmonic voltage exists on bus 1, that is, $V_1 = 0$.

Figure 4-50 shows experimental waveforms when no active filter is connected. Harmonic current propagation makes each seventh-harmonic current between a bus and an adjacent bus larger than 1.8 A.

Figure 4-51 shows experimental waveforms when the active filter with the gain of $K_V = 1.0$ S is installed on bus 2. The seventh-harmonic voltage at bus 2 or at the point of installation of the active filter is the smallest, whereas the seventh-harmonic current flowing between bus 1 and bus 2, that is, I_{1-2} is magnified by two. The seventh-harmonic voltage at bus 4 reaches 3.8 V (3.3%).

Figure 4-52 shows experimental waveforms when the active filter is installed on bus 4, that is, on the end bus of the distribution line. All of I_{1-2}, I_{2-3}, and I_{3-4} are smaller than 1.8 A. In other words, each current-magnifying factor is less than uni-

Table 4.4. Magnifying factors of seventh-harmonic currents when the active filter is disconnected or connected

		Connected	
	Disconnected	Bus 2	Bus 4
I_{1-2}/I_h	3.2	1.9	0.7
I_{2-3}/I_h	2.5	1.9	0.6
I_{3-4}/I_h	1.6	0.9	0.9

ty. The seventh-harmonic voltage at bus 4 is reduced to 1.3 V (1.1%), which is one-third lower than that in the case of installing the active filter on bus 2. In addition, Table 4.3 concludes that installation on bus 4 makes the required current rating of the active filter, I_C, smaller than does installation on bus 2. In this case, the required volt-ampere rating of the active filter installed on bus 2 is 658 VA, whereas that of the active filter installed on bus 4 is 520 VA.

In summary, a shunt active filter based on voltage detection is very effective in damping harmonic propagation. It is controlled in such a way as to present infinite impedance to the external circuit at the fundamental frequency, and to exhibit low resistance ($1/K_V$) for harmonic frequencies. The above experimental results obtained from the laboratory system, along with theoretical results, are summarized as follows.

- Installation of the active filter on the end bus of a power distribution line is more effective in harmonic damping than installation on the beginning bus or in the vicinity of a primary distribution transformer.
- Installation on the end bus makes the required current rating of the active filter smaller than does installation on the beginning bus.
- Harmonic mitigation of voltage and current is a welcome "by-product," when harmonic termination is performed.

In conclusion, the voltage-detection-based active filter intended for "harmonic termination" should be installed not on the beginning bus, but on the end bus of a radial power distribution line subjected to harmonic propagation.

4.2.4.C.4. Adjustment of the Active Filter Gain.
The gain K_V in the controller of the shunt active filter should be adjusted carefully. As suggested before, the shunt active filter for damping of harmonic propagation is very effective if it is installed at the end of the radial feeder and the gain K_V corresponds to the "equivalent characteristic impedance" of the compensated feeder.

As shown in (4.30), the gain K_V is used to determine the compensating current reference of the shunt active filter as $i_C^* = K_V \cdot v_h$. Under the assumption that the actual compensating current i_C is equal to its reference i_C^*, the active filter behaves like a resistor of $1/K_V$ [Ω] for harmonic frequencies. Hence, the active filter acts as a damping resistor or a harmonic terminator to mitigate harmonic resonance in the distribution line. In order to perfectly achieve the harmonic termination of the line, the optimal gain of the active filter should be set as $K_V = 1/Z_0$, where Z_0 is the "equivalent" characteristic impedance of the line, taking into account loads and capacitors for power factor correction along the line. In the previous study of harmonic propagation, the simplified system presented in Fig. 4-41 is considered. In this case, $Z_0 = \sqrt{L_1/C_1}$ [Ω] for a lossless line, whereas it is very difficult to know an actual value of Z_0 in a real system. Strictly speaking, the value is not constant even if it is possible to know it in advance. Therefore, the design of an optimal gain is a challenge in power engineering, and the following analysis is limited to explaining the principles of harmonic propagation.

4.2. THREE-PHASE, THREE-WIRE SHUNT ACTIVE FILTERS

A strange phenomenon related to resonance effects can occur in distribution lines: voltage harmonics are mitigated at the point of installation, whereas they are magnified on other busses where no filter is connected. This phenomenon is referred to as "whack-a-mole." The distributed-constant circuit theory is used to clarify this point [51].

Distributed-Constant Model. Figure 4-53 shows a three-phase power distribution feeder. Table 4.5 shows circuit constants of the feeder for the following analysis. The circuit constants are designed based on a real distribution feeder rated at 6.6 kV and 3 MW in Japan. However, in order to clearly demonstrate the occurrence of the "whack-a-mole" phenomenon, the feeder length and the total capacity of capacitors for power factor correction are modified to 9 km and 3 MVA, which are three times as large as those in the real feeder. It is assumed that each line inductor between a bus and its adjacent bus has the same inductance as L, and each capacitor has the same capacitance as C for the sake of simplicity. Moreover, the line resistance is also neglected. Attention is paid to a radial trunk feeder, so that it is assumed that no branch exists on any bus. No-load conditions are assumed when considering the most serious harmonic propagation.

Figure 4-54 depicts a distributed-constant model of the power distribution feeder, where l is the feeder length and x is the distance from bus 1. A harmonic voltage source exists on bus 1, and a resistor of $1/K_V$ Ω on bus 10. Note that the resistor represents a shunt active filter based on voltage detection when attention is paid to voltage and current harmonics.

The characteristic impedance of the feeder, Z is given by

$$Z = \sqrt{\frac{L}{C}} = 8.9 \text{ Ω} \qquad (4.32)$$

The wavelength λ and the propagation constant γ of the feeder are defined as

$$\lambda = \frac{1}{f\sqrt{LC}} \qquad (4.33)$$

$$\gamma = j\beta = j\omega \sqrt{LC} \qquad (4.34)$$

With focus on the present simplified analysis and the given values of L and C, Table 4.6 summarizes a relationship among harmonic frequency, wavelength, and

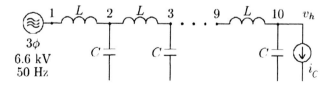

Figure 4-53. Power distribution feeder (lumped-element model).

170 SHUNT ACTIVE FILTERS

Table 4.5. Circuit constants on a 3φ, 50-Hz, 6.6-kV, 3-MVA base

Line voltage	6.6 kV
Line frequency	50 Hz
Feeder length	9 km
Total capacity of capacitors	3 MVA
Number of buses	10
Line inductance, L	1.98 mH/km (4.3%)
Line resistance, R	0.36 Ω/km (2.5%)
Capacitance, C	25 μF (11.1%)

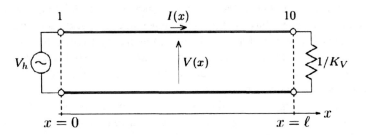

Figure 4-54. Distributed-constant model when a harmonic voltage source exists on bus 1, and the active filter with a gain of K_V is installed on bus 10.

feeder length. A feeder length of 9 km corresponds to approximately half the wavelength for the fifth harmonic and three-fourths the wavelength for the seventh harmonic. As shown in the following discussion, the line behaviors for a half wavelength and three-fourths wavelength are quite different, and can justify the assumption of the "whack-a-mole" phenomenon.

Figure 4-54 shows a harmonic-voltage standing wave $V(x)$ and a harmonic-current standing wave $I(x)$ as a function of distance x, given by

$$V(x) = Ae^{-\gamma x} + Be^{\gamma x} \qquad (4.35)$$

$$I(x) = \frac{1}{Z}(Ae^{-\gamma x} - Be^{\gamma x}) \qquad (4.36)$$

Table 4.6. Relationship among frequency, wavelength, and feeder length

Frequency f	Wavelength λ	Feeder length ℓ
250 Hz (5th)	18.3 km	9 km ($\cong \lambda/2$)
350 Hz (7th)	13.1 km	9 km ($\cong 3\lambda/4$)

The constants A and B can be determined if the boundary conditions at the two terminals; the sending and receiving ends, are known. If a harmonic voltage source V_h is connected to the sending end and a resistor of $1/K_V$ [Ω] is connected to the receiving end, instead of a actual active filter, the constants A and B are given by

$$A = \frac{(\cosh \gamma l + \sinh \gamma l)(K_V Z + 1)}{2(\cosh \gamma l + K_V Z \sinh \gamma l)} V_h \quad (4.37)$$

$$B = \frac{(\sinh \gamma l - \cosh \gamma l)(K_V Z - 1)}{2(\cosh \gamma l + K_V Z \sinh \gamma l)} V_h \quad (4.38)$$

Substituting A and B into (4.35), the following harmonic-voltage standing wave is derived:

$$V(x) = \frac{\cosh \gamma(l - x) + K_V Z \sinh \gamma(l - x)}{\cosh \gamma l + K_V Z \sinh \gamma l} V_h \quad (4.39)$$

Figures 4-55 and 4-56 show voltage standing waves excited by the fifth- and seventh-harmonic voltage sources on bus 1, respectively. The horizontal axis shows the bus number and the distance from bus 1. The vertical axis shows a voltage-magnifying factor, that is, the ratio of a harmonic voltage on a bus to that on bus 1. Marks "o," "x," and "●" indicate digital simulation results, and lines "—," "– –," and "- - -" are analytical results obtained from (4.39). In both figures, "$K_V = 0$" corresponds to no installation of active filters, "$K_V = 1/Z$" means that the active filter acts as the characteristic impedance, and "$K_V = \infty$" forms a short circuit across the

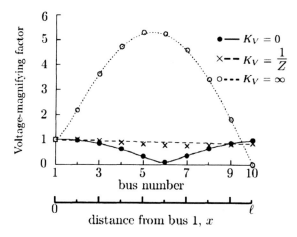

Figure 4-55. A voltage-magnifying factor on each bus when a fifth-harmonic voltage source is connected to bus 1.

172 SHUNT ACTIVE FILTERS

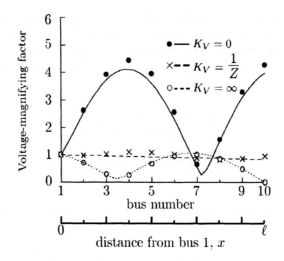

Figure 4-56. A voltage-magnifying factor on each bus when a seventh-harmonic voltage source is connected to bus 1.

end bus for harmonic frequencies. The line resistance, R, in Table 4.5 is considered in achieving both analytical calculation and digital simulation. However, to approximate the results, a feeder length of 9.5 km is considered in (4.39) and the lumped-element model of Fig. 4-53 is implemented in the digital simulator.

Fig. 4-55 shows the fifth-harmonic-voltage standing waves. It shows a typical behavior of a standing wave in a half-wave transmission line. In case of $K_V = 0$ (open-end terminal), the voltage magnifying factor on each bus is less than unity. When the active filter with a gain $K_V = \infty$ is installed (the active filter forces a short-circuit at the end terminal for the fifth-harmonic), no fifth-harmonic voltage appears on the end bus. However, the harmonic voltage on bus 5 is magnified by five times, compared with that on bus 1, thus resulting in the occurrence of the above-denominated "whack-a-mole" phenomenon. On the other hand, the harmonic voltage on each bus is equal to that on bus 1 when the active filter acts as a "harmonic terminator," as a resistor $R = 1/K_V$ that is equal to the characteristic impedance Z of the feeder. This implies that installation of the active filter does not cause any "whack-a-mole" in the feeder.

In contrast to the half-wave behavior of the fifth harmonic, the seventh harmonic imposes a resonant, one-fourth-wave behavior on the feeder. This can be verified in Fig. 4-56 by a high magnifying factor at the end terminal (bus #10), when the circuit is open, that is, no active filter is installed ($K_V = 0$). Therefore, if no active filter is installed, the feeder does not magnify the fifth harmonic, but it does magnify the seventh harmonic. On the other hand, if the active filter has a very high gain ($K_V = \infty$), it does not excite the "whack-a-mole" phenomenon by the seventh harmonic, but it does by the fifth harmonic. Note that the voltage profile for the seventh harmonic is also flat when the active filter gain is $K_V = 1/Z$.

In conclusion, the active filter gain has an optimal value to perform harmonic damping. When the active filter gain K_V is equal to $1/Z$, the active filter acts as a "harmonic terminator" ($R = 1/K_V = Z$). In this case, it performs critical damping such that the harmonic voltage along the feeder is equal to that on bus 1. However, to achieve continuously the critical damping becomes a challenge in a real network, because of its dynamic configuration and varying conditions of loading and compensation.

4.2.5. A Digital Controller

A digital controller using a DSP or a microprocessor is preferable to an analog controller when a shunt active filter is installed on an actual power distribution system. However, the active filter forms a feedback loop in controlling the compensating current, so that time and phase delays in the digital controller may lead to poor damping performance for harmonic propagation or, in the worst case, may induce instability in the control system.

In this section, an example of a digital controller for a shunt active power filter is presented based on the work of Jintakosonwit et al. [52]. To achieve hardware/software implementation of the digital controller, attention is paid to the following items:

- Conversion time of A/D converters
- Data aliasing caused by switching ripples
- Calculation time for signal processing

Generally, the switching frequency of a PWM converter used as a power circuit of the active filter should be around 10 to 20 kHz and, therefore, the sampling rate of 10 to 40 kS/s for voltage and current is required. A low-speed A/D converter with a conversion time of 25 to 100 µs could meet the above sampling rate in a conventional digital controller. However, a digital controller equipped with such a low-speed A/D converter might have two sampling delays:

1. One is due to the conversion of an analog signal to digital signal
2. The other is due to the calculation in the DSP

Such a time delay might contribute to a unsatisfactory frequency response of the current control system. Switching ripples appear in the output voltage and current of the active filter. When the voltage and current including switching ripples are sampled, voltage and current components having frequencies higher than half of the sampling frequency are distorted because of the so-called "aliasing." To avoid this aliasing effect, switching ripples have to be eliminated by an antialiasing low-pass filter with a cutoff frequency lower than, but near to the switching frequency. However, this low-pass filter is accompanied by phase delay that produces a bad effect on the current control system of the active filter.

It is a common procedure to use sinusoidal signals, $\cos \omega t$ and $\sin \omega t$, for calculations in the active filter controllers. A "table look-up" scheme is widely employed in a digital system—the sinusoidal data is calculated and stored in a data table to reduce processing time. At every sampling time, the data is taken out, which refers to the phase of the supply voltage. Unless the sampling time is synchronized with the voltage phase, a nonnegligible phase error might occur; otherwise, a signal process or routine is required for compensation. An effective way of solving all the above problems caused by the time and phase delays inherent in the digital controller is to synchronize signal sampling and processing with the line frequency (here, it is considered to be a line frequency of 60 Hz). This also produces a good "side-effect" in that it is possible to eliminate an antialiasing low-pass filter from the controller.

4.2.5.A. System Configuration of the Digital Controller

Figure 4-57 depicts a fully digital controlled shunt active filter presented in [52]. Table 4.7 summarizes the specifications of its digital controller, in which signal sampling and processing are characterized by synchronization with the line frequency. The digital controller consists of an A/D unit with sampling of voltage and current, a DSP unit for processing of harmonic-voltage extraction, a digital PWM unit, and a phase-locked-loop (PLL) unit. A 16-bit fixed-point ADSP-2101 (Analog Devices) is used as a DSP with a sampling period of 50 µs (20 kS/s). The A/D unit consists of five A/D converters connected in parallel.

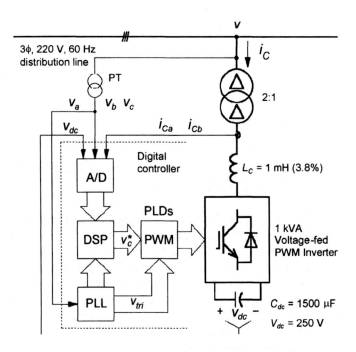

Figure 4-57. Digital controlled active filter.

Table 4.7. Specification of the digital controller

DSP	
Device	ADSP-2101
Bits of resolution	Fixed-point 16 bits
Clock	16.67 MHz
Instruction cycle time	60 ns
Data memory	1 k words (on-chip)
	8 k words (external)
Program memory	2 k words (on-chip)
	32 kB EEPROM
A/D	
Device	ADS-230s
Bits of resolution	12 bits (± 0.24%)
Conversion time	1 µs
PWM	
Device	PLDs (MACH230s)
Blanking time	1 to 4 µ adjustable
Switching frequency	10 kHz (at 60 Hz)
PLL	
Device (PD and VCO)	74HC4046A
System line frequency	60 Hz
VCO frequency	10 MHz
Counters	336-step for sampling
	512-up–down for PWM

In order to reduce the time and phase delays, a particular design for the digital controller is presented. The feature of this digital controller is its operation with synchronism between the system line frequency, the PWM controller, and the sampling of the A/D unit.

4.2.5.A.1. Operating Principle of PLL and PWM Units.

Figure 4-58 shows a block diagram of the PLL and PWM units. The PLL unit consists of a phase detector (PD), a voltage-controlled oscillator (VCO), and two digital counters. Counter I is a 512-step up–down counter that generates the 9-bit carrier signal, v_{tri}, for the PWM unit. Counter II is a 336-step counter that generates the 9-bit phase information, ωt, for calculation of coordinate transformations. In the PLL unit, the 60-Hz line frequency is detected by the PD. After the phase is locked, the VCO frequency is 60 Hz · 336 · 512 which is approximately 10 MHz (exactly 10,321,920 Hz if the line frequency is exactly 60 Hz). When the 512-step up–down counter changes its state from up to down and vice versa, a starting signal with a frequency around 20 kHz (approximately 10 MHz/512) is applied to the A/D unit. This means that the sampling period T is close to 50 µs. Note that the VCO frequency and the sampling frequency are bound up with the resolution of both counters. High resolution produces a low quantization error, whereas both frequencies are high. In this system,

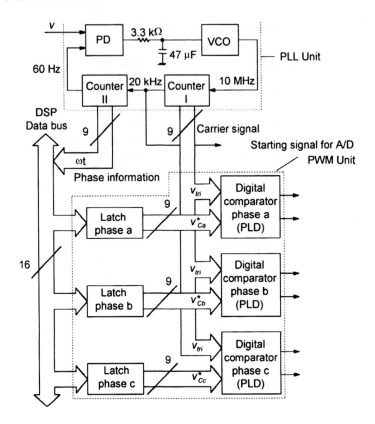

Figure 4-58. Block diagram of the PLL and PWM units.

the sampling frequency is designed around 20 kHz because that is high enough to convert the voltage and current, including the 5th-, 7th-, 11th- and 13th-harmonic frequency components with a low quantization error of 1/4096 (approximately 0.02%). Besides, an integrated circuit of 74HC4046A, used as the PD and VCO, does not work properly over a frequency of 19 MHz. The VCO frequency, therefore, is designed to be 10 MHz.

The PWM unit is used to generate three-phase PWM switching signals. Three digital comparators integrated into the PLDs compare a single triangular-carrier signal, v_{tri}, obtained from the 512-step up–down counter, with three compensating voltage references, v_{Ca}^*, v_{Cb}^*, and v_{Cc}^*, obtained from the DSP. The quantization error in the PWM unit is $1/512 \approx 0.2\%$, and the switching frequency is 10 kHz (10MHz/1024). A three-phase latch circuit is used to hold $v_C^* = [v_{Ca}^*, v_{Cb}^*, v_{Cc}^*]^T$. Thus, v_C^* is updated at every sampling time in the PWM unit. Moreover, the use of the PLDs makes it possible to adjust the so-called "blanking time" in a range of 1 to 4 μs by means of programming.

Figure 4-59 shows a timing diagram of the digital controller. Since synchronous operation is achieved between the line frequency and the sampling in the A/D unit, the number of samples in one cycle of the supply voltage is fixed to be 336 even if the line frequency is slightly changed. No phase error appears in either sampling or processing because of synchronous operation.

4.2.5.A.2. Sampling Operation in the A/D Unit.
Figure 4-60 shows synchronous operation between sampling in the A/D unit and processing in the PWM controller. One phase of three-phase PWM signals is shown for the sake of simplicity. Here, it is assumed that each ac terminal in the active filter is connected to an ideal inductor L_C without supply voltage. The validity of this assumption will be discussed in the following section. In order to generate the PWM signals, the voltage reference, v_C^*, coming from the DSP, is held during one sampling period T, and then it is compared with the triangular-carrier signal v_{tri}, generating the actual output voltage of the active filter v_C. Note that the average value of v_C over one sampling period is equal to v_C^*. The actual output current i_C, depicted by a thick solid line, is calculated by

$$i_C = \int (v_C/L_C)dt \qquad (4.40)$$

and the hypothetical current i_C', depicted by a broken line, is given by

$$i_C' = \int (v_C^*/L_C)dt \qquad (4.41)$$

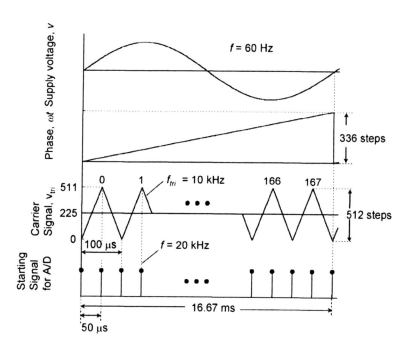

Figure 4-59. Timing diagram of the digital controller.

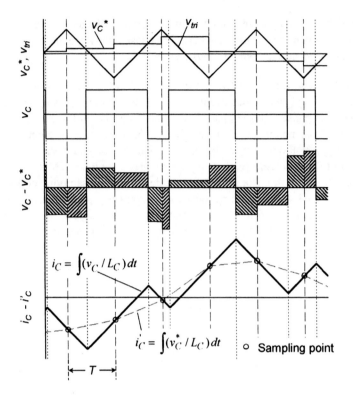

Figure 4-60. Principle of synchronous sampling.

When attention is paid to the single sampling period T in Fig. 4-60, the average value of the harmonic voltage $v_C - v_C^*$ is always zero. Thus, the value of i_C is equal to that of i_C' at each sampling time, so that no aliasing effect occurs in the A/D unit. As a result, the cutoff frequency of a low-pass filter installed at the input terminals of the A/D unit can be set much higher than the sampling frequency. The phase delay caused by the low-pass filter does not influence the current control performance in the active filter.

Another advantage of synchronous operation is that electrical noises resulting from the switching operation of IGBTs do not interfere with the sampled current because the sampling time of current never coincides with the switching time of IGBTs, as shown in Fig. 4-60. The synchronous operation makes a significant contribution to reducing the time and phase delays inherent in the digital controller. Actually, the total delay time in the digital controller implemented was one sampling period of 50 μs (in fact, 49.6 μs for line frequency equal to 60 Hz).

4.2.5.B. Current Control Methods

4.2.5.B.1. Modeling of Digital Current Control. Figure 4-61 shows a control block diagram of the active filter. The supply voltage v on the installation bus and

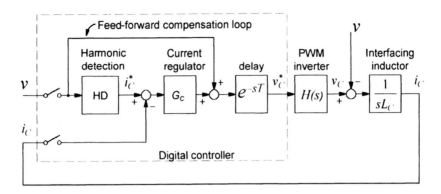

Figure 4-61. Control block diagram of the active filter.

the compensating current i_C are sampled at the same time. Then, the harmonic components contained in v are extracted by the harmonic-detecting circuit, for instance, as shown in Fig. 4-43, thus yielding the compensating current reference i_C^*. Next, the reference current i_C^* and its actual current i_C are compared, and the difference is controlled by the current regulator G_C. The sampled supply voltage v is added to the output signal of the current regulator in order to achieve feed-forward compensation. As a result, it is possible to eliminate the effect of the supply voltage, which is considered as a disturbance from the current controller, as if the load of the PWM converter was only the interfacing inductor L_C, as shown in Fig. 4-61. This sequence yields the voltage reference v_C^* for the PWM converter, accompanied by a time delay of one sampling period T. The effect of the time delay is represented as e^{-sT} in the control block diagram.

Since the harmonic components included in v_C produce no effect on the sampled compensating current, as shown in Fig. 4-60, the PWM converter can be assumed to be a zero-order hold circuit:

$$H(s) = \frac{1 - e^{-sT}}{s} \qquad (4.42)$$

Figure 4-62 shows a current-control block diagram in the z plane. As described above, the supply voltage v can be assumed to be a disturbance signal in the current control. This disturbance signal is canceled by the function of the feedforward compensation loop, so that both disturbance signal and feedforward compensation loop are eliminated from Fig. 4-62. Moreover, combination of the PWM converter and the inductor L_C leads to a single discrete-transfer function.

4.2.5.B.2. Proportional Control.
In the case of the proportional-control method, the current regulator $G_C(z)$ is represented by a feedback gain K_C as

$$G_C(z) = K_C \qquad (4.43)$$

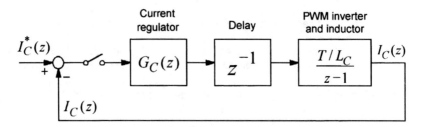

Figure 4-62. The current-control block diagram.

According to the block diagram in Fig. 4-62, the closed-loop, discrete-transfer function $W(z)$ is given by

$$W(z) = \frac{I_C(z)}{I_C^*(z)} = \frac{K_C T/L_C}{z^2 - z + K_C T/L_C} \quad (4.44)$$

The characteristic roots in (3) are expressed as

$$z = \frac{1}{2} \pm \sqrt{\frac{1}{4} - \frac{K_C T}{L_C}} \quad (4.45)$$

To guarantee the stability of current control, the characteristic roots have to lie within the unit circle in the z plain. It is concluded that the current control is stable for

$$0 \le K_C \le \frac{L_C}{T} \quad (4.46)$$

In order to achieve current control without any overshoot or oscillation in the time domain, the critically damped system should be taken in an actual system. Therefore, the feedback gain K_C that offers the same characteristic roots, should be selected as

$$K_C = \frac{L_C}{4T} \quad (4.47)$$

4.2.5.B.3. Deadbeat Control. When the deadbeat-control method is applied to Fig. 4-62, the current control system is subject to a time delay of at least two sampling periods because it takes one sampling period to produce the voltage reference v_C^* from sampling of voltage and current in the digital controller, and it takes the following sampling period in the PWM converter and inductor. Therefore, the closed-loop discrete-transfer function of the current control system based on the deadbeat-control method should be assumed to be

$$W(z) = \frac{I_C(z)}{I_C^*(z)} = z^{-2} \quad (4.48)$$

As a result, the discrete-transfer function of the current regulator $G_C(z)$ can be expressed as

$$G_C(z) = \frac{L_C}{T} \frac{z}{z+1} \quad (4.49)$$

4.2.5.B.4. Frequency Response of Current Control. In the following, the s domain is used to characterize the active filter for ease of explanation. In the case of harmonic damping in the active filter, the relationship in the s domain between the current reference $I_C^*(s)$ and the harmonic voltage $V_h(s)$ at the point of installation is given as $I_C^*(s) = K_V \cdot V_h(s)$. Assuming that no time delay exists in the harmonic-detecting circuit, the transfer function of the active filter $Y_C(s)$ is given by

$$Y_C(s) = \frac{I_C(s)}{V_h(s)} = K_V \cdot W(s) \quad (4.50)$$

The above transfer function implies that the active filter acts as an admittance of $I_C(s)/V_h(s)$ for harmonic frequencies. Moreover, the characteristic of the active filter is dependent on the control gain K_V and the characteristic of the current control $W(s) = I_C(s)/I_C^*(s)$, which has a phase delay. This means that the equivalent circuit of the active filter is the admittance including the phase delay. Here, the characteristic of $W(s)$ is emphasized in the following.

Figure 4-63 shows the frequency response of the current control system based on the proportional-control or deadbeat-control method with the transfer function $W(s)$ as follows. Proportional control is defined as

$$W(s) = \frac{1}{(2e^{sT} - 1)^2}$$

Deadbeat control is defined as

$$W(s) = e^{-2sT}$$

The sampling period of $T = 50$ μs is used for analysis. In the case of the proportional-control method, the frequency response has a low gain and a poor phase lag as the frequency is higher than 1 kHz. On the other hand, the deadbeat-control method significantly improves the frequency response in terms of both gain and phase characteristics.

Equation (4.50) suggests that the performance of the active filter based on voltage detection depends on the current controllability. In a frequency range of 100 Hz to 1 kHz, the current controller has a small amount of phase lag, irrespective of current control methods. Therefore, the active filter acts as a resistor–inductor circuit for harmonic damping of the distribution line in this frequency range. However, the current controller using the proportional-control method has a phase lag of 90° at the 21st-harmonic frequency (1.26 kHz). As a result, the active filter behaves like

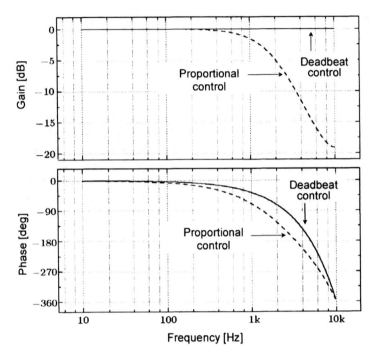

Figure 4-63. Frequency response of the current control.

an inductor rather than a resistor. Moreover, at the 55th-harmonic frequency (3.3 kHz), the phase lag reaches 180°, so that the active filter acts as a negative resistance. The active filter might cause harmonic propagation at that frequency.

When the deadbeat-control method is applied, the phase lag is 90° at the 42nd-harmonic frequency (2.5 kHz). Thus, the deadbeat-control method damps out harmonic propagation in a wider frequency range than the proportional-control method does. At the 83rd-harmonic frequency (5 kHz), the current controller has a phase lag of 180° under a unity gain. As a result, the deadbeat-control method might cause more severe harmonic propagation than the proportional-control method. However, an actual power distribution line exhibits such a high inductive impedance as to attenuate existing harmonic components in a frequency range higher than 3 kHz. Consequently, no significant harmonic propagation would occur in this frequency range.

4.3. THREE-PHASE, FOUR-WIRE SHUNT ACTIVE FILTERS

This section describes shunt active filters intended for grounded three-phase systems or three-phase systems with neutral conductors. These active filters are specially designed for compensating neutral currents (zero-sequence current components) and also have all those compensation characteristics shown in the previous sections for three-phase, three-wire systems.

4.3.1. Converter Topologies for Three-Phase, Four-Wire Systems

Three-phase, four-wire shunt active power filters have been realized using four-leg converters [33,34,35]. Alternatively, an attractive solution that still uses a conventional three-leg converter, is presented here.

Figure 4-64 shows two possible converter topologies that can be used as the power circuit of a three-phase, four-wire shunt active power filter. The fundamental difference between Fig. 4-64(a) and Fig. 4-64(b) is the number of power semiconductor devices. A three-leg conventional converter is used in Fig. 4-64(a) and the ac neutral wire is directly connected to the electrical midpoint of the dc bus. In Fig. 4-64(b), the ac neutral wire connection is provided through the fourth switch leg.

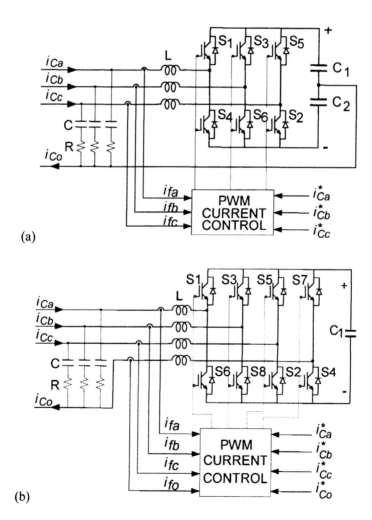

Figure 4-64. Three-phase four-wire PWM converters. (a) Three-leg converter: "split-capacitor" converter topology. (b) Four-leg converter: "four-leg" converter topology.

Other configurations of converters can be employed for realizing three-phase, four-wire shunt active filters. An interesting configuration using a zigzag autotransformer for coupling the active filter to the network is presented in [36]. This can greatly reduce the kVA rating of the PWM converter. However, additional costs arise from the special transformer. Moreover, this approach compensates only for zero-sequence currents, whereas the configurations given in Fig. 4-64 can compensate not only for the neutral current, but also for the harmonics from positive- and negative-sequence components.

The "four-leg" converter topology shown in Fig. 4-64(b) has better controllability than the "split-capacitor" converter topology shown in Fig. 4-64(a) [33,34]. However, the conventional three-leg converter is preferred here because of its lower number of power semiconductor devices [37,38].

There are some problems related to the dc capacitor voltages to be solved in the "split-capacitor" converter topology. This converter topology allows currents to flow through one of the dc capacitors (C_1 or C_2) and to return to the ac neutral wire, causing voltage deviation between the dc capacitors [38]. Once this voltage deviation is controlled, the "split-capacitor" converter topology can become an attractive solution to be generally applied in n-wire systems since it uses a $(n-1)$-leg PWM converter. For example, a 2-leg converter could be used in a three-phase system, where two phases are connected to the converter legs and the third one is connected to the midpoint of the dc bus. An approach to controlling the dc capacitor voltage deviation is introduced in the next section.

4.3.2. Dynamic Hysteresis-Band Current Controller

Hysteresis-based current control is a common PWM (pulse width modulation) control used in voltage-fed converters to force these converters to behave as controlled ac *current* source to the power system. A particular problem arises when controlling the "split-capacitor" converter topology (Fig. 4-64a) with a hysteresis-based PWM current control. If the current references i_{Ck}^* ($k = a, b, c$) are assumed to be from zero-sequence quantities, the converter currents i_{fk} ($k = a, b, c$) will return through the neutral wire. This forces, in the "split-capacitor" converter topology, the current of each phase to flow either through C_1 or through C_2 and to return through the ac neutral wire. Figure 4-65 shows a typical behavior of converter current when controlled by a hysteresis-based PWM control.

The currents can flow in both directions through the switches and capacitors. Table 4.8 summarizes the conditions that cause voltage deviations in the capacitors C_1 and C_2 for a zero-sequence current in the "split-capacitor" converter topology. When $i_{fk} > 0$, V_{dc1} rises and V_{dc2} decreases, but not with equal amplitudes because the positive and negative values of di_{fk}/dt are different, and depend on the instantaneous values of the ac phase voltages. The inverse occurs when $i_{fk} < 0$. The dc voltage variations depend also on the shape of the current reference and the hysteresis bandwidth. Therefore, the total dc voltage, as well as the voltage difference ($V_{dc2}-V_{dc1}$) will oscillate not only at the switching frequency but also at the corresponding frequency of i_0 that is being generated by the converter.

4.3. THREE-PHASE, FOUR-WIRE SHUNT ACTIVE FILTERS

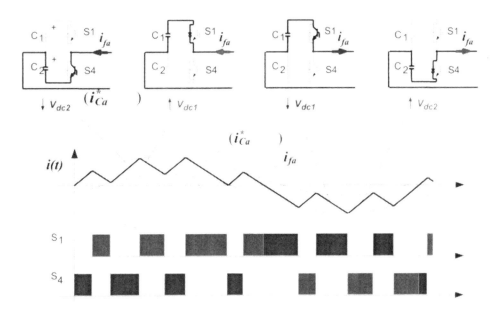

Figure 4-65. Hysteresis-band PWM current control.

In the example given in Fig. 4-65, the phase current i_{fa} causes voltage variations such that the voltage V_{dc1} is higher and V_{dc2} lower at the end of the period. In other words, the positive current integral through the switch S_1 is greater than the negative current integral through S_1, and the contrary happens with the current through the switch S_2. If a dynamic offset level is added to both limits of the hysteresis band, it is possible to control the capacitor *voltage difference* and to keep it within an acceptable tolerance margin. For the example given in Fig. 4-65, a negative offset would counteract the above voltage variation that tends to charge C_1 and discharge C_2.

Therefore, a dynamic offset (ε) should be created from the measurement of the dc capacitor voltages V_{dc1} and V_{dc2}. A scheme for generating the signal ε is present-

Table 4.8. Variation conditions for capacitor voltages V_{C1} and V_{C2}

$i_{fk} > 0$ and $\dfrac{di_{fk}}{dt} < 0$	increase the voltage in C_1
$i_{fk} < 0$ and $\dfrac{di_{fk}}{dt} < 0$	decrease the voltage in C_1
$i_{fk} < 0$ and $\dfrac{di_{fk}}{dt} > 0$	increase the voltage in C_2
$i_{fk} > 0$ and $\dfrac{di_{fk}}{dt} > 0$	decrease the voltage in C_2

ed in the next section. If this signal ε is added to both hysteresis-band limits, a new dynamic hysteresis current control that provides dc voltage equalization in the "split-capacitor" converter topology is given by

$$\begin{cases} \text{upper hysteresis-band limit} = i^*_{Ck} + \Delta(1 + \varepsilon) \\ \text{lower hysteresis-band limit} = i^*_{Ck} - \Delta(1 - \varepsilon) \end{cases} \quad (4.51)$$

where i^*_{Ck} ($k = a, b, c$) are the instantaneous current references provided by an active filter controller and Δ is a fixed half hysteresis band. Thus, the signal ε ($-1 \leq \varepsilon \leq 1$) shifts the hysteresis-band limits around the current references, but does not alter the total hysteresis band (equal to 2Δ), to change the switching times such that

$$\begin{cases} \varepsilon > 0 \Rightarrow \text{rises } V_{dc1} \text{ and lowers } V_{dc2} \\ \varepsilon < 0 \Rightarrow \text{rises } V_{dc2} \text{ and lowers } V_{dc1} \end{cases} \quad (4.52)$$

In the next section, a very simple control circuit is proposed for generating the dynamic offset signal ε. Additionally, another signal to provide total dc bus voltage regulation ($V_{dc1} + V_{dc2}$) should be provided, as it was in the previous sections on three-phase, three-wire active filters by the addition of the signal $\bar{p}_{loss}$ in the active filter controller.

4.3.3. Active Filter dc Voltage Regulator

The voltages in dc capacitors C_1 and C_2 of Fig. 4-64(a) may be controlled by a dc voltage regulator. In this case, both signals $\bar{p}_{loss}$ and ε are generated in this dc voltage regulator, as presented in Fig. 4-66.

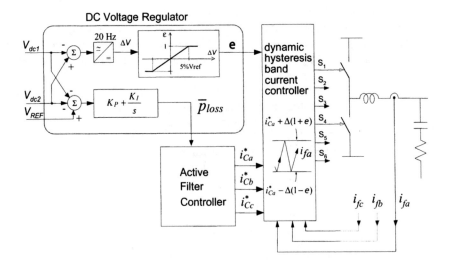

Figure 4-66. DC voltage regulator for the "split-capacitor" topology of Fig. 4-64a.

A low-pass filter with a cutoff frequency of 20 Hz is used in the voltage regulator to render it insensitive to the fundamental frequency (50 Hz) voltage variations. This is the frequency of dc voltage deviation that appears when the active filter compensates for the *fundamental zero-sequence current* of the load, as shown in Fig. 4-65.

The filtered voltage difference $\Delta V = V_{dc2} - V_{dc1}$ produces ε, according to the following limit function generator:

$$\begin{cases} \varepsilon = -1 & \Leftrightarrow \Delta V < -0.05 V_{ref} \\ \varepsilon = \dfrac{\Delta V}{-0.05 V_{ref}} & \Leftrightarrow -0.05 V_{ref} \leq \Delta V \leq 0.05 V_{ref} \\ \varepsilon = 1 & \Leftrightarrow \Delta V > 0.05 V_{ref} \end{cases} \quad (4.53)$$

where V_{ref} is a predefined dc bus voltage reference, and $\pm 5\% V_{ref}$ was arbitrarily chosen as an acceptable tolerance margin of voltage variation.

The signal $\bar{p}_{loss}$ is used in the active filter controller as an *average real power*. It is included in the current reference calculation to force the PWM converter to absorb (deliver) energy from (to) the ac network. If the sum of V_{dc1} and V_{dc2} is smaller than the predefined dc bus voltage reference V_{ref}, the PWM converter should absorb energy from the ac network to charge the dc capacitors. The inverse occurs if $(V_{dc1} + V_{dc2}) > V_{ref}$. The gains in the PI controller of Fig. 4-66 have to be adjusted to provide an adequate dynamic to neutralize the dc bus voltage variations.

The slower feedback loop provided by $\bar{p}_{loss}$ is also useful for correcting voltage variations due to compensation errors that may occur during the transient response of the shunt active filter. This point will be discussed later when the dynamic of the shunt active filter is analyzed. Therefore, the signal $\bar{p}_{loss}$ is very important for providing energy balance inside the active filter.

4.3.4. Optimal Power Flow Conditions

The load powers of the load that can be compensated in terms of the $\alpha\beta 0$ variables are $\tilde{p}$, $\bar{q}$, $\tilde{q}$ and $p_0 = \bar{p}_0 + \tilde{p}_0$. It is important to remember that storage elements are necessary in the active filter if $\tilde{p}$ and p_0 are being compensated. If this *optimal compensation* is done, only the constant power $\bar{p}$ of the load will be supplied by the source. For instance, if a three-phase balanced voltage source is considered, that is, if it comprises only the fundamental positive-sequence voltage $\dot{V}_{+1}$, only the portion of the fundamental positive-sequence load current that is *in phase* with the voltage can produce $\bar{p}$, without generating $\bar{q}$ [see (3.95) and (3.96)].

Equations (3.94) and (3.97), conclude that p_0 never produces a constant power $\tilde{p}_0$ without an associated oscillating power $\bar{p}_0$. Therefore, if an active filter is used for compensating p_0, it has to compensate the total power $p_0 = \bar{p}_0 + \tilde{p}_0$, because it is impossible to produce $\bar{p}_0$ separately from $\tilde{p}_0$. The average zero-sequence power $\bar{p}_0$ exists only in unbalanced systems and, like the real power $\bar{p}$, represents a one-way (not bi-directional) energy flow delivered to the load. Therefore, the active filter needs a power source for supplying the energy related to $\bar{p}_0$ whenever it compensates p_0. To overcome this need for the power supply in the active filter, a new principle of compensation is proposed to allow the use of dc capacitors (energy storage elements) instead of dc sources.

The basic idea consists in keeping the three-phase *average active* power of the shunt active filter equal to zero if the active filter is ideal [41]. In a real implementation, switching and conducting losses exist in the power converter that force the signal $\bar{p}_{loss}$ to be nonzero at steady state. Neglecting the losses allows the active filter to compensate p_0 of the load without the need for an energy source at the dc side of the active filter. Although the energy storage element is still necessary, the three-phase average active power of the shunt active filter should be zero:

$$\bar{p}_{3\phi c} = \frac{1}{T} \int_0^T (v_a i_{Ca} + v_b i_{Cb} + v_c i_{Cc}) \, dt = 0 \tag{4.54}$$

In a real implementation, the energy balance inside the active filter can be met if te active filter takes an average *real power* $\Delta \bar{p}$ from the ac source,

$$\Delta \bar{p} = \bar{p}_0 + \bar{p}_{loss} \tag{4.55}$$

where $\bar{p}_{loss}$ is used to compensate for the losses in the active filter. These losses are represented by the signal $\bar{p}_{loss}$ from the dc voltage regulator (Fig. 4-66). The additional portion of real power, equivalent to the power $\bar{p}_0$ that the shunt active filter is delivering to the load, provides energy balance inside the active filter, when it is compensating the zero-sequence current of the load. Hence, the zero-sequence power $\bar{p}_0$ that the active filter supplies to the load can be taken as a balanced real power $\Delta \bar{p}$ from the source. Note that it is always possible to generate or draw constant real power ($\bar{p} \neq 0$) without generating oscillating power ($\tilde{p} = 0$), even under nonsinusoidal conditions [41,42]. An example was given in the previous section dealing with the constant instantaneous power control strategy (Fig. 4-23) for three-phase, three-wire shunt active filters.

The above ideas are illustrated in Fig. 4-67. It shows a power flow diagram in a circuit in which the *abc* phases were replaced by their equivalent "$\alpha\beta 0$ wires." The active filter compensates all undesirable powers of the load ($\tilde{p}$, $\bar{q}$, $\tilde{q}$, $\bar{p}_0$, and $\tilde{p}_0$) and balances the energy to maintain the dc capacitor voltage around its reference value. The active filter provides optimal power conditions to the source, even under nonsinusoidal conditions. Figure 4-67 shows that the active filter takes real power from the "$\alpha\beta$ wires," represented by $\Delta \bar{p}$, and supplies $\bar{p}_0$ to the load through the "0 wire." In other words, it receives an average positive-sequence power from the source, and supplies an average zero-sequence power to the load. Note that it would also receive a negative-sequence power through $\Delta \bar{p}$ if the source voltage were not balanced.

Equation (3.95) reveals that negative-sequence components simultaneously present in voltage and current can produce constant real power $\bar{p}$. Unfortunately, they

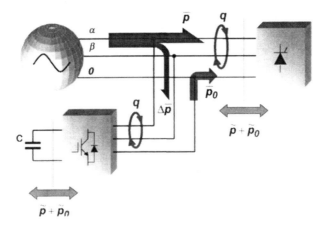

Figure 4-67. Optimal power flow related to the $\alpha\beta 0$ reference frame.

produce $\tilde{p}$ and $\tilde{q}$ if a positive-sequence component is present simultaneously with the negative-sequence component, as shown in (3.98) and (3.99).

The example of Fig. 4-23 shows that the active filter can compensate load currents, even under nonsinusoidal voltages, to provide constant real power $\bar{p}$ to the source. In this case, it was proven that the source current is neither sinusoidal nor proportional to the voltage. Figure 4-32 shows that the shunt active filter can compensate load currents, even under nonsinusoidal voltages, to provide sinusoidal compensated currents to the source. In the next sections, extended control algorithms are presented for realizing the constant instantaneous power control strategy and the sinusoidal current control strategy, for three-phase, four-wire shunt active filters. Both strategies can compensate for the whole zero-sequence current of the load, but it is impossible to satisfy simultaneously both conditions—*sinusoidal currents* and *constant power* to the source—if the voltages are unbalanced and/or distorted.

Figure 4-68 shows a three-phase, four-wire shunt active filter implemented with the "split-capacitor" converter topology. The active filter controller realizes the constant instantaneous power control strategy or the sinusoidal current control strategy, and additionally compensates the load neutral current. Ideally, $i_{Co} = -i_{Lo}$, and the neutral current to the source (i_{So}) is zero, regardless the implemented control strategy.

4.3.5. Constant Instantaneous Power Control Strategy

Figure 4-69 presents the functional control block diagram for three-phase, four-wire shunt active filter that realizes the constant instantaneous power control strategy. The inputs to the control system are the phase voltages and the line currents of the load. Unfortunately, the phase voltages cannot be used directly in the control because of instability problems. It was verified that resonance at a relatively high fre-

190 SHUNT ACTIVE FILTERS

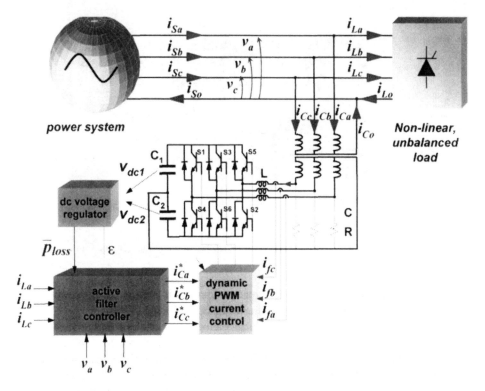

Figure 4-68. A three-phase, four-wire shunt active filter using a three-leg conventional converter.

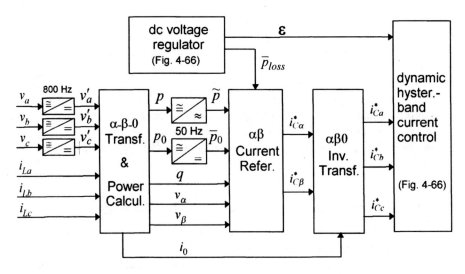

Figure 4-69. Control block diagram of the shunt active filter controller that realizes the constant instantaneous power control strategy.

quency may appear between the source impedance and the small high-pass filter (RC branch circuit) of Fig. 4-68, if the measured phase voltages are directly used in the active filter controller. Therefore, low-pass filters were used (shaded area in Fig. 4-69) with a relatively high cutoff frequency (800 Hz) to attenuate the harmonics in the phase voltages that contribute to the resonant effects at frequencies higher than 800 Hz. In a real implementation, these low-pass filters may be eliminated, depending on system parameters such as a switching frequency of the PWM converter and a frequency response of the electronic circuit for voltage measurement.

The "$\alpha\beta 0$ transformation and power calculation" block in Fig. 4-69 realizes the following equations:

$$\begin{bmatrix} v_0 \\ v_\alpha \\ v_\beta \end{bmatrix} = \sqrt{\frac{2}{3}} \begin{bmatrix} \frac{1}{\sqrt{2}} & \frac{1}{\sqrt{2}} & \frac{1}{\sqrt{2}} \\ 1 & -\frac{1}{2} & -\frac{1}{2} \\ 0 & \frac{\sqrt{3}}{2} & -\frac{\sqrt{3}}{2} \end{bmatrix} \begin{bmatrix} v'_a \\ v'_b \\ v'_c \end{bmatrix} \qquad (4.56)$$

$$\begin{bmatrix} i_0 \\ i_\alpha \\ i_\beta \end{bmatrix} = \sqrt{\frac{2}{3}} \begin{bmatrix} \frac{1}{\sqrt{2}} & \frac{1}{\sqrt{2}} & \frac{1}{\sqrt{2}} \\ 1 & -\frac{1}{2} & -\frac{1}{2} \\ 0 & \frac{\sqrt{3}}{2} & -\frac{\sqrt{3}}{2} \end{bmatrix} \begin{bmatrix} i_{La} \\ i_{Lb} \\ i_{Lc} \end{bmatrix} \qquad (4.57)$$

$$\begin{bmatrix} p_0 \\ p \\ q \end{bmatrix} = \begin{bmatrix} v_0 & 0 & 0 \\ 0 & v_\alpha & v_\beta \\ 0 & v_\beta & -v_\alpha \end{bmatrix} \begin{bmatrix} i_0 \\ i_\alpha \\ i_\beta \end{bmatrix} \qquad (4.58)$$

A high-pass filter with a cutoff frequency of 50 Hz separates the power $\tilde{p}$ from p, and a low-pass filter separates $\bar{p}_0$ from p_0. The powers $\tilde{p}$ and p_0 of the load, together with q, should be compensated to provide optimal power flow to the source. Thus, the $\alpha\beta$ current references are found to be

$$\begin{bmatrix} i'_\alpha \\ i_{C\alpha} \\ i_{C\beta} \end{bmatrix} = \frac{1}{v_\alpha^2 + v_\beta^2} \begin{bmatrix} v_\alpha & v_\beta \\ v_\beta & -v_\alpha \end{bmatrix} \begin{bmatrix} -\tilde{p} + \Delta \bar{p} \\ -q \end{bmatrix} \qquad (4.59)$$

where $\Delta \bar{p}$, given by (4.55), provides energy balance inside the active filter.

Finally, the "$\alpha\beta 0$ inverse transformation" block of Fig. 4-69 calculates the instantaneous current references for the dynamic-hysteresis current control of the VSC:

$$\begin{bmatrix} i^*_{Ca} \\ i^*_{Cb} \\ i^*_{Cc} \end{bmatrix} = \sqrt{\frac{2}{3}} \begin{bmatrix} \frac{1}{\sqrt{2}} & 1 & 0 \\ \frac{1}{\sqrt{2}} & -\frac{1}{2} & \frac{\sqrt{3}}{2} \\ \frac{1}{\sqrt{2}} & -\frac{1}{2} & -\frac{\sqrt{3}}{2} \end{bmatrix} \begin{bmatrix} -i_0 \\ i_{C\alpha} \\ i_{C\beta} \end{bmatrix} \qquad (4.60)$$

With this approach, the active filter supplies the whole i_0 to the load. If no zero-sequence voltage is present, the zero-sequence power p_0 is zero. In this case, the zero-sequence current i_0 of the load is completely compensated without the need for energy balance inside the active filter, since $p_0 = 0$.

The above compensation principle that provides constant source instantaneous power [41,42] is an extended version of the original control strategy proposed by Akagi et al. [3,4,43]. The control method does not use any rms value calculation, although it uses a low and a high-pass filter to separate the powers $\bar{p}_0$ and $\tilde{p}$, which influence the dynamic response of the active filter. It should be noted that the controller for three-phase three-wire systems (Fig. 4-7) can be treated as a simplification of Fig. 4-69, just considering $v_0 = i_0 = p_0 = 0$ and the elimination of signal ε.

4.3.6. Sinusoidal Current Control Strategy

The sinusoidal current control strategy makes the active filter to compensate the current of a nonlinear load to guarantee balanced, sinusoidal current drawn from the network, even under an unbalanced and/or distorted system voltage. Therefore, it compensates also the neutral current of the load, in addition to all the compensation features presented in Fig. 4-26, for active filters in three-phase, three-wire systems.

Now, a positive-sequence detector must replace the 800 Hz cutoff frequency low-pass filter (shaded area in Fig. 4-69) to allow the realization of the sinusoidal current control strategy. The *phase angle* and *frequency* of the fundamental positive-sequence voltage, corresponding to the phasor $\dot{V}_{+1}$, must be accurately determined by the positive-sequence detector. Otherwise, the active filter controller cannot exactly determine the *fundamental* reactive power of the load included in $\bar{q}$, which in turn cannot produce ac currents $\dot{I}_{+1}$ *orthogonal* to the ac voltages $\dot{V}_{+1}$ to produce only $\bar{q}$. The positive-sequence voltage detector as presented in Fig. 4-27, and used in the controller for three-phase three-wire systems, satisfies the above constraints.

If the circuit of Fig. 4-27 replaces the 800 Hz low-pass filters in Fig. 4-69, the compensating powers $\tilde{p}$ and q of the load, calculated in the main part of the active filter controller, will include all fundamental negative-sequence power, the fundamental positive-sequence reactive power, as well as the harmonic powers. In other words, the active filter controller handles the load current as "connected to a sinusoidal balanced voltage source." In this case, only the portion of $\dot{I}_{+1}$ of the load current that is in phase with $\dot{V}_{+1}$ can produce only $\bar{p}$, which is left to be supplied by the source. Therefore, using the present sinusoidal current control strategy makes the source current sinusoidal and balanced if $\tilde{p}$, $\bar{q}$, $\tilde{q}$, and i_0 of the load are compensated. This results in source currents containing only the active portion of fundamental positive-sequence current that is in phase with $\dot{V}_{+1}$.

Therefore, the sinusoidal current control strategy is realized if two changes in Fig. 4-69 are made as follows:

1. Replace the 800 Hz low-pass filters in Fig. 4-69 by the circuit of Fig. 4-27.
2. Remove the 50 Hz low-pass filter that obtains $\bar{p}_0$ in Fig. 4-69, because the new input voltages v'_a, v'_b, and v'_c corresponds to the instantaneous values of $\dot{V}_{+1}$, and contain no zero-sequence components. Thus, p_0 is always zero.

Figure 4-70 shows the complete control block diagram of the shunt active filter that realizes the sinusoidal current control strategy for three-phase, four-wire systems. One simplification was done in the positive-sequence detector shown in Fig. 4-27, and included as part of the controller of the three-phase, four-wire shunt active filter. The voltages v'_α and v'_β given by (4.10) and marked with ② in Fig. 4-27, instead of v'_a, v'_b, and v'_c, given by (4.12), are directly used in the main control.

At this point, it is important to remark that the voltage regulator of Fig. 4-70 that generates the signal $\bar{p}_{loss}$ has received an additional task besides those listed in the last sections: to correct errors in power compensation. This occurs because the feed-forward control circuit is now unable to supervise the zero-sequence power. Since the active filter compensates the whole neutral current of the load in the presence of zero-sequence voltages, the shunt active filter eventually supplies $\bar{p}_0$. Now, (4.55) is no longer used, and $\Delta \bar{p}$ is replaced simply by $\bar{p}_{loss}$. Therefore, if the active filter supplies $\bar{p}_0$ to the load, this causes dc voltage variations, which are sensed by the PI controller of the dc voltage regulator. Hence, an additional amount of *average real power*, numerically equal to $\bar{p}_0$, is automatically added to the signal $\bar{p}_{loss}$ that is mainly used to provide energy to cover for losses in the power circuit of the shunt active filter. Actually, the constant instantaneous power controller presented in Fig. 4-69 would behave in the same manner if $\Delta \bar{p}$ is replaced only by $\bar{p}_{loss}$ in (4.59).

Another kind of error in compensating power calculation may also appear. From (3.95), harmonic components other than the fundamental positive-sequence component can generate average real power $\bar{p}$. Therefore, under unbalanced and/or nonsinusoidal voltage conditions, the fundamental negative-sequence component, as well as harmonic currents, can produce $\bar{p}$. An example follows to better explain this point.

Suppose that the system voltages and load currents comprise only fundamental positive- and negative-sequence components, that is, only $\dot{V}_{+1}$, $\dot{V}_{-1}$, $\dot{I}_{+1}$, and $\dot{I}_{-1}$ are present. The positive-sequence detector extracts the fundamental positive-sequence voltage component $\dot{V}_{+1}$, and delivers it in the form of v'_α and v'_β to the main part of the active filter controller. In this case, the calculated power p and q of the nonlinear load will comprise $\tilde{p}$ and $\tilde{q}$ due to $\dot{I}_{-1}$. Note that $\dot{I}_{-1}$ and $\dot{V}_{+1}$ does not produce any average real and imaginary power.

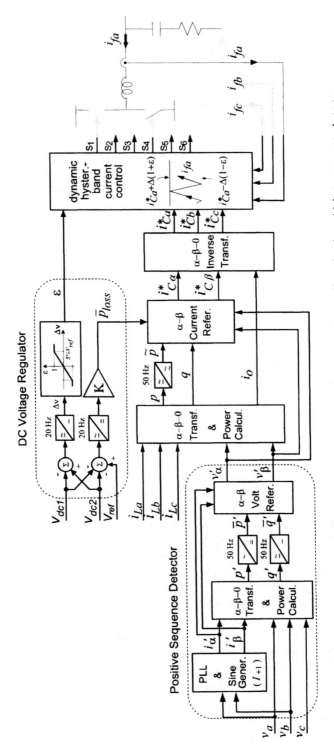

Figure 4-70. Control block diagram of the shunt active filter controller that realizes the sinusoidal current control strategy.

194

Thus, the active filter will supply the whole negative-sequence current to the load, since $\tilde{p}$ and $\tilde{q}$ are being compensated by the shunt active filter. However, the voltages at the load terminal contain a fundamental negative-sequence component. If this component is not orthogonal to $\dot{I}_{-1}$, certainly it is producing average real power, which causes an undesirable continuous flow of energy from or to the dc capacitor of the shunt active filter. This means that the active filter will supply or absorb a nonzero average, negative-sequence power. This causes voltage variations in the dc capacitors. The slower feedback control loop of the dc voltage regulator will sense it, and will properly set $\bar{p}_{loss}$ to make the active filter absorb or supply an additional amount of real power from the source to neutralize the voltage variation. Note that this additional amount of real power is drawn in the form of *fundamental positive-sequence* current, because the compensating current references are being calculated by using only the fundamental positive-sequence component $\dot{V}_{+1}$ of the system voltages.

In conclusion, the control method presented in Fig. 4-70 guarantees sinusoidal and balanced compensated currents drawn from the network. Additionally, the shunt active filter can supply nonzero average negative-sequence and zero-sequence powers to the nonlinear load while it draws the same amount of energy from the network in the form of fundamental, positive-sequence, active power.

4.3.7. Performance Analysis and Parameter Optimization

This section presents some useful aspects when specifying a three-phase, four-wire shunt active filter. Most considerations are also valid for specifying a three-phase, three-wire shunt active filter, since the active filter controller for ungrounded three-phase systems (without neutral conductor) can be implemented by simplifying the three-phase, four-wire controller.

Figure 4-68 shows the basic configuration of shunt active filter that will be considered in the following analysis of performance. As the system is not linear, a digital simulator for mixed control systems, power electronics switches, and electromagnetic circuits is used to optimize the system parameters. The influence of each parameter in the whole system is explained. Afterward, an optimized set of parameters is used to perform the dynamic response analysis. Some economical considerations are summarized. Finally, experimental results from a laboratory prototype are presented.

4.3.7.A. Influence of the System Parameters

The optimization of the system parameters cannot be evaluated separately, because more than one parameter influences a system requisite. For instance, the maximum switching frequency is strongly dependent on the hysteresis band Δ in (4.51) and on the commutation inductance L in Fig. 4-68. Therefore, a complete model of the shunt active filter should be implemented in a digital simulator for optimizing all of the equipment.

Before beginning the optimization, limitations of the power components and the control hardware should be fixed. For instance, the following constraints were considered for a laboratory prototype:

dc bus voltage:
- $V_{dc1} = V_{dc2} < 350$ Vdc

PWM converter:
- switching frequency < 15 kHz
- ac current peak value < 100 A
- ac current rms value < 50 A

Table 4.9 summarizes the trends and goals of the principal system parameters to be optimized.

4.3.7.B. Dynamic Response of the Shunt Active Filter

A comparison between the control strategies, the constant power control strategy, and the sinusoidal current control strategy, given above for three-phase, four-wire systems, has been carried out by digital simulation. Both control strategies presented almost the same dynamic response. The fundamental system frequency is 50 Hz. The source voltages were composed from arbitrarily chosen phasors in terms of symmetrical components. Initially, the amplitudes of the ac voltages were adjusted taking into account the dc capacitor voltage references (2×300 V), such that no coupling transformer was required. The rms amplitudes and phase angles of the voltage phasors are $\dot{V}_{+1} = 110 \angle 0°$, $\dot{V}_{01} = 11 \angle 90°$ (V), $\dot{V}_{+3} = 5 \angle 90°$, $\dot{V}_{-3} = 5 \angle 180°$, and $\dot{V}_{03} = 5 \angle 270°$ (V). The main system parameters shown in Fig. 4-68 are summarized in Table 4.10.

Table 4.9. Trends and goals for parameter optimization

Parameter	Goal	Problems
dc capacitors	Small capacitor to reduce costs	Too small capacitor leads to very high dc voltage variations during transient response
Commutation reactors	Small reactor to obtain faster response (high di/dt)	Too small reactor increases the switching frequency in the hysteresis current control
High-pass filter	No passive filter to reduce costs	Switching harmonics may flow to the ac system
Coupling transformer	No transformer to reduce costs	The magnitude of the ac voltage must be smaller than the dc voltage (boost-type converter)
dc voltage regulator	Slow dynamic to reduce ac transient currents	Too slow dynamic leads to high dc voltage variations
Low-pass and high-pass filters in the controller	Low cutoff frequency to better separate the dc level signals	Lower cutoff frequency improves steady-state response, but deteriorates the dynamics and forces the use of large dc capacitors

4.3. THREE-PHASE, FOUR-WIRE SHUNT ACTIVE FILTERS

Table 4.10. Main parameters of the simulated system

Source impedance	$R = 0.1\ \Omega,\ L = 0.6$ mH
Active filter	$C = 30\ \mu F,\ R = 1\ \Omega$ $L = 2.5$ mH $C_1 = C_2 = 3.0$ mF $V_{ref} = V_{dc} = 600$ V
Load	One three-phase thyristor rectifier, firing angle = 30°, $I_{dc} = 8$ A; commutation inductance = 3 mH One single-phase thyristor rectifier connected between the a-phase and the neutral firing angle = 15°, $I_{dc} = 10$ A, commutation inductance = 3 mH One single-phase diode bridge connected between the b-phase and the neutral, $L_{dc} = 300$ mH, $R_{dc} = 20\ \Omega$, commutation inductance = 3 mH

One simulation for each control strategy with the same optimized set of parameters was performed for comparison between them. The power circuit is the same, and only the control strategy for the active filter is changed (Fig. 4-69 or Fig. 4-70). Both simulations resulted in quite similar phase voltages at the load terminal. Fig. 4-71 shows the phase voltages only for the "constant power" simulation case. The load current is the same for both simulations, and is shown in Fig. 4-72. The diode bridge is connected at $t = 30$ ms, and the controlled rectifiers are connected after $t = 40$ ms, according to their firing angles (see Table 4.10). At $t = 102$ ms, the firing pulses of the three-phase rectifier are blocked, as shown in Fig. 4-72.

The PWM converter of the shunt active filter is blocked until $t = 20$ ms. Figure 4-73 shows the filtered (without switching ripple) compensating currents i_{Ca}, i_{Cb}, and i_{Cc} of the active filter for the "sinusoidal current" simulation case. For the "constant power" simulation case, the current i_{Ca}, i_{Cb}, and i_{Cc} are presented in Fig. 4-74.

Figure 4-75 shows the neutral currents of the load and the source. They are equal for both simulation cases, although the active filter currents in Fig. 4-73 differ from those in Fig. 4-74. The zero-sequence current compensation has the same behavior because both control strategies compensate identically the zero-sequence currents

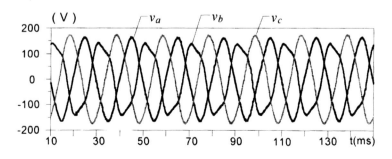

Figure 4-71. Phase voltages at the load terminal.

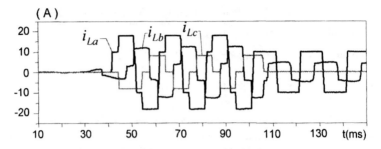

Figure 4-72. Line currents of the load.

of the load, and the same control parameters generating the signal ε for the dynamic-hysteresis control have been used in the dc voltage regulator. The source neutral current has a subharmonic component related to the signal ε [see (4.53) and Fig. 4-66)], which shifts the hysteresis limits to equalize the dc voltages in both dc capacitors.

The amplitude of the subharmonic in the neutral current of the source does not depend on the zero-sequence current of the load, rather it depends mainly on the signal ε, the hysteresis band, and the value of the dc capacitance. Another harmonic component present in the source neutral current is the third harmonic, which is excited by the voltage source component $\dot{V}_{03} = 5 \angle 270°$ V. It is flowing to the passive filter of the shunt active filter ($C = 30$ μF, $R = 1$ Ω). In fact, all those harmonic volt-

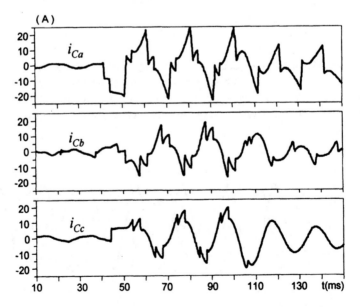

Figure 4-73. Active filter currents for the "sinusoidal current" control strategy.

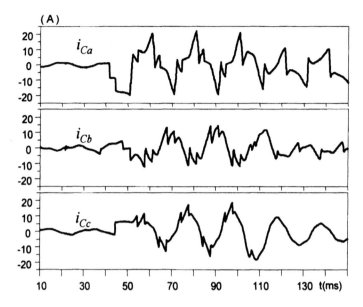

Figure 4-74. Active filter currents for the "constant power" control strategy.

age components of the source excite harmonic currents in the RC-passive filter that are not being compensated by the shunt active filter. The compensated currents, which are drawn from the source are shown in Fig. 4-76(a) for the "constant power" case, and in Fig. 4-76(b) for the "sinusoidal current" case. Although both approaches provide fast response, and equally compensate the neutral current from the harmonic point of view, the "sinusoidal current" control strategy offers a better solution.

Figure 4-77 compares the three-phase instantaneous active power ($p_{3\phi} = p + p_0$) of the load with the powers from the compensated currents shown in Fig. 4-76(a) and Fig. 4-76(b). The "constant power" control strategy ("pconst" curve) should present a perfectly smoothed instantaneous power. Unfortunately, its performance is deteriorated due to the presence of the 800 Hz cutoff frequency low-pass filters in Fig. 4-69. Three fifth-order Butterworth filters were used. Moreover, both control

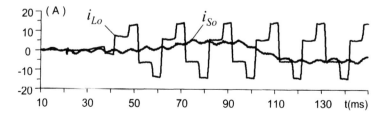

Figure 4-75. Neutral wire currents of the source and the nonlinear load.

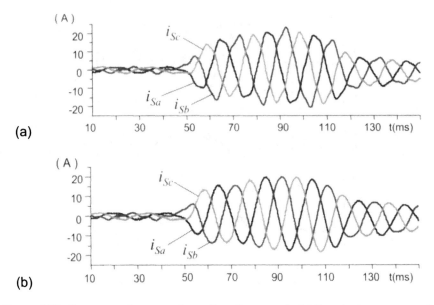

Figure 4-76. Compensated currents drawn from the network: (a) "constant source power" control strategy, (b) "sinusoidal source current" control strategy.

strategies do not compensate the harmonic currents flowing from the source to the high-pass filter ($C = 30$ μF, $R = 1$ Ω), due to the presence of harmonic voltages in the power supply. This characterizes the selective compensation behavior of the active filter. In the next chapter, series active filters are proposed to provide voltage compensation. For the considered system voltages, both control strategies ("pconst" and "isinus" curves) produced compensated currents that draw a small oscillating portion of active power at the same order of magnitude.

The imaginary powers in Fig. 4-78 were calculated from the same currents and voltages as in Fig. 4-77. The "pconst" curve contains a small positive average value (inductive reactive power). This occurs because the 800 Hz cutoff frequency filters introduce a phase delay to the measured voltages, which introduces errors to the

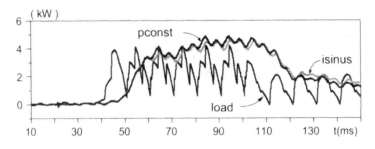

Figure 4-77. Three-phase instantaneous active power.

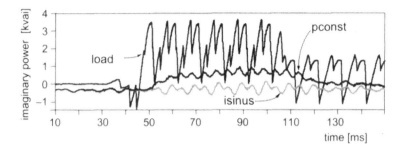

Figure 4-78. Imaginary power.

calculated power $\bar{q}$ of the load, such that the actual value of $\bar{q}$ is not fully compensated. On the other hand, the "constant power" control strategy compensates better the oscillating portion of the load imaginary power.

Both simulation cases presented similar dc voltage variations, as can be seen in the curves in Fig. 4-79. It is possible to see a 50 Hz component in the curves, caused by the compensation of the fundamental zero-sequence current, as explained in Fig. 4-65. Compensation errors occur when the load is connected or disconnected, which tends to discharge or charge the dc capacitors. A time interval is needed for the fifth-order Butterworth filter (50 Hz cutoff frequency) in Fig. 4-69 and in Fig. 4-70 to separate correctly the compensating power $\tilde{p}$ from p. This delay is responsible for the voltage fluctuation during transients. The signal p_{loss} from the voltage regulator contributes to maintaining V_{dc1} and V_{dc2} around their reference value (300 V).

4.3.7.C. Economical Aspects

The simulation results in the previous section have shown that the power rating of the shunt active power filter is of the same order of magnitude as the load. This means that, from an economical point of view, more selective power compensation should be assigned to the shunt active filter. Instead of compensating all undesirable

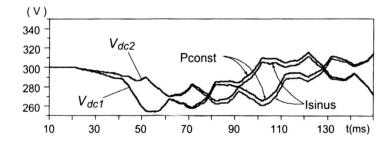

Figure 4-79. dc capacitor voltages.

power portions of the load: $\tilde{p}$, $q = \bar{q} + \tilde{q}$, and the neutral current i_o of the load, some of them might be left uncompensated, or be compensated by other more economical compensators. For instance, if $\bar{q}$ of the load, which includes the fundamental reactive power, is not compensated by the shunt active filter, the peak value of compensating current for the simulation case in the previous section decreases to 15.6 A, instead of 25.1 A as verified above. For the period 65 ms $\leq t \leq$ 85 ms, the rms current calculation results in:

1. With $\bar{q}$ compensation (Fig. 4-73) of $\bar{q}$: $I_{Ca} = 10.6$ A, $I_{Cb} = 7.3$ A, $I_{Cc} = 8.8$ A
2. Without $\bar{q}$ compensation of $\bar{q}$: $I_{Ca} = 5.4$ A, $I_{Cb} = 4.7$ A, $I_{Cc} = 7.8$ A

The aggregate rms current [see (2.46)] result:

1. $\|i_\Sigma\| = 15.6$ A
2. $\|i_\Sigma\| = 10.6$ A

This means a reduction of 32% in the rated power of the shunt active filter.

The imaginary power compensation increases the ac currents of the shunt active filter, although it does not influence the size of the dc capacitors. Theoretically, no energy storage elements are necessary in the active filter to compensate the imaginary power [4]. In the steady state, the size of the dc capacitors is determined by the amount of energy stored/released when the active filter is compensating the powers $\tilde{p}$ and p_0. On the other hand, a nonzero average energy flow arises between the active filter and the ac network, due to compensation errors that appear during transient responses. These compensation errors are mainly induced by the high-pass filter in the controller, because it takes some time to separate correctly $\tilde{p}$ from p after a step change in the load power.

If the voltages at the load terminal do not have zero-sequence components, p_0 is zero. Consequently, now oscillating energy flow related to the power $\tilde{p}_0$, which influences the *total* dc voltage ($V_{dc1}+V_{dc2}$), is present. However, the neutral current compensation still produces dc voltage *deviations* at the corresponding frequency of i_0 in the "split-capacitor" converter topology, as explained in Fig. 4-65 and Table 4.8.

Therefore, the lowest frequency of $\tilde{p}$ and i_0 to be compensated should be carefully taken into account during the specification of the dc capacitors. Some details about the influence of compensation of $\tilde{p}$ on the dc capacitors can be found in [43] and [44].

Additional costs may arise due the need for a coupling transformer between the active filter and the ac network. This will be required to match the ac voltage output of the power system to the dc bus voltage of the PWM converter (see Fig. 4-68). Moreover, the size of the commutation reactors (L in Fig. 4-68) also influences the compensating currents of the shunt active filter. All these parameters should be optimized together to achieve a desired compensation characteristic for the shunt active filter. A useful approach can be found in [45].

4.3.7.D. Experimental Results

A down-scaled prototype of the shunt active filter was built to validate the "split-capacitor" (Fig. 4-68) converter topology and the sinusoidal current control strategy (Fig. 4-70). Six IGBTs (100 A–1200 V) were used in the PWM converter. Several experimental results were recorded, and some of them can be found in [38,39] and [40].

Similar to the simulation cases above, two single-phase converters and one three-phase converter were used as the unbalanced, nonlinear load (see Table 4.10). In order to cause imbalances in the supply voltages, three single-phase transformers are connected in series between the source and the load, as shown in Fig. 4-80. This connection causes imbalances due to negative-sequence components.

Two sets of measurements were prepared, which are shown below. The first one shows the dynamics of the shunt active filter during the start of its PWM converter. The other one shows a transient period caused by the load connection.

For the first case, the series transformers of Fig. 4-80 were bypassed, and the supply voltage was reduced to approximately half (190/110 V). Furthermore, a three-phase Y–Δ transformer was inserted in the circuit to connect the six-pulse thyristor rectifier. Twelve oscilloscope channels were used to record simultaneously the phase voltages (v_a, v_b, v_c), the line currents of the load (i_a, i_b, i_c) and

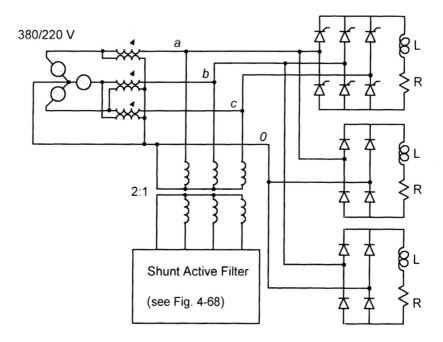

Figure 4-80. Circuit diagram for experimental tests on the prototype three-phase, four-wire shunt active filter.

source (i_{Sa}, i_{Sb}, i_{Sc}), and the shunt active filter (i_{Ca}, i_{Cb}, i_{Cc}). The positive directions of currents are shown in Fig. 4-68. The curves were separated by phase in Fig. 4-81, Fig. 4-82, and Fig. 4-83.

Before the connection of the shunt active filter—while the IGBT's are blocked—the source currents are almost the same as the load ones. They are not only strongly distorted, but also unbalanced. The currents flowing in the neutral wire are shown in Fig. 4-84. This figure shows that the shunt active filter can quickly compensate the neutral current. Compared with the simulation result presented in Fig. 4-75, the sub-harmonic induced by the signal ε in the dynamic hysteresis-band current control was strongly reduced. This was possible mainly due to a fine tuning in the hysteresis band, represented by the constant Δ in (4.51). The firing angle of the three-phase thyristor rectifier was set to 60°, to generate a relatively high inductive fundamental reactive power. The c phase of the load in Fig. 4-83 comprises currents only from this rectifier. After the start of the active filter, the source currents become in phase with the phase voltages, which means that the active filter compensates well the reactive power of the load. Some spikes are present in the source currents. They are caused by a very high di/dt in the load current, which cannot be fully compensated by the shunt active filter. To eliminate the spikes in the source currents, higher di/dt should be allowed for in the compensating currents i_{Ca}, i_{Cb}, and i_{Cc}. One way to do it is to reduce the inductance of the commutation reactors L in Fig. 4-68. However, this change should be carefully carried out, since it implies other changes in the active filter characteristics, as described in Table 4.9.

The regulation of the dc capacitor voltages worked well, because the controller of the active filter was already in operation and reached the steady state at the start of operating the PWM converter. In other words, the high-pass filter in the controller was already correctly separating $\tilde{p}$ from p, so that no compensation errors

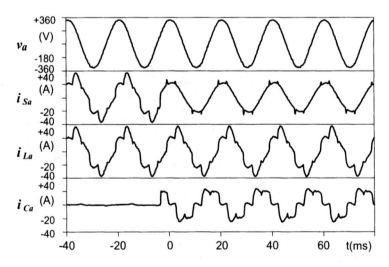

Figure 4-81. Experimental results. Dynamic behavior due to the connection of the shunt active filter (a-phase curves).

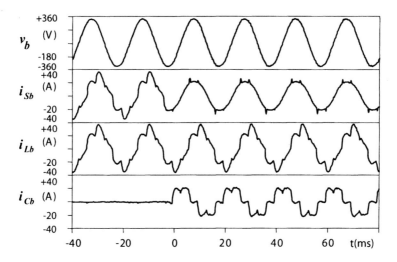

Figure 4-82. Experimental results. Dynamic behavior due to the connection of the shunt active filter (*b*-phase curves).

arose. Another point that should be investigated is the behavior of the shunt active filter under unbalanced supply voltages. These two points are covered in the next set of experimental curves, which involve a load connection transient.

Now, the supply voltage is increased to 380/220 V, and an imbalance around 14% of the fundamental negative-sequence component is superposed, by using the series transformers shown in Fig. 4-80. The new phase voltages at the load termi-

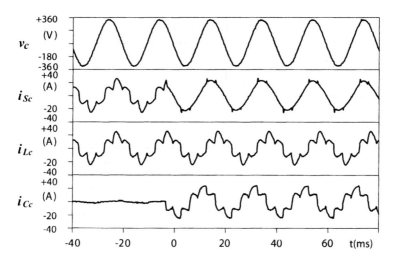

Figure 4-83. Experimental results. Dynamic behavior due to the connection of the shunt active filter (*c*-phase curves).

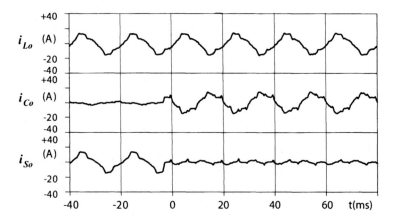

Figure 4-84. Experimental results. Dynamic behavior due to the connection of the shunt active filter (neutral wire currents).

nals are given in Fig. 4-85. The three-phase Y–Δ transformer to connect the six-pulse thyristor rectifier was bypassed, and the new load currents can be seen in Fig. 4-86. The firing angle of the thyristor bridge was set to 40°. The shunt active filter was already in operation when the load was connected. Fig. 4-87 shows the compensated line currents of the source. The high voltages, particularly in the a phase (v_a), causes magnetic saturation in the three-phase transformer (turn ratio 2:1) of the shunt active filter. Consequently, the source currents are distorted, even when the load is disconnected (-20 ms $\leq t \leq 0$ ms). The shunt active filter does not compensate for the current harmonics from the power supply, but only the harmonics from the load. If the currents of Fig. 4-86 and Fig. 4-87 are put over the phase voltages (Fig. 4-85), it is possible to see that the inductive reactive power of the load is well compensated by the shunt active power filter.

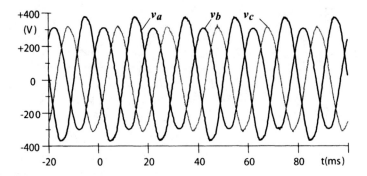

Figure 4-85. Experimental results. Dynamic behavior due to the connection of the load. Phase voltages at load terminal.

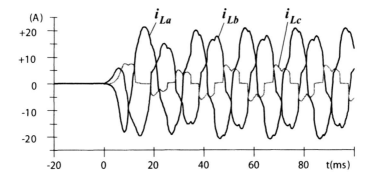

Figure 4-86. Experimental results. Dynamic behavior due to the connection of the load. Line currents of the nonlinear load.

The load is connected at $t = 0$ ms, and the source currents reach their steady state values at $t = 30$ ms. During this period, the shunt active filter is supplying energy to the load, discharging its dc capacitors. This is the time interval during which the fifth-order Butterworth high-pass filter needs to separate correctly the new value of the constant real power $\bar{p}$ of the load. Simultaneously, the dc capacitor voltage regulator are changing the value of the signal $\bar{p}_{loss}$ to conform to the new situation of losses in the PWM converter and to correct the dc voltage variation, as shown in Fig. 4-88.

An oscillation in the amplitude of the source currents is verified, before the steady state is reached at $t = 50$ ms. The average values of the dc capacitor voltages before and after the connection of the load are not the same, as can be seen in Fig. 4-88. In the experimental results, a proportional gain K, instead of a PI controller as suggested in Fig. 4-66, was implemented in the shunt active filter prototype. If zero error is required at the steady state, a PI controller is needed. Figure 4-88 shows that

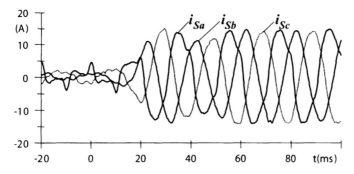

Figure 4-87. Experimental results. Dynamic behavior due to the connection of the load. Compensated currents drawn from the network.

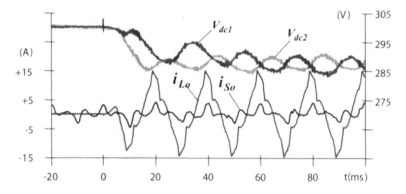

Figure 4-88. Experimental results. Dynamic behavior due to the connection of the load. Neutral currents and the dc capacitor voltages.

the neutral current flowing into the source remains almost the same before and after the connection of the load. Its waveform is similar to the *a*-phase current of Fig. 4-87, before $t = 0$ ms. This means that the magnetization phenomenon in the active filter transformer produces mainly zero-sequence components. As mentioned before, this confirms the selective compensation characteristics of the shunt active filter, which compensate only the currents of the nonlinear load, and cannot compensate the harmonic currents excited by the imbalances and voltage harmonics of the power supply.

4.4. SHUNT SELECTIVE HARMONIC COMPENSATION

The *p-q* Theory can be applied to many different types of harmonic compensation. For instance, Section 3.5.1 showed that different undesirable power components could be selected to be compensated. However, some specific harmonic component or specific set of harmonic components needs to be compensated in some cases. This problem happens in situations where the harmonic filter is designed to filter out only the harmonics that are limited by the regulation, and this filtering is done just to keep these harmonics below the allowed limits and not to eliminate them all [57,58]. With this procedure, an active filter with minimum size and costs is obtained. On the other hand, in some applications, it may be necessary to filter out some given harmonic component completely. Howeverconventional techniques cannot achieve it due to the current synthesis delay by converter PWM control [59].

Figure 4-89 shows the selective filter basic cell (SFBC) block diagram for selective harmonic calculation. In this block diagram, the measured currents i_a, i_b, and i_c, with a given harmonic component at the frequency ω_h, are first transformed from the *abc* coordinates to the $\alpha\beta$ axes. The voltages $v_{\alpha h+}$ and $v_{\beta h+}$ are the positive-sequence components of a harmonic voltage at the frequency ω_h. Then, the the $\alpha\beta$

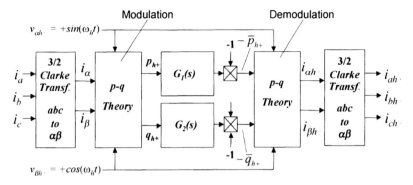

Figure 4-89. Selective filter basic cell (SFBC).

axes current and these harmonic voltages are used to calculate powers p_{h+} and q_{h+} based on the *p-q* Theory. These powers may have average values that are filtered by using low-pass filters $G_1(s)$ and $G_2(s)$. These average values will be present if the positive-sequence current component at the frequency ω_h exists in the load current. Normally, these filters have the same cutoff frequency. The average powers are multiplied by -1, using the same voltage references $v_{\alpha h+}$ and $v_{\beta h+}$, the positive-sequence component of the harmonic current components $i_{\alpha h}$ and $i_{\beta h}$ are obtained at the frequency ω_h. Finally, by transforming back these currents from the $\alpha\beta$ axes to the *abc* axes, the positive-sequence harmonic current given by i_{ah+}, i_{bh+}, and i_{ch+}, are obtained, and can be used as compensation current references for the active filter converter.

For a complete compensation, negative-sequence components at the frequency ω_h have to be measured as well. For this purpose, a similar block diagram as shown in Fig. 4-89 has to be used, with the difference that the reference voltage for the calculation has to be at the frequency ω_h, but with negative sequence. Hence, this complete calculation for each undesired harmonic component, that is, for its positive- and negative-sequence components, allows the implementation of a selective harmonic filter.

In most cases, the converter used for the synthesis of the active filter current presents some delay due to the finite switching frequency and the PWM modulation techniques. This delay causes errors in the active filter compensation characteristics, and it may worsen the harmonic content in some cases if the filter injects currents phase-shifted by an angle close to 180°. If the delay is known, the demodulation in Fig. 4-89 can be made using positive-sequence voltages with a phase shift equal to δ_h [59]. This phase shift should be calculated from the time delay introduced by the PWM converter. The modified selective harmonic calculation block diagram is shown in Fig. 4-90, where the only difference from Fig. 4-89 is the demodulation voltage that is phase-shifted by δ_h.

The above algorithm has to be used for each harmonic frequency that requires

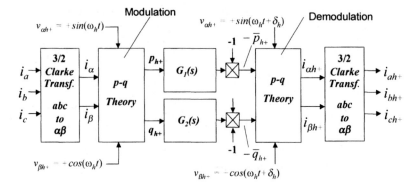

Figure 4-90. Modified selective filter basic cell considering converter delay compensation.

elimination or reduction. This fact makes the algorithm slightly complicated; however, it guarantee better and more precise performance.

It is important to mention that the algorithm shown in Fig. 4-89 and Fig. 4-90 could be employed using the synchronous-reference frames or the d-q transformation instead of the p-q Theory, as presented in [59] and [60]. The results are the same; the difference is in the interpretation of the variables.

The load current compensation scheme for one sequence of a given harmonic component is shown in Fig. 4-91. If various harmonic components, including the positive- and negative-sequence components, are to be compensated, a corresponding number of selective filter basic cells must be used.

In fact, in a generic selective active filter design, it is normal to have to filter not only one harmonic component at one of the sequences, but various components in both positive and negative sequences. These harmonic components may be also close together in the frequency spectrum. The selective detection of these harmonic

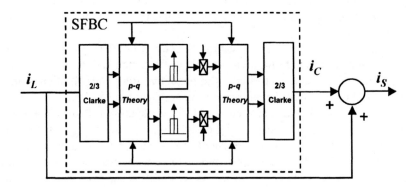

Figure 4-91. Compensation scheme for one harmonic component in a positive or negative sequence.

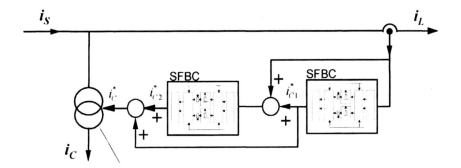

Figure 4-92. Series calculation method for selective filtering of two harmonic components.

components can be made in series or in parallel. Fig. 4-92 shows the series calculation method, whereas Fig. 4-93 shows the parallel calculation method.

In the series calculation method, the load current i_L is measured, and the harmonic component i_{C1} is calculated, using one SFBC. Then, i_{C1} is subtracted from the load current i_L, and applied again to another SFBC to obtain the second harmonic component i_{C2}. This procedure can be done as many times as necessary for the number of harmonic components to be selectively filtered. On the other hand, in the parallel calculation method, the harmonic components i_{C1} and i_{C2} are calculated simultaneously in parallel. If more harmonic components are needed, more parallel SFBCs have to be used.

As has been shown in [61], if the references are calculated from the load current measurement (feedforward), the series method is better than the parallel method because there is less interference between each SFBC. Inversely, if the control input is the line current (feedback), the parallel method gives better results than the series method.

To show the performance of the selective harmonic filter, some simulations were

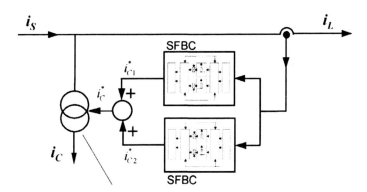

Figure 4-93. Parallel calculation method for selective filtering of two harmonic components.

212 SHUNT ACTIVE FILTERS

performed with measured currents from an arc furnace. The circuit diagram of this arc-furnace system is shown in Fig. 4-94. The point of common coupling (PCC) where the harmonic distortion must comply with the regulations is at 150 kV. If no filter is present in this system, the harmonic content is higher than that allowed at the PCC. To reduce harmonic content at this point, a hybrid shunt filter was installed in place of the 2.2 MVA capacitor bank. The schematic diagram of the new system with the hybrid filter is shown in Fig. 4-95.

In this figure, Z_L is the 150/31.5 kV transformer short-circuit impedance. The transfer function of the passive system is given by

$$\frac{I_L}{I_C} = \frac{1}{1 + \frac{Z_L}{Z_{PF}}} \qquad (4.61)$$

where Z_{PF} is the impedance of the shunt passive filter given by the series connection of L_{PF}, C_{PF}, and R_{PF}. Figure 4-96 shows the gain of the transfer function I_L/I_C for the original case, in which only a capacitive bank was present, and for the case in which the passive filter is connected. It should be noticed that the passive filter (Z_{PF}) is tuned to $8.8f_1$ (where f_1 is the fundamental frequency), and the resonance of Z_{PF} with Z_L is located in the interharmonic $6.5f_1$.

The design of the active filter controller was based on the selective harmonic calculation method considering positive and negative sequences for the following harmonic orders: second, third, fourth, fifth, sixth, and seventh. The gain for each harmonic positive- and negative-sequence component was calculated using an optimization technique presented in [58], with the objective of minimizing these harmonic components up to the level that meets the regulations at the PCC. These

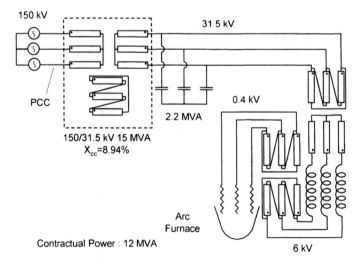

Figure 4-94. Arc furnace system.

4.4. SHUNT SELECTIVE HARMONIC COMPENSATION

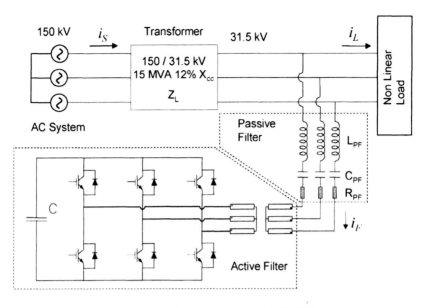

Figure 4-95. Arc furnace system with hybrid filter.

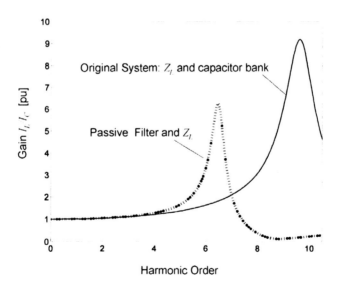

Figure 4-96. Gain of the transfer function I_L/I_C for the original system and with passive filter.

gains are shown in Fig. 4-97 for each positive- and negative-sequence harmonic component separately. The idea is that compensating the harmonics more than specified by the regulations would imply unnecessary costs. The optimization was performed for a THD_p value equal to 12%, where THD_p is the root-mean-square average value of the THD in each phase. In addition, some limits were considered for each harmonic according to the regulations in Argentina [58]. For example, the third-order harmonic to be compensated had a negative sequence that was much higher than the positive one. In this case, through the optimization procedure, the negative sequence is filtered until it is equal to the positive one. Since this is not enough, some additional filtering must be performed in both the positive and negative sequences. In the same way, positive and negative sequences of the sixth harmonic must be compensated almost 80% to oppose to the above-mentioned resonance. Finally, it is not necessary to compensate the eighth, ninth, and tenth harmonics because of the tuned filter effect.

Taking as a reference the calculation time indicated in [59], the calculation time used for compensating the twelve harmonic sequences in an individual and selective way is about 0.14 ms per sample. Thus, in a 50-Hz system, the control can sample at 128 values per cycle, which is good enough to produce satisfactory compensation.

Figure 4-98 shows the actual measurement of the source current at the PCC and the arc-furnace current in the installation shown in Fig. 4-94. Applying a selective active filter with gains as given in Fig. 4-97 to this source current, simulation results shown in Fig. 4-99 are obtained. For this calculation, the series method, in which the output of one SFBC is the input of another SFBC, was used. The current values were normalized to a line voltage of 150 kV. The active filter inverter should work at a lower rated voltage through an adequate transformer, as shown in Fig. 4-95. Figure 4-100 shows that the final source current THD complies with the established value. It also shows that each individual harmonic complies with the corresponding maximum allowed limit. Although the real system has a nonperiodic characteristic, a satisfactory response in real time can be obtained by using the selective harmonic filter.

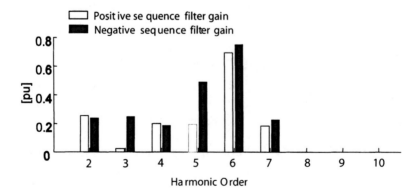

Figure 4-97. Gain for each positive- and negative-sequence harmonic component.

4.4. SHUNT SELECTIVE HARMONIC COMPENSATION

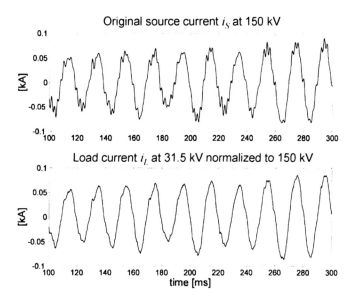

Figure 4-98. Measured source and load currents.

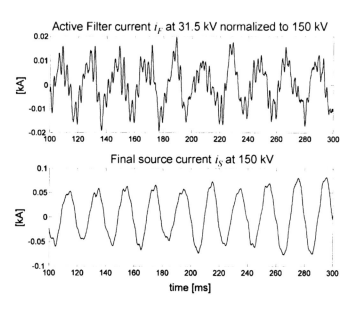

Figure 4-99. Active filter and final source currents.

216 SHUNT ACTIVE FILTERS

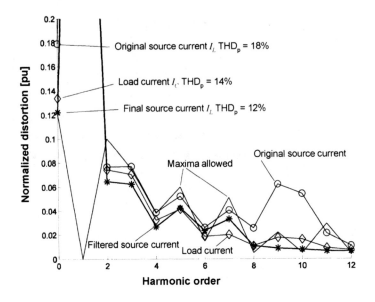

Figure 4-100. Current harmonic spectra.

This method of optimized calculation of the gains of the different SFBCs of a selective active filter is effective in complying with a given harmonic emission regulation. Although the method has been developed for steady-state operation, the simulations show that it can be applied in real time to a system with nonperiodic characteristics. The obtained results show fair agreement with those expected.

4.5. SUMMARY

This chapter covered in detail shunt active filters, including applications to three-phase, four-wire systems. Additionally, an interesting solution has been proposed for allowing the use of $(n-1)$-leg PWM converter in n-wire systems. This was enabled by the use of the "split-capacitor" converter topology. Some important results from this chapter should be pointed out:

1. The principle of shunt active filtering using the instantaneous powers defined in the $\alpha\beta 0$-reference frames can be applied advantageously also to three-phase, four-wire systems.
2. The idea that an active filter should have no dc power supply was fulfilled and extended for applications to three-phase, four-wire systems. It was reinforced by the idea that the shunt active filter can supply an *unbalanced zero-sequence power* to the load taking a *balanced real power* from the source, *without* the need of extra dc power supply in the active filter.

3. The selective compensation characteristic of the control algorithms can be useful for minimizing the power rating of the active filter.
4. The harmonic currents flowing from the source, which are related to the harmonic voltages of the power supply, are not compensated by the shunt active filter. In fact, it would be theoretically possible to filter these harmonic currents, although this would bring very high compensating current to the shunt active filter. Therefore, this is not economically interesting. Other configurations of active filters are better adapted for compensating source harmonics, as will be shown in the next chapters.

REFERENCES

[1] L. Gyugyi and E. C. Strycula, "Active ac Power Filters," in *Proceedings of IEEE Industry Applications Annual Meeting*, vol. 19-C, 1976, pp. 529–535.

[2] H. Kawahira, T. Nakamura, and S. Nakazawa, M. Nomura, "Active Power Filter," in *IPEC'83—Int. Power Electronics Conference*, Tokyo, Japan, 1983, pp. 981–992.

[3] H. Akagi, Y. Kanazawa, and A. Nabae, "Generalized Theory of the Instantaneous Reactive Power in Three-Phase Circuits," in *IPEC'83—International Power Electronics Conference*, Tokyo, Japan, 1983, pp. 1375–1386.

[4] H. Akagi, Y. Kanazawa, and A. Nabae, "Instantaneous Reactive Power Compensator Comprising Switching Devices Without Energy Storage Components," *IEEE Transactions on Industrial Applications*, vol. IA-20, no. 3, 1984, pp. 625–630.

[5] L. Malesani, L. Rossetto, and P. Tenti, "Active Filter for Reactive Power and Harmonics Compensation," in *IEEE-PESC'86—Power Electronics Special Conference*, 86CH2310-1, pp. 321–330.

[6] G. Superti-Furga, E. Tironi, and G. Ubezio, "General Purpose Low-Voltage Power Conditioning Equipment," in *IPEC'95—International Power Electronics Conference*, Yokohama, Japan, Apr. 1995, pp. 400–405.

[7] S. D. Simon and R. M. Duke, "Real-time Optimization of an Active Filter's Performance," *IEEE Transactions on Industrial Electronics*, vol. 41, no. 3, 1994, pp. 278-284.

[8] S. Fukuda and T. Endoh, "Control Method for a Combined Active Filter System Employing a Current Source Converter and a High Pass Filter," *IEEE Transactions on Industrial Applications*, vol. 31, no. 3, May/June 1995, pp. 590–595.

[9] S. J. Jeong and M. H. Woo, "DSP-Based Active Power Filter with Predictive Current Control," *IEEE Transactions on Industrial Electronics*, vol. 44, no. 3, 1997, pp. 329–336.

[10] S. Buso, L. Malesani, P. Mantovelli, and R. Veronese, "Design of Fully Digital Control of Parallel Active Filters for Thyristor Rectifiers to Comply with IEC-1000-3-2 Standards," *IEEE Transactions on Industry Applications*, vol. 34, no. 3, 1998, pp. 508–517.

[11] B. Singh, B. N. Singh, A. Chandra, and K. Al-Haddad, "Real Time DSP Based Implementation of a New Control Method of Active Power Filter," in *1998 IEEE Canadian Conference on Electric and Computer Engineering*, vol. 2, 1998, pp. 794–797.

[12] V. Cardenas, N. Vasquez, and C. Hernandez, "Sliding Mode Control Applied to a 3ϕ

Shunt Active Power Filter Using Compensation with Instantaneous Reactive Power Theory," in *IEEE-PESC'98*, vol. 1, 1998, pp. 236–241.

[13] M. Sedighy, S. D. Dewan, and F. P. Dawson, "Internal Model Current Control of VSC-Based Active Power Filters," in *IEEE-PESC'99*, vol. 1, 1999, pp. 155–160.

[14] Y. Hayashi, N. Sato, and K. Takahashi, "A Novel Control of a Current-Source Active Filter for ac Power System Harmonic Compensation," *IEEE Transactions on Industrial Applications*, vol. 27, no. 2, Mar./Apr. 1991, pp. 380–385.

[15] T.-N. Lê, "Kompensation schnell veränderlicher Blindströme eines Drehstromverbrauchers," *etzArchiv Elekt.*, vol. 11, no. 8, 1989, pp. 249–253.

[16] M. Kohata, T. Shiota, and S. Atoh, "Compensator for Harmonics and Reactive Power Using Static Induction Thyristors," *EPE'87—European Conference Power Electronics Applications*, Genoble, France, 1987, pp. 1265–1270.

[17] S.-Y. Choe and K. Heumann, "Harmonic Current Compensation Using Three-Phase Current Source Converter," in *EPE'91—European Conference Power Electronics Applications*, vol. 3, Firenze, Italy, 1991, pp. 3.006–3.011.

[18] H. Akagi, "Trends in Active Power Line Conditioners," *IEEE Transactions on Power Electronics*, vol. 9, no. 3, May 1994, pp. 263–268.

[19] H. Akagi, "New Trends in Active Filters," in *EPE'95—European Conference Power Electronics Appl.*, vol. 0, Sevilla, Spain, Sep. 1995, pp. 0.017–0.026.

[20] H. Akagi, "Active Harmonic Filters," *Proceedings of the IEEE*, vol. 93, no. 12, December 2005, pp. 2128–2141.

[21]. M. Routimo, M. Salo, and H. Tuusa, "Comparison of Voltage Source and Current-Source Shunt Active Power Filters," in *Conference Records IEEE-PESC 2005*, pp. 2571–2577.

[22] H. Akagi, H. Fujita, and K. Wada, "A Shunt Active Filter Based on Voltage Detection for Harmonic Termination of Radial Power Distribution Line," *IEEE Transactions on Industrial Applications*, vol. 35, no. 3, 1999, pp. 638–645.

[23] P. Jintakosonwit, H. Fujita, H. Akagi, "Performance of a DSP-Controlled Shunt Active Filter for Harmonic Damping on a Power Distribution System," in *IPEC-Tokyo 2000—International Power Electronics Conference*, vol. 1, Tokyo, Japan, 3-7 April 2000, pp. 27–32

[24] M. Aredes, J. Häfner, and K. Heumann, "Three-Phase Four-Wire Shunt Active Filter Control Strategies," *IEEE Transactions on Power Electronics*, vol. 12, no. 2, March 1997, pp. 311–318.

[25] T. Sezi, "Ein Beitrag zur Wirk- und Blindleistungssteuerrung eines Zwischenkreisumrichters," *Dr.-Ing. Thesis*, Technische Universität Berlin, Germany, 1985, pp. 54-60.

[26] N. G. Hingorani, "Power Electronics in Electric Utilities: Role of Power Electronics in Future Power Systems," *Proceedings of the IEEE*, vol. 76, no. 4, April, 1988.

[27] L. Gyugyi, "Solid-state control of ac power transmission," in Workshop on the Future in High-Voltage Transmission: Flexible AC Transmission Systems (FACTS), Cincinnati, Ohio, Nov. 1990.

[28] N. G. Hingorani, "High Power Electronics and Flexible AC Transmission System," *IEEE Power Engineering Reviews*, July 1988.

[29] N. G. Hingorani, "Introducing Custom Power," *IEEE Spectrum*, pp. 41–48, June, 1995.

[30] N. Hingorani and L. Gyugyi, *Understanding Facts: Concepts and Technology of Flexible AC Transmission Systems*, IEEE Press, 1999.

[31] M. Aredes, K. Heumann, and J. Häfner, "A Three-Phase Four-Wire Shunt Active Filter Employing a Conventional Three-Leg Converter," *EPE Journal*, vol. 6, no. 3-4, Dec. 1996, pp. 54–59.

[32] M. Aredes, K. Heumann, and E. H. Watanabe, "A Universal Active Power Line Conditioner," *IEEE Transactions on Power Delivery*, vol. 13, no. 2, April 1998, pp. 545–551.

[33] C. A. Quinn and N. Mohan, "Active Filtering of Harmonic Currents in Three-Phase, Four-Wire Systems with Three-Phase and Single-Phase Non-Linear Loads," in *IEEE-APEC'92 Appl. Power Electronics Conference*, 1992, pp. 829-836.

[34] C. A. Quinn, N. Mohan, and H. Mehta, "A Four-Wire, Current-Controlled Converter Provides Harmonic Neutralization in Three-Phase, Four-Wire Systems," *IEEE-APEC'93 Applications of Power Electronics Conference*, 1993, pp. 841–846.

[35] D. Sutanto and M. Bou-Rabee, "Active Power Filters with Reactive Power Compensation Capability," in *International Power Engineering Conference*, Singapore, March 1993, pp. 73-78.

[36] G. Kamath, N. Mohan, and V. D. Albertson, "A Transformer-Coupled Active Filter for 3-Phase, 4-Wire Systems," in *IEEE/KTH Stockholm Power Technology Conference*, vol. Power Electronics, Sweden, June 1995, pp. 253–255.

[37] G. Superti-Furga, E. Tironi, and G. Ubezio, "General Purpose Low-Voltage Power Conditioning Equipment," in *IPEC'95—International Power Electronics Conference*, Yokohama, Japan, April 1995, pp. 400–405.

[38] M. Aredes, J. Häfner, and K. Heumann, "A Three-Phase Four-Wire Shunt Active Filter Using Six IGBT's," *EPE'95—European Conference Power Electronics Applications*, vol. 1, Sevilla, Spain, Sep. 1995, pp. 1.874–1.879.

[39] M. Aredes, K. Heumann, E. H. Watanabe, and J. Häfner, "A Three-Phase Four-Wire Shunt Active Filter Using Six IGBT's," in *COBEP'95—3rd Brazilian Power Electronics Conference*, Dec. 1995, São Paulo SP, Brazil, pp. 21–26.

[40] B. Rizanto, "Aufbau und Inbetriebnahme eines 3-Phasen 4-Leiter Shunt Active Power Filters mit einem herkömmlichen 3-Zweige Umrichter," *Studienarbeit*, TU Berlin—Institut für Allgemeine Elektrotechnik, Germany, 1995, pp. 42–49.

[41] M. Aredes and E. H. Watanabe, "New Control Algorithms for Series and Shunt Three-Phase Four-Wire Active Power Filters," *IEEE Transactions on Power Delivery*, vol. 10, no. 3, July 1995, pp. 1649–1656.

[42] M. Aredes, "Novos Conceitos de Potência e Aplicações em Filtros Ativos," M.Sc. Thesis, COPPE—Federal University of Rio de Janeiro, Brazil, Nov. 1991 (in Portuguese).

[43] H. Akagi, A. Nabae, and S. Atoh, "Control Strategy of Active Power Filters Using Multiple Voltage-Source PWM Converters," *IEEE Transactions on Industrial Applications*, vol. 22, no. 3, 1986, pp. 460–465.

[44] F.-Z. Peng, H. Akagi, and A. Nabae, "A Study of Active Power Filters Using Quad-Series Voltage-Source PWM Converters for Harmonic Compensation," *IEEE Transactions on Power Electronics*, vol. 5, no. 1, Jan. 1990, pp. 9–15.

[45] H. Akagi, Y. Tsukamoto, and A. Nabae, "Analysis and Design of an Active Power Filter Using Quad-Series Voltage-Source PWM Converters," *IEEE Transactions on Industrial Applications*, vol. 26, no. 1, Jan./Feb. 1990, pp. 93–98.

[46] H. Akagi, "New Trends in Active Filters for Power Conditioning," *IEEE Transactions on Industrial Applications*, vol. 32, no. 6, 1996, pp. 1312–1322.

[47] H. Akagi, "Control Strategy and Site Selection of a Shunt Active Filter for Damping of

Harmonic Propagation in Power Distribution Systems," *IEEE Transactions on Power Delivery*, vol. 12, no. 1, January 1997, pp. 354–363.

[48] P. Jintakosonwit, H. Fujita, H. Akagi, and S. Ogasawara, "Implementation and Performance of Cooperative Control of Shunt Active Filters for Harmonic Damping Throughout a Power Distribution System," *IEEE Transactions on Industrial Applications*, vol. 39, no. 2, Mar./Apr. 2003, pp. 556–564.

[49] T.-N. Lê, M. Pereira, K. Renz, and G. Vaupel, "Active Damping of Resonance in Power Systems," *IEEE Transactions on Power Delivery*, vol. 9, no. 2, April 1994, pp. 1001–1008.

[50] H. Akagi, H. Fujita, and K. Wada, "A Shunt Active Filter Based on Voltage Detection for Harmonic Termination of a Radial Power Distribution Line," *IEEE Transactions on Industrial Applications*, vol. 35, no. 3, 1999, pp. 638-645.

[51] K. Wada, H. Fujita, and H. Akagi, "Considerations of a Shunt Active Filter Based on Voltage Detection for Installation on a Long Distribution Feeder," in *Proceedings of The IEEE Industry Applications Annual Meeting 2001*, vol. 1, 2001, pp. 157–163.

[52] P. Jintakosonwit, H. Fuijita, and H. Akagi, "Control and Performance of a Fully-Digital-Controlled Shunt Active Filter for Installation on a Power Distribution System," *IEEE Transactions on Power Electronics*, vol. 17, no. 1, January 2002, pp. 132–140.

[53] E. J. Currence, J. E. Plizga, and H. N. Nelson, "Harmonic Resonance At a Medium-sized Industrial Plant," *IEEE Transactions on Industrial Applications*, vol. 31, no. 4, 1995, pp. 682–690.

[54] D. Andrews, M. T. Bishop, and J. F. Witte, "Harmonic Measurements, Analysis, and Power Factor Correction in a Modern Steel Manufacturing Cacility," *IEEE Transactions on Industrial Applications*, vol. 32, no. 3, 1996, pp. 617–624.

[55] K. Oku, O. Nakamura, and K. Uemura, "Measurement and Analysis of Harmonics in Power Distribution Systems, and Development of a Harmonic Suppression Method," *IEE of Japan Transactions*, vol. 114-B, no. 3, 1994, pp. 234–241 (in Japanese).

[56] K. Oku, O. Nakamura, J. Inoue, and M. Kohata, "Suppression Effects of Active Filter on Harmonics in a Power Distribution System Including Capacitors," *IEE of Japan Trans.*, vol. 115-B, no. 9, 1995, pp. 1023–1028 (in Japanese).

[57] G. Gonzalo, A. Salvia, C. Briozzo, and E. H. Watanabe, "Control Strategy of Selective Harmonic Current Shunt Active Filter," *IEEE Proceedings of Generation Transmission and Distribribution*, vol. 149, no. 2, Dec. 2002, pp. 689–694.

[58] G. Gonzalo, A. Salvia, C. Briozzo, and E. H. Watanabe, "Selective Filter with Optimum Remote Harmonic Distortion Control," *IEEE Transactions on Power Delivery*, vol. 19, no. 4, Oct. 2004, pp. 1990–1997.

[59] P. Matavelli, "A Closed-Loop Selective Harmonic Compensation for Active Filters," *IEEE Transactions on Industrial Applications*, vol. 37, No. 1, Jan./Feb. 2001, pp. 81–89.

[60] S. Bhattacharya, P.-T. Cheng, and D. Divan, "Hybrid Solutions for Improving Passive Filter Performance in High Power Applications," *IEEE Transactions on Industrial Applications*, vol. 33, no. 3, May/Jun. 1997, pp. 732–747.

[61] G. Casaravilla, A. Salvia, C. Briozzo, and E. H. Watanabe, "Selective Active Filter Applied to an Arc Furnace Adjusted to Harmonic Emission Limitations," in *IEEE/PES Transmission and Distribution Conference—Latin America*, 2002.

CHAPTER 5

HYBRID AND SERIES ACTIVE FILTERS

Considering the duality of its circuits, a series active filter should be a power electronics device that, in principle, would block harmonic voltages in the load from those appearing in the source. In fact, this would be a dual device of a shunt active filter. In the case of the shunt active filter, the source is represented by a voltage source and the load by a current source, including harmonic currents that have to be compensated by the shunt active filter. Therefore, the shunt active filter has to generate harmonic currents to cancel load harmonic currents. On the other hand, the series active filter should generate harmonic voltages to cancel the load harmonic voltages. However, since duality is being considered, in the case of the series active filter, the source should be a current source. This is not the case normally. In some cases, a combination of the source and its impedance can be considered as a current source. As a result, the series active filter operating as a dual circuit of the shunt active filter is rather a theoretical situation [1]. In this chapter, first this theoretical filter will be explained. Then, following the technological developments based on the p-q Theory, a combined system of a shunt passive filter and a series active filter [2] will be presented. The combination of a passive filter and an active filter is generally called a hybrid filter in this book. Also, a series active filter integrated with a diode rectifier will be presented in detail. Naturally, it will not be possible to cover all possible cases, but the idea is to give a basis for different filter designs. Finally, some comparisons between hybrid and pure shunt active filters will be presented.

5.1. BASIC SERIES ACTIVE FILTER

Figure 5-1 shows an example of a basic series active filter. For simplicity, system "A" represents the source side, which has been represented by balanced sinusoidal

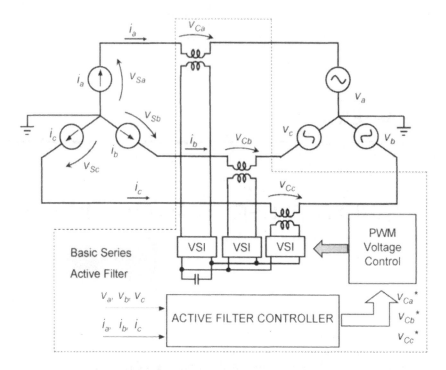

Figure 5-1. Basic series active filter.

current sources, and system "B" represents the load side with its voltage sources containing harmonic components. Moreover, it is assumed that there is no zero-sequence current. The voltages on the current sources are given by v_{Sa}, v_{Sb}, and v_{Sc}. On the other hand, the relation among the source voltage, the load voltage, and the active filter voltage is given by

$$\begin{bmatrix} v_{Sa} \\ v_{Sb} \\ v_{Sc} \end{bmatrix} = \begin{bmatrix} v_a \\ v_b \\ v_c \end{bmatrix} - \begin{bmatrix} v_{Ca} \\ v_{Cb} \\ v_{Cc} \end{bmatrix} \qquad (5.1)$$

The basic series active filter voltages are synthesized by three single-phase converters with a common dc capacitor. The reference voltage for these converters is calculated by the "active filter controller" shown in Fig. 5-1, which has as input signals the load voltages and currents (equal to the source currents) [1]. In this case, the calculation of these voltages can be performed by using the dual p-q Theory, as given in (3.62):

$$\begin{bmatrix} p_0 \\ p \\ q \end{bmatrix} = \begin{bmatrix} i_0 & 0 & 0 \\ 0 & i_\alpha & i_\beta \\ 0 & -i_\beta & i_\alpha \end{bmatrix} \begin{bmatrix} v_0 \\ v_\alpha \\ v_\beta \end{bmatrix} \qquad (5.2)$$

The instantaneous real, imaginary and zero-sequence powers calculated enable us to calculate the oscillating real power $\tilde{p}$ and the oscillating imaginary power $\tilde{q}$, where the zero-sequence powers $\bar{p}_0$ and $\tilde{p}_0$ are assumed to be zero because it was assumed that there is no zero-sequence current. With these oscillating powers, it is possible to calculate the instantaneous voltages that have to be injected by the series active filter to compensate for the load voltage harmonics by using:

$$\begin{bmatrix} v_{Ca}^* \\ v_{Cb}^* \end{bmatrix} = \frac{1}{i_\alpha^2 + i_\beta^2} \begin{bmatrix} i_\alpha & -i_\beta \\ i_\beta & i_\alpha \end{bmatrix} \begin{bmatrix} \tilde{p} \\ \tilde{q} \end{bmatrix} \quad (5.3)$$

Of course, it is considered that the converters are ideal, without power losses; therefore, it is not necessary to control dc capacitor voltage. However, in actual cases, a certain amount of $\Delta \bar{p}$ should be added to $\tilde{p}$ with the objective of compensating for the losses, as was done in the case of a shunt active filter. The reference voltages $v_{C\alpha}^*$ and $v_{C\beta}^*$ can be converted to the *abc*-reference frames by

$$\begin{bmatrix} v_{Ca}^* \\ v_{Cb}^* \\ v_{Cc}^* \end{bmatrix} = \sqrt{\frac{2}{3}} \begin{bmatrix} 1 & 0 \\ -\frac{1}{2} & \frac{\sqrt{3}}{2} \\ -\frac{1}{2} & -\frac{\sqrt{3}}{2} \end{bmatrix} \begin{bmatrix} v_{C\alpha}^* \\ v_{C\beta}^* \end{bmatrix} \quad (5.4)$$

These are the voltages that the series active filter has to generate to compensate for the harmonic voltage components in the load, which produce oscillating real power $\tilde{p}$ and oscillating imaginary power $\tilde{q}$ at the load side. This approach guarantees that the voltages on the current sources have purely sinusoidal waveforms. However, a displacement angle may be present because $\bar{q}$ was not being considered in (5.3). To eliminate this displacement angle, $\bar{q}$, if existing, should be added to that equation. Doing so, the voltages v_{Sa}, v_{Sb} and v_{Sc} on the current sources will be purely sinusoidal and in phase with the currents. Since this is a rather theoretical situation, only these concepts of this basic series active filter are presented.

5.2. COMBINED SERIES ACTIVE FILTER AND SHUNT PASSIVE FILTER

One traditional solution to mitigating harmonics problems is the use of passive filters. Here, a system with a passive filter will be analyzed.

Figure 5-2 shows a nonlinear load and a shunt passive filter connected to the grid. The objective of this shunt passive filter is to eliminate harmonic currents generated by the nonlinear load. In fact, this is a quite traditional solution to preventing load harmonic currents from flowing into the source or other loads. For the design of the passive filter, it is necessary to know the source impedance, which is not con-

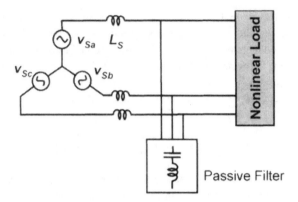

Figure 5-2. Nonlinear load with passive filter.

stant and depends directly on the system configuration. This means that the filtering characteristic of the passive filter is strongly influenced by this impedance. In addition, the following two problems may occur.

The first problem is that the shunt passive filter is designed to have low impedance at the tuned frequency, say ω_h, and may serve as a current sink at another frequency ω_i if the source voltage includes such a harmonic component. This means that the harmonic components coming from other sources distort the voltage source waveform. Consequently, a harmonic current flows into the shunt passive filter. In the worst case, series resonance between the source impedance $\omega_i L_S$ and the shunt filter impedance may occur, possibly with an overcurrent flowing into the passive filter. This resonance situation is shown in Fig. 5-3.

The second problem is that antiresonance (parallel resonance) between the source impedance ωL_S and the shunt filter impedance occurs at the frequency ω_i. Therefore, a given load harmonic current i_h flowing through the circuit causes a high impedance voltage as a result of parallel connection of the shunt passive filter

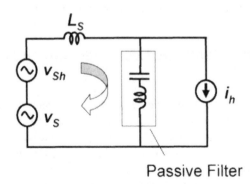

Figure 5-3. Series resonance phenomena producing large source current.

and the source impedance, producing a harmonic amplifying phenomenon. Hence, high harmonic voltage v_h appears on the load, as shown in Fig. 5-4.

The question that arises here is whether to use a shunt active filter in this situation. In some cases, this kind of filter may be impractical for the following reasons:

1. It may not be technically feasible to realize a large-rating PWM converter with a current response fast enough to compensate for the harmonic components with high efficiency.
2. The initial cost of the shunt active filter is high, compared with that of a shunt passive filter.

Although the use of the shunt passive filter is justified, it is possible to say that:

1. Its filtering characteristics depend on the source impedance, which may be not well defined, or it may vary as the grid topology changes.
2. The larger the source impedance at one given frequency, the better its ability to attenuate the corresponding harmonic component. However, the source impedance should be as small as possible at the fundamental frequency to keep negligible the fundamental voltage drop.

To avoid the above-mentioned problems regarding the applications of passive filters, it would be interesting to combine the shunt passive filter with a series active filter to adjust the equivalent system impedance as a function of the frequency. Therefore, as shown in Fig. 5-5(a), the single-phase equivalent circuit at the fundamental frequency for Fig. 5-2 should present the source impedance as it is. On the other hand, for harmonic frequencies, the source impedance should be increased as shown in Fig. 5-5(b). In this figure, the block K, generically, is a resistance higher than the source impedance at the harmonic frequencies. To achieve this objective, this block K should function as an "active impedance" inserted between the source and the passive filter. In addition, this "active impedance" should have the "job" of eliminating the series or parallel resonances. The problem here is how to synthesize

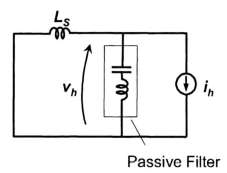

Figure 5-4. Antiresonance phenomena producing large v_{ll}.

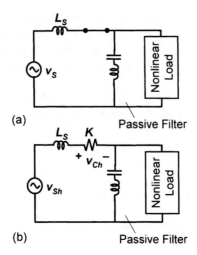

Figure 5-5. Equivalent circuit for the "active impedance". (a) For the fundamental frequency, and (b) for the harmonic components.

this "active impedance." This can be easily done by PWM converters. The basic structure of the system shown in Fig. 5-2 is repeated in Fig. 5-6 with the insertion of three voltage source-converters in series with each phase. The converter and its respective control block have the basic objective of synthesizing the above-mentioned "active impedance."

5.2.1. Example of an Experimental System

Figure 5-7 shows a block diagram of a combined series active filter and shunt passive filter for an experimental system tested in a laboratory [2]. The nonlinear load

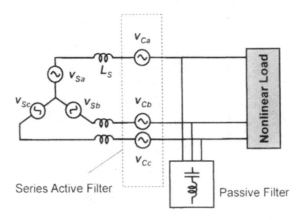

Figure 5-6. Combined series active filter and shunt passive filter.

5.2. COMBINED SERIES ACTIVE FILTER AND SHUNT PASSIVE FILTER

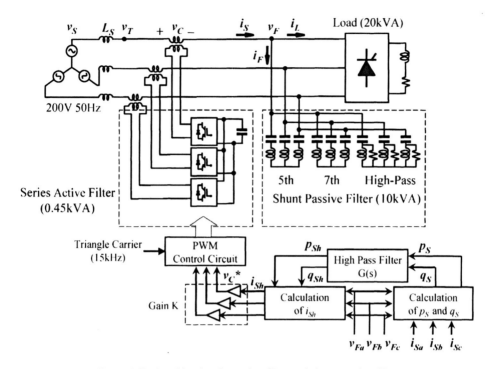

Figure 5-7. Combined series active filter and shunt passive filter.

is a 20-kVA thyristor rectifier. The 10-kVA shunt passive filter consists of a fifth-order- and a seventh-order tuned harmonic filters, plus a high-pass filter. The shunt passive filter parameters are given in Table 5.1. One converter is connected in series in each phase through a current transformer with a turn ratio equal to 1:20. The transformer is necessary to match the voltage and current of the converter with that of the system, as well as to guarantee galvanic isolation. The total rated power of the converters is 0.45 kVA, where the dc side capacitor is common for all three converters. A PWM control block is designed so that the ac output voltage of the converters follows the references that are calculated by the control block, based on the *p-q* Theory. One single-phase diode rectifier, not shown in Fig. 5-7, rated at 50 VA, is connected to the dc-side capacitor, to supply the power losses of the converters.

Table 5.1. Parameters of the shunt passive filters

	Inductance [mH]	Capacitance [μF]	
Fifth-order filter	1.2	340	$Q = 14$
Seventh-order filter	1.2	170	$Q = 14$
High-pass filter	0.26	300	$R = 3\,\Omega$

In this hybrid filter, the function of the converter is not to compensate for the harmonics voltages directly as shown in Fig. 5-1, but to enhance the shunt passive filter characteristics by providing harmonic isolation. In other words, the converters should provide high resistance for the harmonic frequencies and, of course, zero resistance for the fundamental frequency. One important result of this approach is that the series converters are much smaller in capacity or volt-ampere rating than a conventional shunt active filter designed to filter directly the load harmonic currents.

5.2.1.A. Compensation Principle

Taking, for simplicity, only one phase of the circuit in Fig. 5-7 enables us to draw the circuit shown in Fig. 5-8. In this figure, the thyristor rectifier is being represented by a current source; the voltage source is split into its fundamental voltage source and harmonic voltage source. The series active filter is represented by the controlled voltage source $\dot{V}_C$ and the shunt passive filter is represented by the equivalent impedance Z_F.

The basic compensation principle follows the concepts presented in Fig. 5-5—the series active filter should synthesize the "active impedance" presenting zero impedance at the fundamental frequency and a high resistance K at the source or load harmonic frequencies. In fact, this "active impedance" is synthesized by the PWM converter shown in Fig. 5-7, and redrawn in detail in Fig. 5-9. The inductance L_r and capacitance C_r in this circuit form a small-rated passive filter to suppress switching ripples. To explain the synthesis of the "active impedance" that is the basis of the compensation principle, some equations have to be derived from Fig. 5-8.

5.2.1.A.1. Source Harmonic Current $\dot{I}_{Sh}$.
The harmonic current flowing in the source is dependent on both the load harmonic current $\dot{I}_{Lh}$ and the source harmonic voltage $\dot{V}_{Sh}$. It is given by

$$\dot{I}_{Sh} = \frac{Z_F}{Z_S + Z_F + K} \dot{I}_{Lh} + \frac{\dot{V}_{Sh}}{Z_S + Z_F + K} \qquad (5.5)$$

where

$$\dot{I}_{Sh} \cong 0 \quad \text{if} \quad K \gg Z_S, Z_F \qquad (5.6)$$

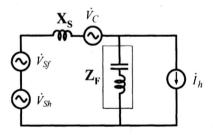

Figure 5-8. Single-phase equivalent circuit for the series active filter and shunt passive filter.

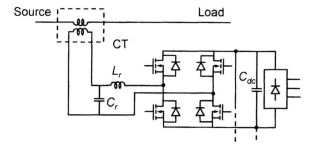

Figure 5-9. Single-phase detailed circuit of the voltage source converter.

with Z_S being the magnitude of $\mathbf{Z_S}$ and Z_F the magnitude of $\mathbf{Z_F}$. The first term on the right-hand side of (5.5) shows that the series active filter operates as an "active impedance." In fact, it acts as a "damping resistance," which can eliminate the parallel resonance between the shunt passive filter and the source impedance. The second term means that the series active filter acts as a "blocking resistance," which can prevent the harmonic current produced by the source harmonic voltage from flowing into the shunt passive filter. It is important to point out that if the equivalent resistance K is much larger than the source impedance, variations in this impedance produce no effect on the filtering characteristics of the shunt passive filter. Consequently, this reduces the source harmonic current to zero, as shown in (5.6).

5.2.1.A.2. Output Voltage of Series Active Filter $\dot{V}_C$.

The output voltage of the series active filter, which corresponds to the harmonic voltage across the equivalent resistance K in Fig. 5-5, is given by

$$\dot{V}_C = K\dot{I}_{Sh} = K\frac{\mathbf{Z_F}\dot{I}_{Lh} + \dot{V}_{Sh}}{\mathbf{Z_S} + \mathbf{Z_F} + K} \tag{5.7}$$

$$\dot{V}_C \cong \mathbf{Z_F}\dot{I}_{Lh} + \dot{V}_{Sh} \quad \text{if} \quad K \gg Z_S, Z_F \tag{5.8}$$

Equation (5.8) shows clearly that the voltage rating of the series active filter $\dot{V}_C$ is given by two factors: the first term in the right hand side of this equation, which is inversely proportional to the quality factor of the shunt passive filter, and the second term that is equal to the source harmonic voltage.

5.2.1.A.3. Shunt Passive Filter Harmonic Voltage $\dot{V}_{Fh}$.

The harmonic voltage on the shunt passive filter is given by

$$\dot{V}_{Fh} = \frac{\mathbf{Z_S} + K}{\mathbf{Z_S} + \mathbf{Z_F} + K}\mathbf{Z_F}\dot{I}_{Lh} + \frac{\mathbf{Z_F}}{\mathbf{Z_S} + \mathbf{Z_F} + K}\dot{V}_{Sh} \tag{5.9}$$

$$\dot{V}_{Fh} \cong -\mathbf{Z_F}\dot{I}_{Lh} \quad \text{if} \quad K \gg Z_S, Z_F \tag{5.10}$$

The above equations show that by choosing $K \gg Z_S, Z_F$, no source harmonic voltage appears across the shunt passive filter.

5.2.1.B. Filtering Characteristics

Taking into account the equivalent circuit given in Fig. 5-8 and applying the superposition law enables us to analyze the filtering characteristics [3]. Two harmonic-current flow should be investigated: one from the load to the source and the other from the source to the load.

5.2.1.B.1. Harmonic Current Flowing from the Load to the Source. From Fig. 5-8, together with neglecting the harmonic voltage source, it is easy to see that the load harmonics are divided between the shunt passive filter and the source, proportionally to the admittance of the parallel branches. Therefore, the ratio of the source harmonic current to the load harmonic current for the case of $\dot{V}_{Sh}$ is given by

$$\frac{\dot{I}_{Sh}}{\dot{I}_{Lh}} = \frac{Z_F}{Z_S + Z_F + K}, \quad \text{for } \dot{V}_{Sh} = 0 \tag{5.11}$$

The ratio given above can be called the "distribution factor," and is shown for various values of K in Fig. 5-10(a) for the case of source inductance L_S equal to 0.02 pu, and in Fig. 5-10(b) for the case of $L_S = 0.056$ pu. The inductance calculation takes into account the system base of $V_@ = 200$ V, $S_@ = 20$ kVA, and $f_@ = 50$ Hz. The source resistance is not considered because it is much smaller than the source reactance. In these figures, harmonic amplification occurs whenever the ratio I_{Sh}/I_{Lh} is greater than zero dB, whereas harmonic attenuation occurs if the ratio is smaller than zero dB.

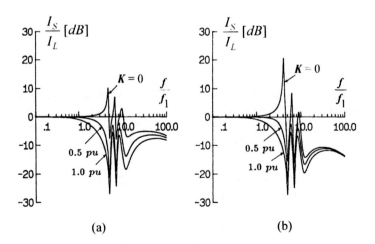

Figure 5-10. Ratio of source current to load current in the case of (a) $L_S = 0.02$ pu, and (b) $L_S = 0.056$ pu.

Referring to Fig. 5-10, the case of $K = 0$ means that no series active filter is present, that is, the shunt passive filter is operating alone. In this situation, it is possible to see that resonance may occur at some frequencies. For instance, the fourth-order harmonic is one example of a resonant frequency that, in the case of $L_S = 0.056$ pu, produces an amplification factor of 22 dB (13 times). Naturally, one may not be worried because this is not a characteristic frequency in conventional thyristor rectifiers. However, if the load is a cycloconverter or even a thyristor rectifier during transient operation, this harmonic frequency may appear and cause significant distortion in the source current. The fourth-harmonic current may arise in an actual thyristor rectifier operating under unbalanced voltages, and the harmonic amplification phenomenon will be shown later in an experimental measurement. When the equivalent resistance K is made equal to 1 or 2 Ω (0.5 and 1.0 pu, respectively), the series active filter can reduce considerably the distribution factor over the frequency range. There is no amplification problem because the filter acts as a damping resistance. The best result is obtained for $K = 2\ \Omega$.

5.2.1.B.2. Harmonic Current Flowing from the Source to the Shunt Passive Filter.
From Fig. 5-8, with the assumption that the load harmonic current is $I_{Lh} = 0$, the source harmonic voltage produces a harmonic current flowing into the shunt passive filter that corresponds to the second term in the right-hand side of (5.5). This harmonic current is repeated below:

$$\dot{I}_{Sh} = \frac{\dot{V}_{Sh}}{\mathbf{Z}_S + \mathbf{Z}_F + K} = \frac{\dot{V}_{Sh}}{\mathbf{Z}_1}, \quad \text{for } \dot{I}_{Lh} = 0 \tag{5.12}$$

Considering that Z_0 is the rated impedance of the circuit, which in this case is equal to 2 Ω on a 200-V, 20-kVA, and 50-Hz basis, the ratio Z_1/Z_0 as a function of frequency, where Z_1 is the magnitude of $\mathbf{Z}_1$, is shown in Fig. 5-11(a) for the case of source inductance L_S equal to 0.02 pu, and in Fig. 5-11(b) for the case of $L_S = 0.056$ pu. For the case of $K = 0$, that is, the case of no series active filter, the ratio Z_1/Z_0 is almost zero around the fifth-order harmonic in Fig. 5-11(a) and at the fourth-order harmonic in Fig. 5-11(b). In this last case, for example, even a small amount of harmonic voltage included in the source voltage at this frequency produces an excessive current through the shunt passive filter. This situation should be avoided to prevent overheating in the filter, which might put it at risk. In the case of $K = 1$ or 2 Ω, the ratio Z_1/Z_0 is increased over the frequency range as compared to the previous case of $K = 0$. In these two cases, the series active filter acts as a blocking resistance, so that the harmonic current flowing into the source or the shunt passive filter is negligible.

5.2.1.C. Control Circuit
The control of the series active filter is based on a very simple equation:

$$v_C^* = K i_{Sh} \tag{5.13}$$

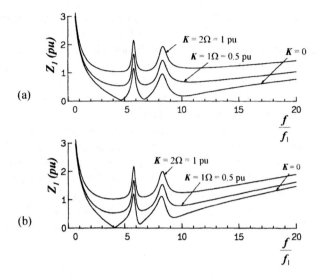

Figure 5-11. Source current due to the load current in the case of (a) $L_S = 0.02$ pu and (b) $L_S = 0.056$ pu.

where i_{Sh} is the source harmonic current, which can be calculated based on the instantaneous power theory (p-q Theory). For this calculation, the voltages in the load side (v_a, v_b, v_c) and the currents in the source sides (i_{Sa}, i_{Sb}, i_{Sc}) have to be transformed into the $\alpha\beta$-orthogonal coordinates by a simplified Clark transformation (v_0 is not considered):

$$\begin{bmatrix} v_\alpha \\ v_\beta \end{bmatrix} = \sqrt{\frac{2}{3}} \begin{bmatrix} 1 & -\frac{1}{2} & -\frac{1}{2} \\ 0 & \frac{\sqrt{3}}{2} & -\frac{\sqrt{3}}{2} \end{bmatrix} \begin{bmatrix} v_a \\ v_b \\ v_c \end{bmatrix} \qquad (5.14)$$

$$\begin{bmatrix} i_{S\alpha} \\ i_{S\beta} \end{bmatrix} = \sqrt{\frac{2}{3}} \begin{bmatrix} 1 & -\frac{1}{2} & -\frac{1}{2} \\ 0 & \frac{\sqrt{3}}{2} & -\frac{\sqrt{3}}{2} \end{bmatrix} \begin{bmatrix} i_{Sa} \\ i_{Sb} \\ i_{Sc} \end{bmatrix} \qquad (5.15)$$

The real and imaginary power can be calculated by

$$\begin{bmatrix} p \\ q \end{bmatrix} = \begin{bmatrix} v_\alpha & v_\beta \\ v_\beta & -v_\alpha \end{bmatrix} \begin{bmatrix} i_{S\alpha} \\ i_{S\beta} \end{bmatrix} \qquad (5.16)$$

In this book, the unity for q is vai, meaning volt-ampere-imaginary. By using a high-pass filter, the oscillating components $\tilde{p}$ and $\tilde{q}$ can be extracted from p and q.

5.2. COMBINED SERIES ACTIVE FILTER AND SHUNT PASSIVE FILTER

In this case, a first-order, high-pass filter with a cutoff frequency of 35 Hz is used. Then, the source harmonic current is calculated by

$$\begin{bmatrix} i_{Sah} \\ i_{Sbh} \\ i_{Sch} \end{bmatrix} = \sqrt{\frac{2}{3}} \begin{bmatrix} 1 & 0 \\ -\frac{1}{2} & \frac{\sqrt{3}}{2} \\ -\frac{1}{2} & -\frac{\sqrt{3}}{2} \end{bmatrix} \begin{bmatrix} v_\alpha & v_\beta \\ -v_\beta & v_\alpha \end{bmatrix}^{-1} \begin{bmatrix} \tilde{p} \\ \tilde{q} \end{bmatrix} \quad (5.17)$$

The reference voltage for the series active filter given by (5.13) is compared with a triangular carrier waveform to produce the PWM switching pattern. The carrier frequency is 15 kHz. With this control, the series active filter operates as a controlled voltage source. This is the dual situation as compared with shunt active filter, in which it operates as a controlled current source. Naturally, a current-source PWM converter would not be suitable for this application as it is the voltage-source PWM for this case of series active filter.

The voltage-source PWM converter used to synthesize the series active filter can be protected against overvoltage and overcurrent by adopting the following approach:

1. All the upper power MOSFETs in each converter leg are turned off to disconnect dc capacitors from the secondary of the current transformers (CT's).
2. All the lower power MOSFETs are turned on to short the secondary of the current transformers through the on-state MOSFETs and the diodes.

5.2.1.D. Filter to Suppress Switching Ripples

Generally, it is very difficult to design a passive filter to suppress harmonics due to the PWM switching because the impedance of the power system is not well defined. Here, it is much simpler because the PWM converter is connected to the system through a current transformer with a large turn ratio. As mentioned before, the inductor L_r (= 1 mH) and capacitor C_r (= 0.33 µF) that appear in Fig. 5-9 are used to suppress the ripples due to the switching of the PWM converters, which is done at a frequency f_s of 15 kHz. The equivalent circuit seen from the PWM converter is shown in Fig. 5-12(a), and the equivalent circuit for the ripple voltage source is shown in Fig. 5-12(b). The impedance Z_{PWM} is equal to the sum of Z_S and Z_F, which are seen from the secondary of the current transformer and can be expressed as

$$Z_{PWM} = (n_2/n_1)^2 |Z_S + Z_F| \quad (5.18)$$

Considering the parameters given in Table 5.1, the switching frequency f_s = 15 kHz, n_1/n_2 = 20, and L_s = 2%, Z_{PWM} = 9.6 kΩ. On the other hand, the reactance giv-

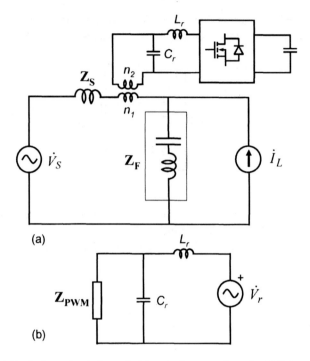

Figure 5-12. (a) Equivalent circuit for switching ripples seen from the PWM converter and (b) equivalent circuit for the ripple voltage source v_r.

en by the capacitor C_r {= $1/[2\pi(2f_s)C_s]$} and the reactor L_r [= $2\pi(2f_s)L_r$] at the frequency of the switching ripple, which is equal to $2f_s$ are equal to 16 Ω and 200 Ω, respectively. This means that the harmonic current due to the voltage ripple does not go to the system, but flows through the capacitor C_r. Since the reactance of the capacitor is much smaller than the reactance of the inductor, the ripple voltage becomes very small in the secondary of the transformer. Moreover, considering that the turn ratio of the transformer is 1:20, the voltage ripple reflected to the primary side is reduced further.

5.2.1.E. Experimental Results

The experimental results for the combined series active filter and shunt passive filter was obtained for the following conditions: $K = 2$ Ω, dc capacitor voltage equal to 120 V, and capacitance value equal to 1500 μF.

Figure 5-13 shows the experimental waveforms for the situation before and after the series active filter was started. The waveforms are for the line voltage, the source current, the shunt passive filter current, the load current, and the series active filter voltage. It is clear that before the series active filter was started, there was a reasonable amount of harmonic current in the source current, showing that the passive filter used alone was not able to eliminate the harmonic current from the

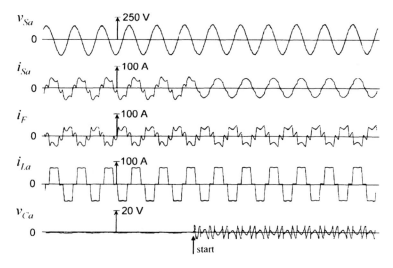

Figure 5-13. Experimental waveforms for $L_S = 2\%$. From the top: line voltage, source current, shunt passive filter current, load current, and series active filter voltage.

source. After the series active filter was started, the source current turned almost purely sinusoidal. The compensating voltage generated by the series active filter presents an rms value of only 2.5 V. This means that the volt-ampere (VA) rating of the series active filter is only 2.5V × 60A × 3 = 450 VA. This figure represents only 2.3% of the load rating of 20 kVA.

Figure 5-14 shows the frequency spectra for the source current in the case of Fig. 5-13, when the series active filter was off and after it was started. Almost no harmonic component was seen after the filter was started, confirming the effectiveness of the series active filter.

Figure 5-15 shows the experimental result for the case of $L_S = 2\%$, as in Fig. 5-13, but without load. Although the source harmonic voltage was only 1%, the source and shunt passive filter harmonic currents were as high as 10% before the series active filter was started. After the series active filter was started, these harmonic components simply disappear. In fact, the series active filter blocks the source harmonic voltage from appearing on the shunt passive filter terminals.

Figure 5-16 shows a similar waveform to that in Fig. 5-13 with the difference that the source inductance L_s was made equal to 5.6%. In this situation, before the series active filter was started, a parallel resonance appeared at the fourth-harmonic frequency. This happened because a small amount of the fourth-order harmonic in the load current was largely amplified, distorting severely the source current i_S and the source voltage v_S. After the series active filter was turned on, no parallel resonance appeared.

The measured loss in the series active filter was smaller than 40 W, showing that this combined series active filter and shunt passive filter is much more efficient than conventional shunt active filters.

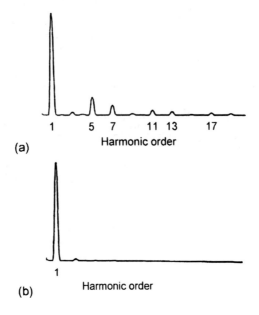

Figure 5-14. Frequency spectra of source current i_S, (a) before the series active filter started, and (b) after it started.

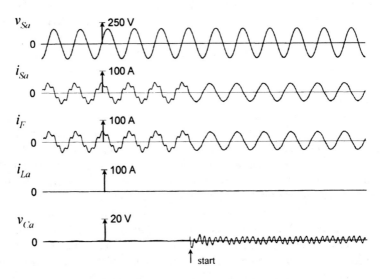

Figure 5-15. Experimental waveforms for $L_S = 2\%$ and no-load condition. From the top: line voltage, source current, shunt passive filter current, load current and series active filter voltage.

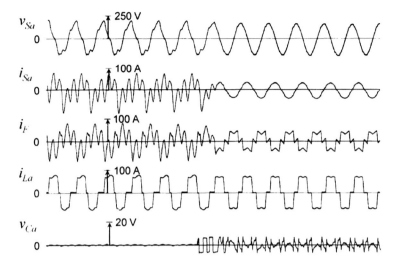

Figure 5-16. Experimental waveforms for $L_S = 5.6\%$. From the top: line voltage, source current, shunt passive filter current, load current, and series active filter voltage.

5.2.2. Some Remarks about the Hybrid Filters

The combination of a shunt passive filter and a low-rated series active filter resulted in a very practical and economical way to filter harmonic currents. The concept adopted in this approach differs from conventional shunt active filters or pure series active filters in terms of guaranteeing better filtering characteristics as well as lower initial costs. The required rating of the series active filter is mainly determined by the quality factor of the shunt passive filter and the amount of background harmonic voltage existing in the power system.

Although the quality factor of the shunt passive filter used in the experiments was equal to 14, it may be in the range of 50 to 80 in actual cases. Consequently, the rated power of the series active filter may be as low as 1% of the rated power of the nonlinear load (thyristor rectifier or cycloconverter) if no background harmonic voltage exists in the source. This leads to the important conclusion that this combined series active filter and shunt passive filter is one of the most suitable solutions to high-power thyristor rectifiers, like those used in high-voltage dc transmission systems (HVDC) or cycloconverters.

An interesting alternative to the hybrid filter is presented in [4]. In this approach, the active filter is connected in series with the passive filter, as shown in Fig. 5-17.

The composed branch of passive and active filters is connected in parallel, as closed as possible to the harmonic-producing load. This new approach provides almost the same results as that discussed before and shown in Fig. 5-7. Although these hybrid filters are slightly different in circuit configuration, they are almost the same in operating principle and filtering performance. Such a combination with the passive filter makes it possible to significantly reduce the rating of the active filter.

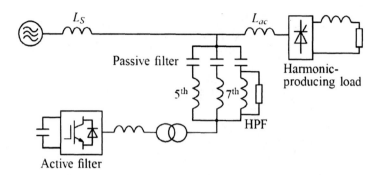

Figure 5-17. Series connection of an active filter and a passive filter.

The task of the active filter is not to compensate for harmonic currents produced by the thyristor rectifier, but to achieve "harmonic isolation" between the supply and the load [2]. As a result, no harmonic resonance occurs, and no harmonic current flows in the supply.

The harmonic distortion that appears in the voltage at the load terminal is different in Fig. 5-7 and Fig. 5-17. In case of Fig. 5-7, the harmonic voltage at load terminal is produced by the harmonic current of the load flowing into the passive filter, as shown in (5.10). Contrarily, in case of the configuration used in Fig. 5-17, this component is eliminated, but the harmonic voltage that may be present in the power system propagates, and appears at the load terminal [4]. To overcome both problems, a combined solution was presented in [5]. In this book, it is called the hybrid unified power quality conditioner, and is described in detail in Chapter 6.

5.3. SERIES ACTIVE FILTER INTEGRATED WITH A DOUBLE-SERIES DIODE RECTIFIER

Harmonic-current-free rectifiers capable of operating at the unity power factor are required as utility interfaces for converter-based industrial loads such as adjustable-speed motor drives and uninterruptible power supplies in a range of 1–10 MW. Pulse-width-modulated (PWM) rectifiers have shown promise in meeting the guidelines for harmonic mitigation [6–9]. The increased cost and switching loss caused by pulse width modulation, however, would make a high-power PWM rectifier rated at 1–10 MW economically impractical. The reason is that the GTO thyristor or IGBT used for the PWM rectifier is subjected to high-frequency switching of the full amount of active power.

An ac/dc power conversion system providing a solution to the above-mentioned problems hidden in the PWM rectifier was proposed in [10]. In this conversion system, a low-rated series active filter for the purpose of achieving harmonic compensation [11,12] is integrated with a high-rated, double-series diode rectifier for the purpose of performing ac/dc power conversion. The active filter enables the diode rectifier to draw three-phase sinusoidal currents from the utility. In addition, the active filter based on supply current detection is characterized by closed-loop control with a relatively high feedback gain. The stability analysis of this kind of active filter was presented in [14].

This section presents the series active filter integrated with a double-series diode rectifier, and analyzes its stability. Special attention is paid to the delay time inherent in the harmonic-current-extracting circuit. The analysis points out that even a delay time as short as 40 μs may make the active filter unstable when the ac/dc power conversion system is installed on a "stiff" power system. The major time delay is caused by coordinate transformation in the harmonic-current-extracting circuit. Here, two control circuits are analyzed: the first-generation control circuit was proposed by Fujita et al. [10], and the second-generation control circuit was proposed by Srianthumrong et al. [14]. It will be shown that although the first-generation control circuit can work well in some situations, it may be unstable in others. The second-generation control circuit, on the other hand, is capable of eliminating the effect of the delay time on the system stability. This second-generation circuit configuration reduces the phase delay in the extracted harmonic current even when it has the same delay time in the coordinate transformation as that in the first-generation control circuit. The theoretical analysis shows that the second-generation control circuit makes it possible to significantly improve the system stability. Furthermore, a switching-ripple filter is designed and tested. The effect of the filter on the system stability is also discussed. Experimental waveforms obtained from a 20-kW laboratory system show stable operation even though the coordinate-transformation circuit has a delay time of 40 μs.

Figure 5-18 shows a harmonic-free, ac/dc power conversion system discussed in this section. It consists of a combination of a double-series diode rectifier rated at

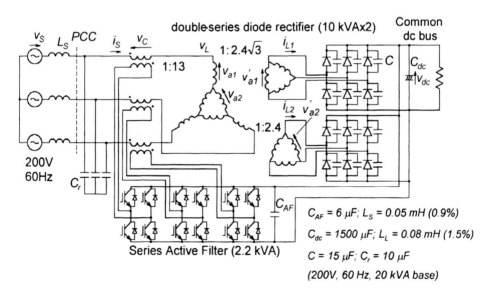

Figure 5-18. Harmonic-current-free, ac/dc power conversion system.

20 kW and a series active filter with a peak voltage and current rating of 2.2 kVA (see Table 5.2). The series active filter consists of three single-phase matching transformers and three single-phase full-bridge, voltage-source PWM converters. Three low-rated capacitors installed at the utility-consumer point of common coupling (PCC) form a passive filter for eliminating switching ripples. The ac terminals of each PWM converter are connected in series with a power line through a single-phase matching transformer. The double-series diode rectifier consists of three-phase Y–Δ and Δ–Δ connected transformers and two three-phase diode rectifiers. The primary windings of the two transformers are connected in series with each other. The dc terminals of the diode rectifiers and the active filter form a common dc bus equipped with an electrolytic capacitor.

The double-series diode rectifier shown in Fig. 5-18 is represented as a series connection of a leakage inductor L_L of the transformers with an ac voltage source v_L. The reason for providing the ac voltage source to the equivalent model of the diode rectifier is that the electrolytic capacitor C_{dc} is directly connected to the dc terminals of the diode rectifier, as shown in Fig. 5-18. The active filter is controlled in such a way as to present zero impedance for the fundamental frequency and to act as a resistor with high resistance of $K\,\Omega$ for harmonic frequencies [13]. The ac voltage of the active filter, which is applied to a power line through the matching transformer, is given by

$$v_C = K i_{Sh} \qquad (5.19)$$

where i_{Sh} is the source harmonic current drawn from the utility. Note that v_C and i_{Sh} are instantaneous values. Figure 5-19 shows an equivalent circuit of the system considering only current and voltage harmonics. Referring to Fig. 5-19, if $K \gg \omega_h (L_S + L_L)$, the supply harmonic current and the active filter voltage are approximated by

$$I_{Sh} = \frac{V_{Sh} - V_{Lh}}{K + j\omega_h(L_S + L_L)} \approx \frac{V_{Sh} - V_{Lh}}{K} \qquad (5.20)$$

$$V_C = \frac{K}{K + j\omega_h(L_S + L_L)} (V_{Sh} - V_{Lh}) \approx (V_{Sh} - V_{Lh}) \qquad (5.21)$$

If the gain K is high enough, (5.20) implies that almost purely sinusoidal three-phase currents are drawn from the utility. Equation (5.21) suggests that no harmonic voltage appears upstream of the active filter or at the utility-consumer point of

Table 5.2. Specification of the series active filter shown in Fig. 5-18

Voltage rating (grid side)	12.5 V
Current rating (grid side)	57.7 A
dc bus voltage	300 V

5.3. SERIES ACTIVE FILTER INTEGRATED WITH A DOUBLE-SERIES DIODE RECTIFIER

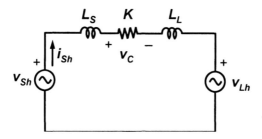

Figure 5-19. Single-phase equivalent circuit considering only of harmonics.

common coupling (PCC) as long as $V_{Sh} = 0$. The 15-µF commutation capacitor C, which is connected across each diode, contributes to a significant reduction of the required rating of the active filter [10]. If the commutation capacitor were removed from each diode, the ac voltage source v_L would be a twelve-step waveform, and so

$$v_L = \sqrt{2} V_S (\sin \omega t + \frac{1}{11} \sin 11\omega t + \frac{1}{13} \sin 13\omega t + \ldots) \qquad (5.22)$$

The 11th-harmonic voltage is the most dominant component in v_L from a theoretical point of view. Equations (5.20) and (5.22) suggest that the gain K should be more than 5 Ω (250% on a 200-V, 60-Hz, 20-kVA base) to reduce the 11th-harmonic current contained in i_{Sh} to 3 % of the fundamental current.

5.3.1. The First-Generation Control Circuit

5.3.1.A. Circuit Configuration and Delay Time

Figure 5-20 shows a block diagram of the first-generation control circuit proposed in [10]. The p-q transformation circuit converts three-phase supply currents i_{Sa}, i_{Sb}, and i_{Sc} into the instantaneous active current i_p and the instantaneous reactive current

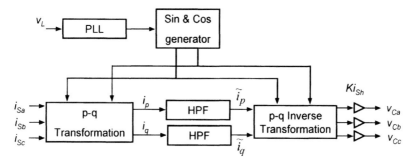

Figure 5-20. First-generation control circuit block diagram [10] (HPF = high-pass filter).

i_q. The PLL circuit generates together with the "sin and cos generator" two signals equal to $\sin(\omega t)$ and $\cos(\omega t)$, where ω is the line angular frequency and ϕ is a generic phase angle guaranteeing that the generated $\sin(\omega t + \phi)$ is in phase with the a-phase voltage. Therefore, after the transformation, the fundamental components in i_{Sa}, i_{Sb}, and i_{Sc} correspond to dc components in i_p and i_q, and the harmonics to the ac components. Two first-order, high-pass filters (HPFs) with a cutoff frequency of 2 Hz extract the ac components $\tilde{i}_p$ and $\tilde{i}_q$ from i_p and i_q, respectively. Then, the p-q inverse transformation produces their supply harmonic currents. Each harmonic current is amplified by the gain K, and then it is applied to the gate control circuit for each PWM converter as a voltage reference v_c^*. Each transformation circuit in Fig. 5-20 has four multiplying D/A converters.

Fig. 5-21 shows the circuit configuration of the multiplying D/A converter (Analog Devices: AD7545). The combination of a resistor for feedback (RFB) and a capacitor (C) behaves like a low-pass filter for eliminating high-frequency components that result from the so-called "zero-order hold" of the 12-bit digital "sin and cos" generators, thus making the output signal continuous. However, it produces a delay time that is dependent on the time constant determined by the RFB and C. Furthermore, a sampling time of 4 µs in the 12-bit digital signal processing also causes an additional delay time. Therefore, the control circuit produces a delay time as large as 35 µs. Moreover, the PWM converters using IGBTs have a blanking time of 5 µs, so that the overall delay time of the system reaches more than 40 µs.

2.3.1.B. Stability of the Active Filter

When attention is paid to harmonic components to be eliminated, the system behaves like a closed-loop control system, as shown in Fig. 5-22. Here, the system inductance L is the sum of the supply inductance L_S and the leakage inductance of the double-series transformer, L_l. For the sake of simplicity, the overall delay time of the system is assumed to be a constant value τ. The open-loop transfer function in Fig. 5-22 is given by

$$G(s) = \frac{K}{sL} e^{-s\tau} \qquad (5.23)$$

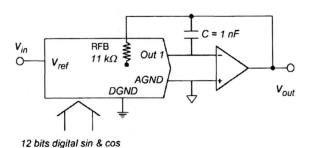

Figure 5-21. Multiplying D/A converter.

5.3. SERIES ACTIVE FILTER INTEGRATED WITH A DOUBLE-SERIES DIODE RECTIFIER

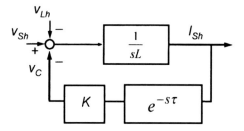

Figure 5-22. Control diagram of the system with a constant delay time τ.

From the Nyquist stability criterion, the stable operation of the system must meet the following condition

$$\frac{\tau K}{L} < \frac{\pi}{2}$$

Equation (5.24) suggests that the system is always stable if $\tau = 0$. In this ideal situation, K can be any value and the filter will operate as an ideal harmonic blocker. However, the existence of a delay time may cause instability, so that a compromise or trade-off exists between the gain K and the delay time τ. A large delay time limits the gain K and, therefore, it would be difficult to obtain satisfactory characteristics if the system inductance L (= $L_S + L_L$) is low. Fig. 5-23 shows relationships between the system inductance and the critical delay time as a parameter of the gain. In this figure, stable operation is only achievable if, for a given inductance L, the delay time is below the line for a given gain K. For example, if $L = 6\%$ and $K = 250\%$, the system will be stable if the time delay is bellow 100 µs.

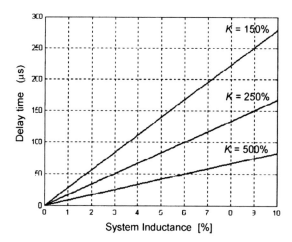

Figure 5-23. Critical delay time of the series active filter.

The 5-kW ac/dc power conversion system described in [10] has the following circuit constants: $L_S = 0.14$ mH, $L_L = 1.2$ mH, and $K = 27$ Ω. If $\tau = 40$ μs, then

$$\frac{\tau K}{L_S + L_L} = \frac{40 \cdot 10^{-6} \cdot 27}{1.34 \cdot 10^{-3}} = 0.81 < \frac{\pi}{2} \tag{5.25}$$

This indicates that the 5-kW system is stable enough. In fact, experimental results have confirmed stable operation, as shown in [10]. The 20-kW system described in Fig. 5-18 has the following circuit constants: $L_S = 0.05$ mH, $L_L = 0.08$ mH, and $K = 5$ Ω. Like the 5-kW system, if $\tau = 40$ μs, then

$$\frac{\tau K}{L_S + L_L} = \frac{40 \cdot 10^{-6} \cdot 5}{0.13 \cdot 10^{-3}} = 1.54 \approx \frac{\pi}{2} \tag{5.26}$$

In this case, even a delay time as short as 40 μs may cause a stability problem. A fully digital control circuit would have a delay time of more than 50 μs even if the leading-edge microprocessors and/or digital signal processors were applicable.

Fig. 5-24 shows experimental waveforms of the 20-kW system after the series active filter was started, in which the first-generation control circuit was applied. It shows that serious oscillation around 3.5 kHz occurred in voltage and current. Fortunately or unfortunately, this experimental result agrees with the analytical result obtained from (5.26). The control circuit with a delay time as short as 40 μs causes the stability problem, and it is not applicable to the 20-kW system, which has such a low system inductance. Thus, it is necessary to solve the stability problem.

5.3.2. The Second-Generation Control Circuit

Figure 5-25 shows a block diagram of the proposed control circuit capable of improving the system stability. The control circuit extracts the fundamental components from detected supply currents by using the p-q and inverse transformations as follows. The dc components, $\bar{i}_p$ and $\bar{i}_q$, are extracted from i_p and i_q by two first-order, low-pass filters (LPFs). Then, the p-q inverse transformation produces the supply fundamental currents i_{S1a}, i_{S1b}, and i_{S1c}. They are subtracted from the supply currents i_{Sa}, i_{Sb}, and i_{Sc}, thus resulting in the supply harmonic currents.

The conventional and proposed control circuits have almost the same characteristics except for the effect of the delay time existing in the p-q and inverse transformation circuits. In the first-generation control circuit, all the frequency components flow through the two transformation circuits, so that the delay time hidden in the transformation circuits causes the instability shown in Fig. 5-24. On the other hand, in the second-generation control circuit, the fundamental-frequency component flows through the transformation circuit, whereas high-frequency components are bypassed from them. Since the fundamental-frequency component is in a low-frequency range, a delay time as short as 40 μs makes no contribution to causing instability in a high-frequency range.

5.3. SERIES ACTIVE FILTER INTEGRATED WITH A DOUBLE-SERIES DIODE RECTIFIER

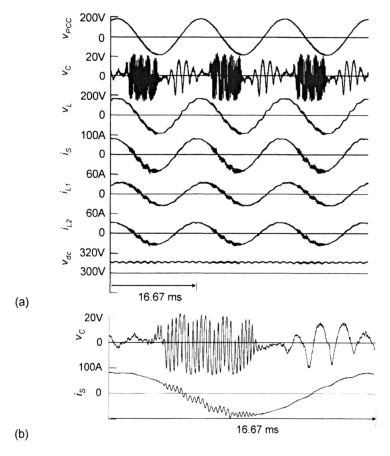

Figure 5-24. Experimental waveforms for a 20-kW system after starting the active filter with the first-generation control circuit.

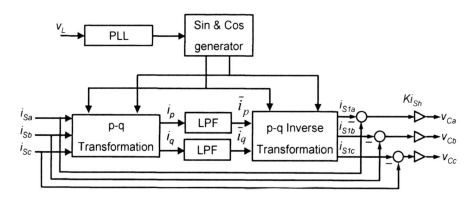

Figure 5-25. Second-generation control circuit.

5.3.3. Stability Analysis and Characteristics Comparison

In this section, the delay time existing in the control circuit is discussed. The system stability is investigated via frequency responses of the open-loop transfer function.

5.3.3.A. Transfer Function of the Control Circuits

Figure 5-26 shows a closed-loop control system when the first-generation control circuit explained in Section 5.3.1 is used. Here, G_{d1} and G_{d2} are transfer functions corresponding to the delay caused by the multiplying D/A converters in the p-q and inverse transformation circuits. G_{d3} corresponds to the delay coming from the PWM converters. These transfer functions are assumed to be first-order delays:

$$G_{d1}(s) = \frac{\omega_{cr}}{s + \omega_{cr}} \quad (5.27)$$

$$G_{d2}(s) = \frac{\omega_{cr}}{s + \omega_{cr}} \quad (5.28)$$

$$G_{d3}(s) = \frac{\omega_{cp}}{s + \omega_{cp}} \quad (5.29)$$

The transfer function of the harmonic-current-extracting circuit including a high-pass filter can be expressed as

$$D(s) = \frac{s - \omega_1}{s - j\omega_1 + \omega_c} G_{d1}(s) G_{d2}(s) \quad (5.30)$$

where ω_1 ($= 2\pi 60$ rad/s) is the fundamental frequency, and ω_c ($=10$ rad/s) is the cut-off frequency of the high-pass filter. In this system, instability occurs in a high-frequency range as shown in Fig. 5-24. Since ω_1 and ω_c are in a low-frequency range,

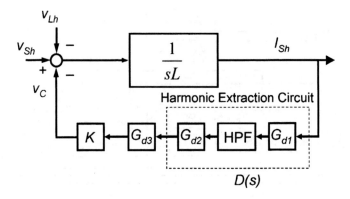

Figure 5-26. Control system diagram when the first-generation control circuit is used.

5.3. SERIES ACTIVE FILTER INTEGRATED WITH A DOUBLE-SERIES DIODE RECTIFIER

they do not influence the system stability, and so they can be neglected in the following analysis. The transfer function of the high-pass filter is unity in a high-frequency range of $\omega \gg \omega_c$. Thus, it is reasonable to approximate $D(s) = G_{d1}(s)G_{d2}(s)$, so that the open-loop transfer function $G(s)$ is given by

$$G(s) = \frac{K}{sL} G_{d1}(s)G_{d2}(s)G_{d3}(s) \quad (5.31)$$

Equation (5.31) shows that three delay components exist in the open-loop transfer function of the first-generation control circuit.

Applying the second-generation control circuit makes $G_{d1}(s)$ and $G_{d2}(s)$ produce no effect on stability, as explained in the following discussion. Figure 5-27 shows the closed-loop control system, in which the transfer function of the harmonic-current-extracting circuit is given by

$$D(s) = 1 - \frac{\omega_c}{s - j\omega_1 + \omega_c} G_{d1}(s)G_{d2}(s) \quad (5.32)$$

Substituting s for $j\omega$ in (5.32) yields frequency responses of $D(\omega)$. The transfer function $D(\omega)$ approaches unity when ω is much larger than ω_c. This implies that $G_{d1}(s)$ and $G_{d2}(s)$ are negligible as long as attention is paid to the system stability. Hence, the open-loop transfer function of the system approximates

$$G(s) = \frac{K}{sL} G_{d3}(s) \quad (5.33)$$

5.3.3.B. Characteristics Comparisons

Figures 5-28 and 5-29 show the Bode diagrams for the open-loop transfer functions presented in (5.31) and (5.33). The control gain is set to $K = 5\,\Omega$, $\omega_{cr}/2\pi = 8.8$ kHz,

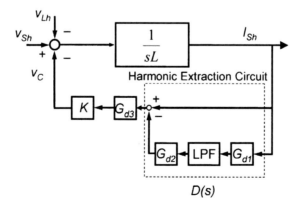

Figure 5-27. Control system diagram when the second-generation control circuit is used.

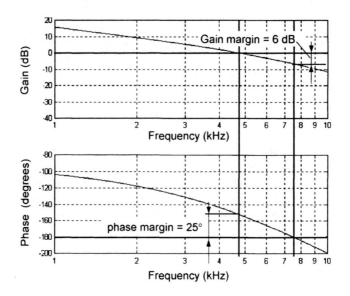

Figure 5-28. Open-loop frequency responses in the case of using the first-generation control circuit with $K = 5\ \Omega$.

and $\omega_{cp}/2\pi = 44.2$ kHz for calculation. The system using the first-generation control circuit has a gain margin as small as 6 dB and a phase margin as small as 25°. Although the system is theoretically stable, such a small phase margin causes poor damping performance and, therefore, oscillation may occur at a specific frequency in a range of 3 to 5 kHz.

On the other hand, Fig. 5-29 shows the Bode diagram of the system using the second-generation control circuit. The phase margin of the system is as large as 83°. This results from having eliminated the effect of G_{d1} and G_{d2} from the harmonic extraction circuit. The closed-loop system is a second-order system with a damping factor of $\xi = 1.43$, which corresponds to an overdamping condition. As a result, no overshoot occurs in the supply current. For this reason, a feedback gain as high as 5 Ω (250% on a 200-V, 60-Hz, 20-kVA base) is applicable, thus making the system performance satisfactory. The proposed control circuit is superior to the first-generation control circuit in terms of both stability and filtering characteristics.

5.3.4. Design of a Switching-Ripple Filter

5.3.4.A. Design Principle

In Fig. 5-18, three Y-connected capacitors C_r are installed at the utility-consumer point of common coupling (PCC) to reduce switching voltage and current ripples generated from pulse-width modulation (PWM). Fig. 5-30 shows an equivalent circuit of the system with focus on switching ripples, where the double-series diode rectifier is represented as a leakage inductor L_L and the supply is represented as an

5.3. SERIES ACTIVE FILTER INTEGRATED WITH A DOUBLE-SERIES DIODE RECTIFIER

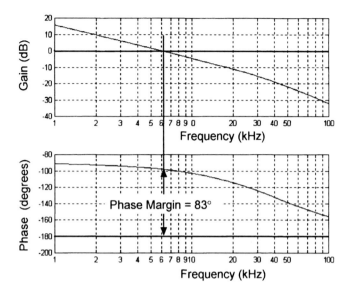

Figure 5-29. Open-loop frequency responses in the case of using the second-generation control circuit with $K = 5\,\Omega$.

inductor L_S. The capacitors C_r are connected in parallel with the supply to reduce the supply impedance seen upstream of the active filter in a switching-frequency range. Therefore, the amount of switching-voltage ripple appearing at the PCC can be given by

$$\frac{V_{PCSW}(s)}{V_{CSW}(s)} = \frac{1}{s^2 L_L C_r + 1 + \frac{L_L}{L_s}} \approx \frac{1}{s^2 L_L C_r} \qquad (5.34)$$

Under the operating conditions of $L_L = 0.08$ mH and $\omega/2\pi = 20$ kHz, C_r can be designed to obtain $V_{PCSW}/V_{CSW} \approx 0.1$, thus leading to 10 μF. If the peak value of

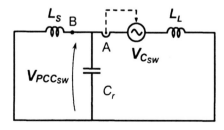

Figure 5-30. Single-phase equivalent circuit with focus on switching voltage and current ripples.

250 HYBRID AND SERIES-ACTIVE FILTERS

the ripple voltage produced by the active filter is 22.5 V, it is reduced to only 2 V at PCC.

5.3.4.B. Effect on the System Stability

When the switching-ripple filter is installed, the system stability depends significantly on the point of supply current detection. If the supply current is detected at point A in Fig. 5-30, the open-loop transfer function of the system in (5.33) becomes

$$G(s) = \frac{K}{sL_L + \dfrac{sL_s}{s^2 L_s C_r + 1}} G_{d3}(s) \qquad (5.35)$$

The only difference between (5.33) and (5.35) is the system impedance seen upstream of the active filter. The filter capacitor C_r reduces the supply impedance seen upstream of its installation site when attention is paid to an equivalent switching-ripple frequency of 20 kHz. Thus, the effective line inductance in a high-frequency range also decreases from $L_L + L_S$ to L_L. However, the inductance L_L may be still large enough to maintain the system stability, because (5.35) tells us that the system remains stable with a phase margin as large as 76°. Note that a parallel resonance seen from the active filter may occur between C_r and the supply inductance L_S. However, this parallel resonance results in high impedance at the resonant frequency, thus producing no bad effect on the system stability. On the other hand, a set of C_r and L_L also appears to be forming a series resonance in Fig. 5-30. However, the resonant frequency of C_r and L_S around 5.6 kHz is lower than that of C_r and L_L around 7.1 kHz. This means that the inductance L_S is more dominant than C_r at the series resonant frequency of C_r and L_L. Therefore, this series resonance is not a serious problem from a practical point of view. Although the gain of the open-loop transfer function in (5.35) is still high in some frequency ranges around the resonant frequency, its phase ranges from −90° to 90°, leading to a stable operation. Moreover, the series-resonant frequency around 5.6 kHz is much lower than the switching frequency of 20 kHz, so that the active filter generates no ripple current around 5.6 kHz.

On the other hand, if the supply current is detected at point B, the open-loop transfer function of the system becomes

$$G(s) = \frac{K}{sL_S} F(s) G_{d3}(s) \qquad (5.36)$$

where $F(s)$ is the transfer function from the active filter voltage to the PCC voltage,

$$F(s) = \frac{K}{s^2 L_L C_r + 1 + \dfrac{L_L}{L_s}} \qquad (5.37)$$

5.3. SERIES ACTIVE FILTER INTEGRATED WITH A DOUBLE-SERIES DIODE RECTIFIER

It is clearly seen from (5.37) that the phase characteristic of $F(s)$ is very poor, especially in a high-frequency range. Thus, detecting of the supply current at point B would make the system unstable.

5.3.4.C. Experimental Testing

In order to make the effect of the ripple filter more clear, the supply inductance L_S in Fig. 5-18 was changed from 0.9% to 5% ($L_S = 0.258$ mH) in this experiment. Figures 5-31 and 5-32 show experimental waveforms under the operation of the series active filter before and after the switching-ripple filter is installed. Figures 5-33 and 5-34 are frequency spectra of the waveforms in Fig. 5-32 and Fig. 5-33. Before installing the switching-ripple filter, the supply impedance is much higher than that of the double-series transformer at the equivalent switching frequency of 20 kHz. Thus, a nonnegligible amount of switching voltage appears at the PCC. After installing the ripple filter, neither switching voltages nor currents appear at the PCC. However, the 11th- and 13th-harmonic voltage and current are not influenced by installing the ripple filter, as shown in Fig. 5-33 and Fig. 5-34.

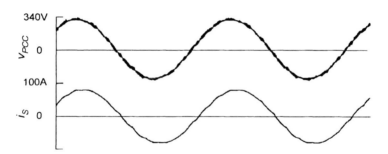

Figure 5-31. Experimental waveforms before installing the ripple filter, in which v_{PCC} is the line-to-line voltage.

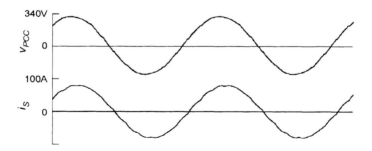

Figure 5-32. Experimental waveforms after installing the ripple filter, in which v_{PCC} is the line-to-line voltage.

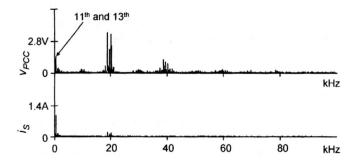

Figure 5-33. Spectra in rms value before installing the ripple filter.

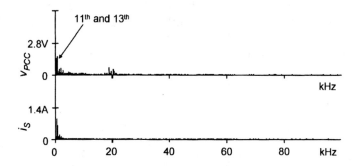

Figure 5-34. Spectra in rms value after installing the ripple filter.

5.3.5. Experimental Results

Figures 5-35 and 5-36 show experimental waveforms under $L_S = 0.05$ mH (0.9%) before and after the series active filter is started. Figures 5-37 and 5-38 are frequency spectra of i_S in Fig. 5-35 and Fig. 5-36.

The proposed control circuit is implemented by a hybrid digital and analog circuit using multiplying D/A converters, and its control gain is set to $K = 5$ Ω in this experiment. A digital filter with a cutoff frequency of 4 kHz is used in taking the waveforms of v_C and v_L, in order to reject the switching ripples and high-frequency noises. Before starting the active filter, a large amount of 11th-harmonic current is included in i_S. Since the rectifier currents i_{L1} and i_{L2} are also distorted, a ripple voltage of 3 V appears in the dc-link voltage v_{dc}. After starting the active filter, the supply current looks sinusoidal because the 11th-harmonic current in i_S is reduced to 1.6 %. Table 5.3 shows the total harmonic distortion (THD) of i_S and the ratio of each harmonic current with respect to the fundamental current contained in i_S. The third-, fifth- and seventh-harmonic currents are smaller than 1%, and the THD is decreased from 31% to 2.0%. Experimental results verify that the series active filter using the second-generation control circuit can effectively reduce supply current harmonics without any instability problem.

5.4. COMPARISONS BETWEEN HYBRID AND PURE ACTIVE FILTERS   253

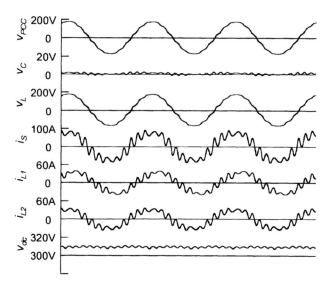

Figure 5-35. Experimental waveforms before starting the series active filter.

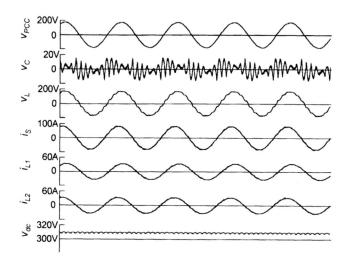

Figure 5-36. Experimental waveforms after starting the series active filter.

5.4. COMPARISONS BETWEEN HYBRID AND PURE ACTIVE FILTERS

Active filters based on leading-edge power electronics technology can be classified into pure active filters and hybrid active filters. The reader may have the following simple question in mind, "Which is preferred, a pure active filter or a hybrid active filter?" Fortunately or unfortunately, engineering has no versatile techniques in terms

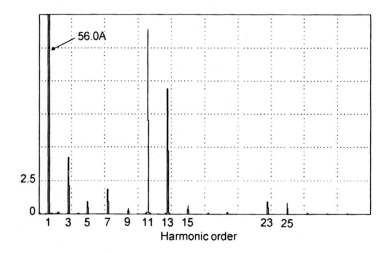

Figure 5-37. Spectrum of supply current in rms value before starting the active filter.

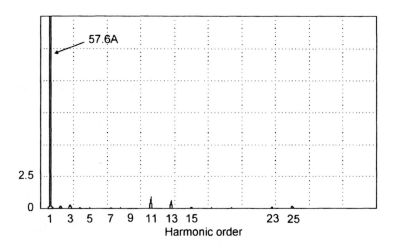

Figure 5-38. Spectrum of supply current in rms value after starting the active filter.

Table 5.3. Supply current harmonics and THD before and after starting the active filter, expressed as the harmonic-to-fundamental current ratio (%)

	3rd	5th	7th	11th	13th	23rd	25th	THD
Before	7.5	1.7	3.3	24.8	16.8	1.7	1.5	31.2
After	0.5	0.1	0.1	1.6	1.1	0.2	0.3	2.0

of cost and performance. It is based on a compromise or a trade-off between cost and performance. Therefore, a comprehensive answer of the authors to the question depends strongly on the function(s) of active filters intended for installation.

A pure active filter provides multiple functions such as harmonic filtering, damping, isolation and termination, load balancing, reactive-power control for power-factor correction and voltage regulation, voltage-flicker reduction, and/or their combinations. A cluster of the above functions can be represented by "power conditioning." Hence, the pure active filter is well suited to "power conditioning" of nonlinear loads such as electric ac arc furnaces and utility/industrial distribution feeders. On the other hand, a hybrid active filter consists of an active filter and a single-tuned filter that are directly connected in series without a transformer. This hybrid filter is exclusively devoted to "harmonic filtering" of three-phase diode rectifiers, because it has no capability of reactive-power control from a practical point of view, although it has from a theoretical point of view.

This section deals with a low-cost transformerless hybrid active filter, comparing it with a pure active filter [16,17].

5.4.1. Low-Voltage Transformerless Hybrid Active Filter

Figure 5-39 shows the circuit configuration of a hybrid active filter connected in parallel with a three-phase diode rectifier rated at 480 V and 20 kW. The hybrid filter is directly connected to the 480-V industrial distribution feeder without any transformer. It is designed to reduce the total harmonic distortion (THD) of i_S below

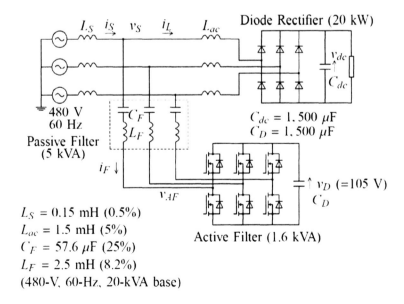

Figure 5-39. Circuit configuration of the 480-V hybrid active filter.

5%. The hybrid filter consists of an active filter based on a 1.6-kVA voltage-source PWM converter with a carrier frequency of 10 kHz, and a 5-kVA passive filter. The passive filter is a three-phase, single-tuned filter tuned to the seventh-harmonic frequency with a quality factor of $Q = 22$. The passive filter exhibits poor filtering performance in a range of low-order harmonic frequencies, except around the seventh-harmonic frequency.

The passive filter and the active filter are directly connected in series with each other. This "hybrid" configuration results in a dc voltage as low as 105 V across the dc bus of the active filter. Moreover, no switching-ripple filter is required for the hybrid filter because the passive filter presents high impedance around 10 kHz. The diode rectifier has an ac inductor of $L_{ac} = 5\%$ at its ac side. This ac inductor is indispensable to achieve proper operation of the hybrid filter because no inductor is installed on the dc side of the diode rectifier.

Figure 5-40 shows the control system of the hybrid filter. The control system has the following three control functions: feedback control, feedforward control, and dc-voltage control. The feedback control forces all the harmonic currents contained in i_L to flow into the hybrid filter, whereas it forces no harmonic current to flow from the power system into the hybrid filter. This improves the filtering performance of the passive filter, and prevents the passive filter from being overloaded and ineffective. Moreover, the feedback control makes the active filter act as a

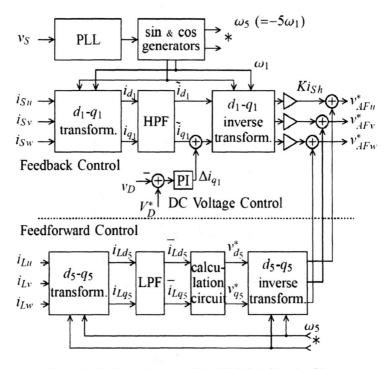

Figure 5-40. Control system of the 480-V hybrid active filter.

damping resistor for low-order harmonic frequencies, so that no harmonic resonance occurs between the passive filter and the power system inductance L_S. As a result, this hybrid filter does not require elaborate investigations into the possibility of harmonic resonance and overloading before installing it.

The feedforward control forces the fifth-harmonic current contained in i_L to flow actively into the hybrid filter. This control part is accomplished, following a principle of selective harmonic compensation, similar to that used in Chapter 4. The dc voltage control makes the active filter build up and regulate its dc capacitor voltage by itself without any external power supply. A proportional-plus-integral (PI) controller is used to do it [16].

The volt-ampere rating required for the active filter in the 480-V hybrid filter is

$$P_{HF} = \sqrt{3} \times \frac{V_{dc}}{\sqrt{2}} \times \frac{I_{Fmax}}{\sqrt{2}}$$
$$= \sqrt{3} \times 74^V \times 13.0^A \quad (5.38)$$
$$= 1.6 \text{ kVA}$$

where I_{Fmax} is a maximum value of the filter current I_F.

The resonant frequency of the passive filter in the hybrid filter is given by

$$f = \frac{1}{2\pi \sqrt{L_F C_F}} \quad (5.39)$$

It is a well-known fact that the passive filter presents good filtering characteristics around the resonant frequency. The seventh-harmonic frequency was selected as the resonant frequency, instead of the fifth-harmonic frequency, for the following lucid reasons:

- The passive filter tuned to the seventh-harmonic frequency is less expensive and less bulky than that tuned to the fifth-harmonic frequency as long as both filters have the same filter inductor L_F.
- The passive filter tuned to the seventh-harmonic frequency offers less impedance to the 11th- and 13th-harmonic components, compared to that tuned to the fifth-harmonic frequency.
- The feedforward control combined with the feedback control makes a significant contribution to improving the filtering performance at the most dominant fifth-harmonic frequency.

The characteristic impedance of the passive filter, Z, is given by

$$Z = \sqrt{\frac{L_F}{C_F}} \quad (5.40)$$

This impedance determines filtering performance at harmonic frequencies except for the resonant frequency. Generally speaking, the characteristic impedance should be as low as possible to obtain better filtering performance. This implies that the capacitance value of C_F should be as large as possible, and the inductance value of L_F should be as small as possible. A lower characteristic impedance reflects a lower dc capacitor voltage as well as lower EMI emissions by the hybrid filter. This allows the hybrid filter to use low-voltage MOSFETs that are less expensive and more efficient than high-voltage ones. On the other hand, a low characteristic impedance has the following disadvantages:

- A large capacitance value of C_F makes it bulky and expensive.
- A large amount of leading reactive current flows into the hybrid filter.
- A smaller inductance value of L_F increases switching ripples. The ratio of the switching-ripple voltage contained in the supply voltage, v_{Ssw}, with respect to that at the ac side of the active filter, v_{AFsw}, can be calculated under an assumption of $L_S \ll L_{ac}$ as follows:

$$\frac{v_{Ssw}}{v_{AFsw}} \approx \frac{L_S}{L_S + L_F}$$

Hence, a trade-off or a compromise exists in the design of the characteristic impedance. In other words, the above-mentioned criteria should also be considered when selecting the value of L_F.

5.4.2. Low-Voltage, Transformerless, Pure Shunt Active Filter

Figure 5-41 shows a pure active filter integrated into the same diode rectifier as that in Fig. 5-39. The pure active filter consists of an inductor L_F and a PWM converter that are directly connected in series. Strictly speaking, this inductor is not a filter inductor but a commutation inductor. However, it has the same inductance value as the filter inductor L_F in Fig. 5-39. The hybrid filter in Fig. 5-39 can be divided into the following two parts connected with each other: the capacitor C_F and the pure active filter consisting of the inductor L_F and the PWM converter. This means that the hybrid filter in Fig. 5-39 can be considered as a series connection of the capacitor and the pure active filter. The pure filter is designed to reduce the total harmonic distortion of i_S below 5%, like the hybrid filter. Unfortunately, this "pure" configuration results in a dc voltage as high as 750 V across the dc bus of the active filter, so the active filter requires the 1.2-kV or higher voltage IGBT as a power device. The carrier frequency of the pure filter is 10 kHz, which is the same as that of the hybrid filter. The other parameters of the pure filter are the same as those of the hybrid filter. The required rating of the 480-V pure filter, P_{PF} is given by

$$\begin{aligned} P_{PF} &= \sqrt{3} \times 530^V \times 13.7^A \\ &= 12.6 \text{ kVA} \end{aligned} \quad (5.41)$$

Fig. 5-42 shows the control system of the pure filter. The control system is almost the same as that of the hybrid filter. The pure filter has the same proportional

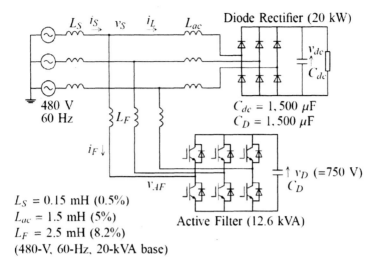

Figure 5-41. Circuit configuration of the 480-V pure shunt active filter.

and integral gains as the hybrid filter. However, the following differences exist, compared to the control system of the hybrid filter:

- The supply voltage v_S is detected and added to the voltage reference of the pure filter, v_{AF}^*, in order to compensate for an effect of v_S on current controllability.
- The electrical quantity controlled in the dc voltage control is not Δi_{q1} but Δi_{d1}, because precise adjustment of a small amount of active power enables to regulate the dc voltage of the pure filter.

The feedforward control in the pure filter has the same task as that in the hybrid filter. It calculates the voltage appearing across the ac inductor L_F, assuming that all of the fifth-harmonic current included in the load current flows into the pure filter.

5.4.3. Comparisons Through Simulation Results

Figure 5-43 shows simulated waveforms of the 480-V hybrid filter in a steady state. The feedback gain of the active filter, K is set to 39 (340%) so that the hybrid filter provides good stability. The supply current i_S becomes nearly sinusoidal. The dc capacitor voltage of the active filter is set to 105 V. The low-voltage MOSFETs used here are easily available from the market at low cost.

Figure 5-44 shows simulated waveforms of the 480-V pure filter under the same conditions as Fig. 5-43. The waveform of i_S is also nearly sinusoidal. Note that the dc capacitor voltage for the 480-V pure filter is observed to be 750 V from computer simulation. This means that the 1.2-kV or higher voltage IGBTs are required as the power devices in the pure-filter configuration.

260 HYBRID AND SERIES-ACTIVE FILTERS

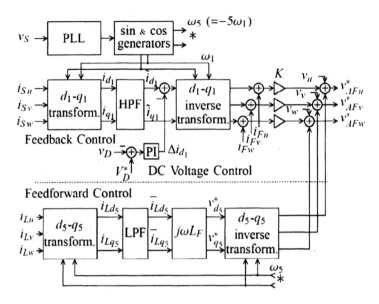

Figure 5-42. Control system of the 480-V pure shunt active filter.

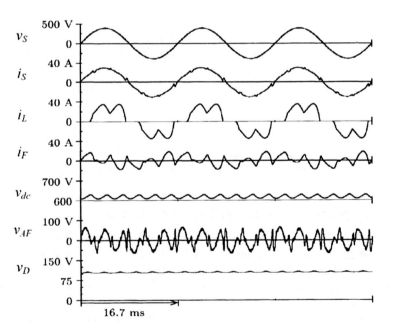

Figure 5-43. Simulated waveforms of the 480-V hybrid active filter.

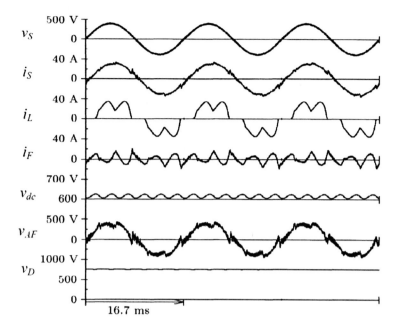

Figure 5-44. Simulated waveforms of the 480-V pure shunt active filter.

A first-order, low-pass filter with a cutoff frequency of 2 kHz is used to eliminate the switching ripples from v_{AF}, thus making the waveform clear. Note that the low-pass filter for signal processing is integrated into the software package. The following significant difference exists in the waveforms of v_{AF} between Fig. 5-43 and Fig. 5-44: no fundamental voltage appears across the ac terminals of the active filter in the hybrid filter because the supply voltage v_S is applied across the filter capacitor C_F. On the other hand, the supply-line-to-neutral voltage as high as 277 V ($= 480/\sqrt{3}$) appears in the waveform of v_{AF} in the case of the pure filter. This is an essential difference in operating principle and performance between the hybrid filter and the pure filter.

When attention is paid to switching ripples contained in v_S and i_S of Fig. 5-43 and Fig. 5-44, it is clear that the switching ripples in the hybrid filter are much smaller than those in the pure filter. Note that no additional switching-ripple filter is installed in both cases. The 10-kHz ripple voltage contained in v_S is 0.38 V (0.14%) in the hybrid filter, whereas it reaches 4.0 V (1.4%) in the pure filter. The 10-kHz current ripple present in i_S is small enough to be neglected in the hybrid filter, whereas it reaches 0.4 A (1.8%) in the pure filter.

5.5. CONCLUSIONS

The first section of this chapter presented the basic principles of a series active filter. In the second section a hybrid series active filter and shunt passive filter were analyzed. This hybrid approach has shown that it is possible to improve significantly the filtering performance of a shunt passive filter when combined with a very small series active filter.

Section 5.3 discussed the stability of the ac/dc power conversion system characterized by integration of a series active filter with a double-series diode rectifier, taking into account the delay time inherent in the harmonic current-extracting circuit. The analysis and experimental results have pointed out that even a delay time as short as 40 μs can cause a stability problem if the power conversion system is installed in a high-power system with a low system inductance. Therefore, a second-generation control circuit of the harmonic-extraction circuit, capable of eliminating the effect of the delay time, was analyzed. An increased gain and phase margin clearly show that the second-generation control circuit significantly improves the system stability, compared with the first-generation configuration. Moreover, a switching-ripple filter has been designed and evaluated. Experimental results obtained from a 20-kW laboratory system verify the viability and effectiveness of the series active filter, which can effectively reduce supply current harmonics without any instability problem.

REFERENCES

[1] M. Aredes and E. H. Watanabe, "New Control Algorithms for Series and Shunt Three-Phase Four-Wire Active Power Filters," *IEEE Transactions on Power Delivery*, vol. 10, no. 3, July 1995, pp. 1649–1656.

[2] F. Z. Peng, H. Akagi, and A. Nabae, "New Approach to Harmonic Compensation in Power Systems—A Combined System of Shunt Passive and Series Active Filters," in *Conference Record IEEE-IAS Ann. Meeting,* 1988, pp. 874–880. (*IEEE Transactions on Industry Applications,* vol. 26, No. 6, Nov/Dec, 1990, pp. 983–990.)

[3] F. Z. Peng, H. Akagi, and A. Nabae, "Compensation Characteristics of the Combined System of Shunt Passive and Series Active Filters," *IEEE Transactions on Industry Applications,* vol. 29, no. 1, pp. 144–152, 1993.

[4] H. Fujita and H. Akagi, "A Practical Approach to Harmonic Compensation in Power Systems: Series Connection of Passive and Active Filters," in *Conference Record IEEE-IAS Annual Meeting,* 1990, pp. 1107-1112. (*IEEE Transactions on Industry Applications,* vol. 27, no. 6, pp. 1020–1025, 1991.)

[5] H. Akagi and H. Fujita, "A New Power Line Conditioner for Harmonic Compensation in Power Systems," *IEEE Transactions on Power Delivery,* vol. 10, no. 3, July 1995, pp. 1570–1575.

[6] T. Okuyama, H. Nagase, and Y. Kubota, "High Performance Ac Motor Speed Control System Using GTO Converters," in *Proceedings of the 1983 International Power Electronics Conference,* Tokyo, Japan, pp. 720–731, 1983.

[7] J. W. Dixon, A. B. Kulkarni, M. Nishimoto, and B. T. Ooi, "Characteristics of a Controlled-Current PWM Rectifier-Inverter Link," in *Proceedings of the 1986 IEEE/IAS Annual Meeting,* pp. 685–691, 1986.

[8] R. Wu, S. B. Dewan, and G. R. Slemon, "A PWM ac to dc Converter with Fixed Switching Frequency," in *Proceedings of the 1988 IEEE/IAS Annual Meeting,* pp. 706–711, 1988.

[9] T. G. Habetler, "A Space Vector-Based Rectifier Regulator for ac/dc/ac Converters," *IEEE Transactions on Power Electronics*, vol. 8, no. 1, pp. 30–36, 1993.

[10] H. Fujita and H. Akagi, "An Approach to Harmonic Current-free ac/dc Power Conversion for Large Industrial Loads: The integration of a series active filter with a double-series diode rectifier," *IEEE Transactions on Industry Applications*, vol. 33, no. 5, pp. 1233–1240, 1997.

[11] H. Akagi, "New Trends in Active Filters for Power Conditioning," *IEEE Transactions on Industry Applications.*, vol. 32, no. 6, pp. 1312–1322, 1996.

[12] F. Z. Peng, "Application Issues of Active Power Filters," *IEEE IAS Magazine*, vol. 4, no. 5, pp. 21–30, 1998.

[13] F. Z. Peng, H. Akagi, and A. Nabae, "A New Approach to Harmonic Compensation in Power Systems—A Combined System of Shunt Passive and Series Active Filters," *IEEE Transactions on Industry Applications*, vol. 26, no. 6, pp. 983–990, 1990.

[14] S. Srianthumrong, H. Fujita, and H. Akagi, "Stability Analysis of a Series Active Filter Integrated with a Double-Series Diode Rectifier," *IEEE Transactions on Power Electronics*, vol. 17, no. 1, pp. 117–124, 2002.

[15] H. Akagi, "Active Harmonic Filters," *Proceedings of the IEEE*, vol. 93, no. 12, pp. 2128–2141, December 2005.

[16] S. Srianthumrong and H. Akagi, "A Medium-Voltage Transformerless ac/dc Power Conversion System Consisting of a Diode Rectifier and a Shunt Hybrid Filter," *IEEE Transactions on Industry Applications*, vol. 39, no. 3, pp. 874–882, 2003.

[17] H. Akagi, S. Srianthumrong, and Y. Tamai, "Comparisons in Circuit Configuration and Filtering Performance between Hybrid and Pure Shunt Active Filters," in *Conference Record IEEE-IAS Annual Meeting*, pp. 1195–1202, 2003.

CHAPTER 6

COMBINED SERIES AND SHUNT POWER CONDITIONERS

Combined series and shunt power conditioners are formed by two power converters connected back-to-back through a common dc link. Figure 6-1 shows the basic configuration of a generic series and shunt controller. The ac output of one converter is inserted in series with the power system, whereas the ac output of the second one is connected in parallel. Although two current-source converters (CSCs) could be employed in the basic configuration shown in Fig. 6-1, it is common to use two voltage-source converters (VSCs) connected back-to-back through a common dc voltage link. Note that the terms "voltage-source converters" and "voltage-fed converters" are synonymous, as are the terms "current-source converters" and "current-fed converters." This book uses "voltage-source converters" and "current-source converters," along with their abbreviations VSCs and CSCs, instead of "voltage-fed converters" and "current-fed converters," respectively. VSC technology is widely used and dominates the market. This chapter presents some combined series/shunt active power conditioners, all based on VSCs.

An important issue related to VSCs is the switching technique of the converter. A common switching technique is switching at the fundamental frequency, which is used in multipulse converters for very high power applications. Another common switching technique is pulse-width modulation (PWM), based on high-frequency switching [1]. However, PWM techniques generate higher switching losses in the power semiconductor devices than the multipulse converter approach.

In Chapter 4, converters with PWM current control were used in shunt active filter applications. The hysteresis-band control with minor current feedback loop is a PWM technique that forces the converter to act as a controlled *current* source. Another well-known PWM controller is the carrier-based sine-PWM control. This

Instantaneous Power Theory and Applications to Power Conditioning. By Akagi, Watanabe, & Aredes
Copyright © 2007 the Institute of Electrical and Electronics Engineers, Inc.

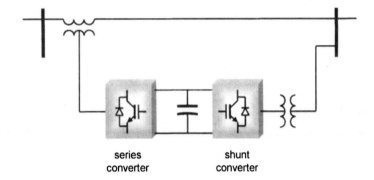

Figure 6-1. Combined series and shunt power conditioner.

switching technique is based on the comparison of a sinusoidal reference signal with a triangular carrier wave. If applied to a VSC, it makes the converter act as a controlled *voltage* source. Thus, PWM converters can represent ideally controlled voltage or current sources, which can drain controlled real (active) power and controlled imaginary (reactive) power independently.

On the other hand, switching at the fundamental frequency forces the converter to generate square-wave outputs, which have a fundamental component with an amplitude directly proportional to the dc voltage of the VSC. Thus, the frequency and phase angle of the outputs can be quickly changed by adjusting the switching instants, whereas their amplitude is changed only if the dc voltage varies. For a specific application, if feasible, PWM converters are preferred, since they can control the generated active and reactive power independently and faster than multipulse converters. The possibility of using multipulse converters or PWM converters will be addressed in the following sections, where series and shunt converters are combined in an integrated series/shunt power conditioner (compensator).

In Fig. 6-1, a dc capacitor is used in the common dc link as an energy storage element. Indeed, the main function of this dc capacitor is to impose dc *voltage* source characteristics on the dc link. Hence, for a long term, the power flowing into the dc link through the series converter must be balanced with the power flowing out through the shunt converter, and vice-versa. If desired, other types of energy storage systems or energy sources, or even dc loads, could be connected to this common dc link. Moreover, the PWM converters could also be separated and interconnected by dc cables or overhead dc lines, to form a dc transmission system. Such applications will not be covered in this book. The active power line conditioners presented in this chapter use the back-to-back arrangement as in Fig. 6-1.

In this chapter, equipment for compensation at the system frequency, known as the *unified power flow controller* (UPFC), which performs power flow control, reactive power compensation, and voltage regulation, is first presented. Then, the same back-to-back arrangement, controlled to perform voltage and current harmonic compensation, is presented. This new equipment, called the *unified power quality conditioner* (UPQC), is a combination of a series active filter for harmonic-voltage compensation and a shunt active filter for harmonic-current compensation. Finally, a universal compensator, the *universal active power line conditioner* (UPLC), aggregates all functions of the UPFC and the UPQC into a single power conditioner, fully based on the *p-q* Theory, is presented.

6.1. THE UNIFIED POWER FLOW CONTROLLER (UPFC)

The unified power flow controller (UPFC) is a FACTS device (flexible ac transmission systems) [2]. The FACTS concept is understood in such a general way as to comprise applications of power-electronics-based equipment to power flow control in a transmission line. Prior to the FACTS concept, electromechanical devices, like transformers or phase shifters, had been the only way to perform power flow control [3]. The phase shifting transformer is perhaps the oldest power flow controller [4]. However, it can regulate only the steady-state power flow, since it has a slow time response due to the inertia of its moving parts in the on-load-tap changer.

The use of FACTS devices for controlling load flow dynamically, damping sub-synchronous oscillations, regulating system voltage, and enhancing transient and dynamic stability, has been made possible by the emergence of high power thyristors [5,6,7,8]. A review of various approaches to static phase shifters is given in [9]. Different FACTS devices can be assembled by using thyristors [10]. The static var compensator (SVC) based on the thyristor-controlled reactor (TCR) and the thyristor-switched capacitor (TSC) or fixed capacitors has been widely applied in power systems for more than 25 years. Series compensation using the thyristor-controlled series capacitor (TCSC) [11], also known as the advanced series compensator [12, 13], has been in use for almost 15 years.

Three principal parameters—the *terminal voltage*, the *series impedance*, and the *phase angle displacement* (power angle)—determine the power flow through a transmission line ($P = V_1 V_2 \sin \delta / X_L$). It is a common sense to restrict the use of the name FACTS device only for equipment that can control one or more such parameters in *real time* [2]. The static var compensator (SVC) can control continuously the reactive power at the bus where it is connected. Hence, it can control continuously the amplitude of the voltage at the controlled ac bus. On the other hand, the thyristor-controlled series capacitor (TCSC) can control the equivalent impedance and static phase shifters can control the power angle. The main goal of these devices is to increase the usable power transmission capacity up to its thermal limit [14].

The second generation of FACTS device exploit the self-commutated power semiconductor devices and their use in PWM converters [15,16,17,18]. For instance, high power gate-turn-off thyristors (GTOs) have led to the development of a ±100 Mvar static synchronous compensator (STATCOM) that is equivalent to an *ideal* synchronous condenser [19]. A ±75 Mvar STATCOM based on integrated gate-commutated thyristors (IGCTs) has been operating since 2003 [20]. The convertible static converter (CSC) with two ±100 Mvar shunt and/or series compensation capability based on GTOs has been in operation since 2004 [21].

The unified power flow controller (UPFC) is a compensator formed by the combination of shunt and series VSC converters, as shown in Fig. 6-1. It can control the active (real) and reactive (imaginary) power through a transmission line and simultaneously regulate the voltage at an ac bus, as suggested in Fig. 6-2. It is a very fast acting device with high performance and flexibility. Thus, the UPFC is an advanced compensator under the FACTS concepts that offer new control capabilities to transmission systems [14,17,23,24,25,26].

6.1.1. FACTS and UPFC Principles

Both FACTS and UPFC concepts were first introduced by Hingorani and Gyugyi [2,14]. Generally, the FACTS concept involves modern power electronic devices for power flow control *in real time*. In other words, it increases system flexibility by means of using fast-acting power conditioners to adjust properly the power flow in the transmission system. Hence, a gain in stability margin is obtained, allowing the system to be operated safely under higher loading conditions. Ideally, FACTS devices could be installed in critical points of the transmission system, providing voltage support and reactive power compensation, as well as power flow control. The purpose is to increase the stability margin and to allow increasing power dispatch up to the thermal limits of line conductors and power ratings of generators and transformers.

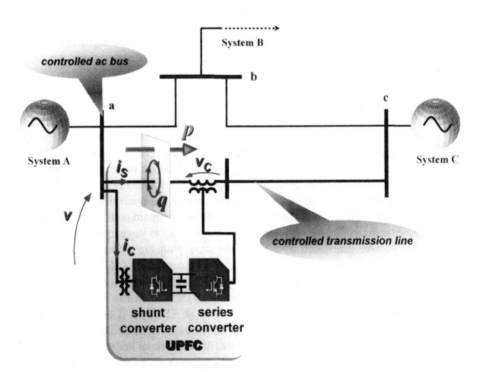

Figure 6-2. Combined series and shunt converters for FACTS application.

In contrast to thyristor-based compensators controlling only their firing angles, compensators based on self-commutated semiconductor devices can control switching patterns to behave as controlled current or voltage sources. Moreover, the VSC can drain real and imaginary powers, which are controlled independently of each other. Thus, each PWM converter in Fig. 6-2 has two degrees of freedom. The compensating current i_C of the shunt converter produces real (p_1) and imaginary (q_1) powers with the controlled bus voltage v, independent of each other. On the other hand, the compensating voltage v_C of the series converter produces, with the controlled transmission line current i_S, real (p_2) and imaginary (q_2) powers, which are also independent of each other. However, the limited energy-storage element, characterized by the capacitor in the common dc link, imposes a restriction on the four degrees of freedom (p_1, q_1, p_2, q_2) in the back-to-back configuration. In order to provide energy balance in the common dc link of the UPFC, the average energy flowing from the shunt converter must be equal to the energy flowing out through the series converter or vice versa. In other words, according to the conventions of currents and voltages adopted in Fig. 6-2, the average real power $\bar{p}_1$ of the shunt converter must be set equal to the average real power of the series converter in the opposite direction, that is, $\bar{p}_1 = -\bar{p}_2$.

The following control principles for the UPFC use q_1 of the shunt converter to regulate the voltage v of the controlled ac bus. In this case, the control strategy is similar to that implemented in the static synchronous compensator (STATCOM) that behaves as an *ideal* synchronous condenser [19]. The real (p_2) and imaginary (q_2) powers of the series converter vary independently in order to allow v_C to be adjusted properly to control the transmission line current i_S. The limitations for q_1, q_2, and p_2 are given by the power ratings of the shunt and series converters. As shown in Fig. 6-2, the current i_S of the controlled transmission line may be continuously adjusted to produce the desired real power p and imaginary power q at the transmission line. The last degree of freedom in terms of controlled powers is the real power of converter #1 (shunt converter), p_1. As mentioned before, this degree of freedom is lost, since no other functionality can be assigned other than to force $\bar{p}_1 = -\bar{p}_2$ to provide energy balance inside the common dc link. In fact, in a real implementation, $\bar{p}_1$ must be slightly different from $-\bar{p}_2$ to compensate for losses in the converters of the UPFC, in a similar manner as was done in the previous chapters, where real implementations of active filters were shown. The principles of voltage regulation and power flow control are explained in the following section.

6.1.1.A. Voltage Regulation Principle

The shunt converter of a UPFC can regulate the voltage v at the controlled ac bus, corresponding to bus "a" in Fig. 6-2. It can generate a controlled reactive current component contained in i_C, composed of a fundamental positive-sequence component ($\dot{I}_{+1}$), which leads ($q_1 < 0$) or lags ($q_1 > 0$) by 90° with respect to the fundamental positive-sequence component $\dot{V}_{+1}$ in the controlled voltage v. If i_S is the current of the controlled transmission line that interconnects bus "a" to bus "c" in Fig. 6-2, it will be controlled by the series converter of the UPFC to produce the desired powers p and q. Hence, variations in the controlled reactive current i_C will cause changes only in the current i through the power system "A." For simplicity, this system is represented by a voltage source behind an equivalent inductance L.

Figure 6-3 shows part of the circuit in Fig. 6-2 and, for simplicity, considers the effect of varying i_C only on the current i flowing through the equivalent inductance L of the power system "A." Fig. 6-4 shows phasor diagrams for the circuit of Fig. 6-3. It suggests a regulation of voltage v by injecting a fundamental current $\dot{I}_C$ orthogonal to the fundamental component $\dot{V}$ of the controlled voltage v. If the compensating current $\dot{I}_C$ produces only imaginary power with voltage $\dot{V}$, the phasors $\dot{V}$ and $\dot{V}_G$ of voltages v and v_G are in phase, whereas the voltage $\dot{V}_L$ across the series inductance L depends on the difference in magnitude between those voltage phasors. Figure 6-4 shows that $\dot{V}_L$ is in phase with $\dot{V}$, if $|\dot{V}| < |\dot{V}_G|$ and an inductive current i_C is drawn from power system "A." If $|\dot{V}| > |\dot{V}_G|$, $\dot{V}_L$ is in counterphase and a capacitive current i_C is drawn from power system "A." Generally, inductive current i_C produces positive imaginary power ($q_1 > 0$) and forces a reduction in voltage magnitude on the controlled ac bus "a." Contrarily, if the shunt converter draws a capacitive current ($q_1 < 0$), the controlled bus voltage is increased. Later, an example is given to show how the above principles of voltage regulation can be implemented in a controller for UPFC using the p-q Theory.

6.1.1.B. Power Flow Control Principle

The series converter of a UPFC can control the instantaneous real and imaginary powers, which are produced by the voltage v of the controlled ac bus and the controlled current i_S of the transmission line. The series converter inserts the compensating voltage v_C, as shown in Fig. 6-5, to change the voltage v_S at the right side of the UPFC, such that $v_S = v - v_C$. Thus, it varies the terminal voltage v_S of the controlled transmission line and controls the current i_S to match the desired loading

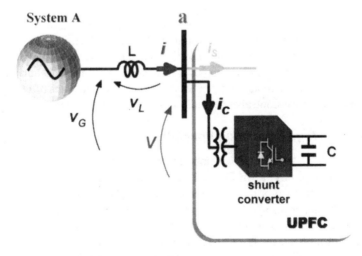

Figure 6-3. Voltage regulation principle realized by the shunt converter of the UPFC.

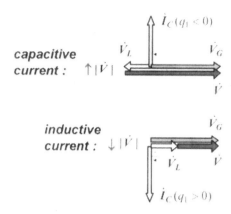

Figure 6-4. Phasor diagrams corresponding to Fig.6-3.

conditions for this line. These desired loading conditions, here referred to a real power order P_{REF} and a imaginary power order Q_{REF}, could be locally fixed or commanded remotely by a power dispatch control center. The goal is to force the powers p and q to match the power orders P_{REF} and Q_{REF}, respectively.

The basic idea consists in inserting a fundamental voltage component $\dot{V}_C$ through the series converter of the UPFC to change the phase angle and the amplitude of terminal voltage $\dot{V}_S$. Here, the reference is the phasor $\dot{V}$ of the controlled ac bus voltage v. The compensating voltage phasor $\dot{V}_C$ consists of two components $\dot{V}_c(p_C)$ and $\dot{V}_c(q_C)$. The first one is parallel to the phasor $\dot{V}$ and the second one is orthogonal, as shown in Fig. 6-6. The variables p_C and q_C will be explained later.

If the series converter is bypassed or operating with $v_C = 0$, then $v_S = v$. The current i_S of the controlled transmission line, here simply represented by a series induc-

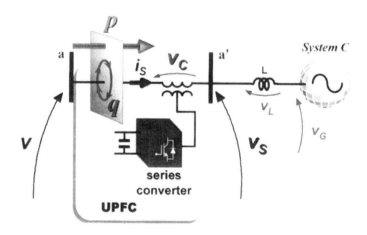

Figure 6-5. Power flow control by the series converter of the UPFC.

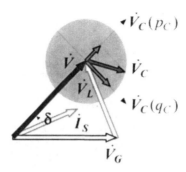

Figure 6-6. Composition of the compensating voltage of the series converter of the UPFC as a function of a parallel and an orthogonal voltage component.

tance, lags by 90° the voltage v_L on this transmission line. Note that $v_L = v_S - v_G$, as shown in Fig. 6-5. In Chapter 3, it was shown that a shunt active filter produces a compensating current that is *orthogonal,* leading (capacitive) or lagging (inductive) current, if it is instantaneously calculated as a function of an imaginary power q_C and the system voltage at the point of common coupling (PCC). On the other hand, a compensating current that is in phase or in counterphase with this system voltage can be calculated instantaneously as a function of a compensating real power p_C. However, as for series compensation, compensating voltages are required instead of compensating currents.

The same kernel as that in the control algorithm of the shunt active filter can be employed here, to generate orthogonal and parallel compensating *voltage* components. To clarify this point, a design of the UPFC controller based on the p-q Theory is presented in the next section. For now, Fig. 6-6 suggests the use of an "imaginary control power" signal q_C to build up orthogonal components for $\dot{V}_C$ and a "real control power" signal p_C to build up parallel components for $\dot{V}_C$. Note that p_C and q_C are merely control signals without any relation, neither to the real (p) and imaginary (q) powers of the controlled transmission line of Fig. 6-5, which are calculated from v and i_S, nor to the powers p_2 and q_2 produced by the series converter, which are calculated from v_C and i_S.

Figure 6-7(a) and (b) show that the orthogonal component of $\dot{V}_C$, that is, $\dot{V}_C$ as a function of q_C and noted as $\dot{V}_C(q_C)$ modifies more the phase angle than the magnitude of terminal voltage $\dot{V}_S$. If the voltage $\dot{V}_G$ on the other terminal of the controlled transmission line cannot change so quickly, the orthogonal component of the series compensating voltage, $\dot{V}_C(q_C)$, affects more the real power p passing through the line than the imaginary power q produced by v and i_S. In other words, the orthogonal component $\dot{V}_C(q_C)$ has more controllability over the real power p of the controlled transmission line than over the imaginary power q. In contrast, Fig. 6-7(c) and (d) show that parallel components of $\dot{V}_C$, that is, $\dot{V}_C(p_C)$, modifies more the magnitude than the phase angle of the terminal voltage $\dot{V}_S$ of the controlled transmission line. Hence, for a voltage $\dot{V}_G$ on the other terminal of the transmission line that cannot change quickly, this means that the parallel component $\dot{V}_C(p_C)$ causes more variations in the imaginary power q produced by v and i_S than in the real pow-

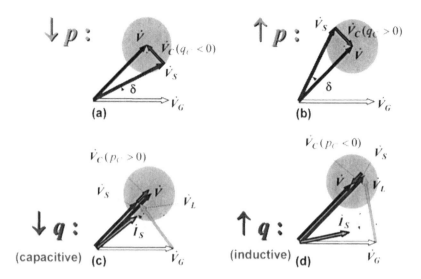

Figure 6-7. Phasors diagrams for power flow control by the series converter of the UPFC.

er passing through the line. In other words, $\dot{V}_C(p_C)$ controls basically the imaginary power q on the terminal of the controlled transmission line where the UPFC is installed.

Of course, the control of the real power p through $\dot{V}_C(q_C)$ is not perfectly decoupled from the control of the imaginary power q. Cross effects in the imaginary power q appear when $\dot{V}_C(q_C)$ is changed to control p. There are also cross effects in the real power p when $\dot{V}_C(p_C)$ is varied to control the imaginary power q. Thus, this control principle does not realize fully independent controls of p and q. In other words, a UPFC controller based on the above principle of power flow control has to adjust simultaneously $\dot{V}_C(q_C)$ and $\dot{V}_C(p_C)$ to control p and q to track accurately their reference values, given by the real and imaginary power orders, P_{REF} and Q_{REF}, respectively.

Fig. 6-7(a) shows that the power angle δ between $\dot{V}_S$ and $\dot{V}_G$ decreases when an orthogonal component leading $\dot{V}$ by 90°, that is, $\dot{V}_C(q_C < 0)$, is inserted in series with the transmission line. This forces a reduction in the real power p. In fact, if this voltage component $\dot{V}_C(q_C < 0)$ increases further and is large enough, the power angle δ can become negative and a power flow reversion will be imposed on the controlled transmission line. The real power increases when lagging component $\dot{V}_C(q_C > 0)$ is inserted, as shown in Fig. 6-7(b).

On the other hand, a compensating component in phase with $\dot{V}$, that is, $\dot{V}_C(p_C > 0)$, forces the current phasor $\dot{I}_S$ to be more capacitive, as shown in Fig. 6-7(c). This decreases the imaginary power q, which means that it forces the imaginary power q to eventually become negative. Contrarily, the imaginary power increases and the current $\dot{I}_S$ becomes more inductive when a compensating voltage component in counterphase, $\dot{V}_C(p_C < 0)$, is generated by the series converter of the UPFC, as shown in Fig. 6-7(d).

The compensating voltage $\dot{V}_C$ can vary inside a limited circle, restricted by the rated voltage of the series converter, to force the instantaneous powers p and q of the controlled transmission line, as shown in Fig. 6-5, to match the power orders P_{REF} and Q_{REF}, respectively.

Figure 6-8 summarizes the above compensation tendencies for the powers in the controlled transmission line, by the insertion of a compensating phasor $\dot{V}_C$ in the four quadrants of a coordinate system aligned to the voltage phasor $\dot{V}$ (fundamental frequency component) of the controlled ac bus.

6.1.2. A Controller Design for the UPFC

A controller for the UPFC that realizes voltage regulation and power flow control can be designed based on the p-q Theory and the concept of instantaneous aggregate voltage. Moreover, it is an integrated design that generates the instantaneous values of the compensating current i_C, shown in Fig. 6-3, for the shunt converter, and the compensating voltage v_C, shown in Fig. 6-5, for the series converter. As suggested in those figures, the shunt converter acts as a controlled current source and the series converter as a controlled voltage source. Here, it is assumed that both converters have a PWM (pulse-width modulation) switching control, although a multipulse switching control could be used in the shunt converter. Later, an alternative design for the shunt converter is presented for the case of multipulse converters.

Figure 6-9 shows the principal compensation functions of the UPFC. By generating controllable reactive current i_C, the magnitude of voltage v of the controlled bus is kept constant, around its reference value. The product of controlled current i_S and controlled voltage v produces the real power p and imaginary power q. They are continuously compared with their reference values (real power order and imaginary power order), to determine properly the compensating voltage of the UPFC series converter. Thus, the voltage v at the left-hand side of the UPFC and the current i_S

Figure 6-8. Real power control by the orthogonal component and imaginary power control by the parallel component.

6.1. THE UNIFIED POWER FLOW CONTROLLER (UPFC) 275

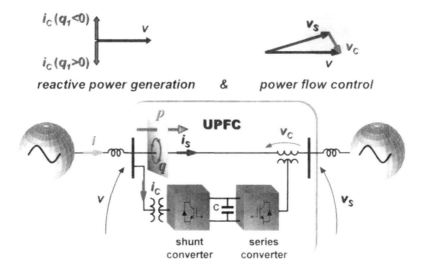

Figure 6-9. Functionality of the UPFC.

through the controlled transmission line are the principal measurements for designing a UPFC controller.

Figure 6-10 shows all measurements that are needed to realize the UPFC controller. As mentioned in the previous sections, the shunt converter is responsible for keeping the dc link voltage constant. This is accomplished by adding a controlled *active* current component to i_C, which produces energy flow into (from) the

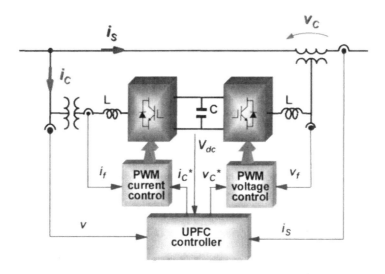

Figure 6-10. Basic control block diagram of the UPFC and converters.

dc link, to counteract to the energy flow going out of (into) the UPFC series converter. To realize this function of energy balance, the UPFC controller needs a measurement of the dc link voltage. The measured compensating current i_f and compensating voltage v_f, as shown in Fig. 6-10, are used as minor feedback loops in the PWM control circuits. Here, the PWM current and voltage control circuits are considered as part of the power converters, instead of part of the UPFC controller.

At this point, it should be emphasized that the power flow control functionality is performed by the series converter of the UPFC, by setting properly a compensating voltage v_C to control the current i_S. The instantaneous real and imaginary powers p and q in Fig. 6-9, produced by i_S and voltage v of the controlled bus, are the powers that are compared to the reference values P_{REF} and Q_{REF}. The real and imaginary powers produced by the series converter of the UPFC, given by the product of i_S and v_C, take the consequences of that power flow control. Thus, the shunt converter has to generate the active-current component that produces a real-power flow in the opposite direction to that at the series converter, in order to provide energy balance inside the dc link and keep the dc voltage constant.

A UPFC controller can be designed using only the concepts learned from the p-q Theory and the concept of instantaneous aggregate voltage, as given in (2.47). The functional control block diagram of this UPFC controller is illustrated in Fig. 6-11(a) for the UPFC series converter, and in Fig. 6-11(b) for the UPFC shunt converter. The UPFC control variables agree with the symbols, phasors diagrams, and voltage and current conventions shown in Fig. 6-3 to Fig. 6-10. Moreover, the p-q Theory gives positive values of imaginary power for inductive currents. Positive values of real power means that energy is flowing into the converter, from the ac side to the dc side, in case of using voltage and current directions as the so-called "load convention." Otherwise, the converter supplies energy if the "generator convention" is used and the real power has positive values.

As mentioned before, only three measurements are needed as inputs to the UPFC controller:

1. The phase voltages of the controlled ac bus (v_a, v_b, v_c)
2. The voltage of the common dc link (V_{dc})
3. The currents of the controlled transmission line (i_{Sa}, i_{Sb}, i_{Sc})

Four reference values are needed:

1. The real power order of the controlled transmission line (P_{REF})
2. The imaginary power order of the controlled transmission line (Q_{REF})
3. The voltage magnitude of the controlled ac bus (V_{REF})
4. The rated voltage of the common dc link (V_{dcREF})

The entire control algorithm consists of expressions that are extensively used in the previous chapters. For instance, the control block of the UPFC series converter

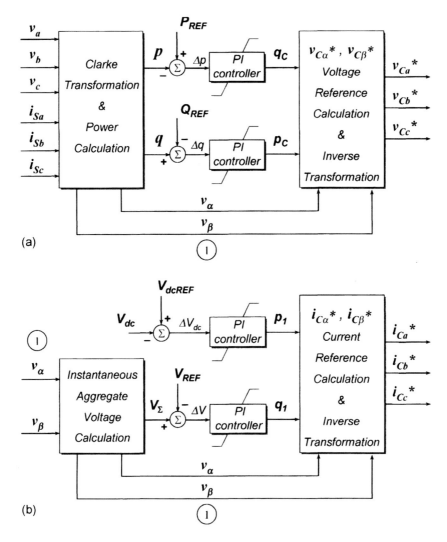

Figure 6-11. A UPFC control block diagram based on the p-q Theory. (a) Control block diagram of the series converter, (b) control block diagram of the shunt converter.

shown in Fig. 6-11(a) contains the direct Clarke transformation (abc to $\alpha\beta$ coordinates) for voltages and currents, and the real and imaginary power calculations. Note that a three-phase system without a neutral conductor is being considered, and zero-sequence components are neglected.

When designing controllers for power conditioners, it is valuable to achieve simplifications in order to reduce computation efforts. Since a three-phase system without a neutral wire is being considered, all calculations related to zero-sequence components are eliminated and only two measurements, instead of three, are sufficient for each three-phase point of measurement, as shown in the following discussion.

If zero-sequence components are neglected, the following relations are valid:

$$v_a + v_b + v_c = 0$$
$$i_a + i_b + i_c = 0 \qquad (6.1)$$

Under this constraint, it is possible to simplify the Clarke transformation in Fig. 6-11(a). Recalling that line voltages do not contain any zero-sequence components, that is,

$$v_{ab} + v_{bc} + v_{ca} = 0 \qquad (6.2)$$

where $v_{ab} = v_a - v_b$, $v_{bc} = v_b - v_c$ and $v_{ca} = v_c - v_a$. From (6.1), it is possible to determine phase voltages as functions of line voltages, that is,

$$\begin{cases} v_a = (v_{ab} - v_{ca})/3 \\ v_b = (v_{bc} - v_{ab})/3 \\ v_c = (v_{ca} - v_{bc})/3 \end{cases} \qquad (6.3)$$

Further simplification can be made, since $v_{ca} = -v_{ab} - v_{bc}$, from (6.2). Thus, phase voltages can be determined from the measurements of only two line voltages:

$$\begin{bmatrix} v_a \\ v_b \\ v_c \end{bmatrix} = \frac{1}{3} \begin{bmatrix} 2 & 1 \\ -1 & 1 \\ -1 & -2 \end{bmatrix} \begin{bmatrix} v_{ab} \\ v_{bc} \end{bmatrix} \qquad (6.4)$$

Neglecting zero-sequence components, the Clarke transformation is given by

$$\begin{bmatrix} v_\alpha \\ v_\beta \end{bmatrix} = \sqrt{\frac{2}{3}} \begin{bmatrix} 1 & -\frac{1}{2} & -\frac{1}{2} \\ 0 & \frac{\sqrt{3}}{2} & -\frac{\sqrt{3}}{2} \end{bmatrix} \begin{bmatrix} v_a \\ v_b \\ v_c \end{bmatrix} \qquad (6.5)$$

From (6.4) and (6.5), the $\alpha\beta$-voltage components can be determined directly from the two measured line voltages:

$$\begin{bmatrix} v_\alpha \\ v_\beta \end{bmatrix} = \sqrt{\frac{2}{3}} \begin{bmatrix} 1 & \frac{1}{2} \\ 0 & \frac{\sqrt{3}}{2} \end{bmatrix} \begin{bmatrix} v_{ab} \\ v_{bc} \end{bmatrix} \qquad (6.6)$$

Therefore, the two line voltages in (6.6) can replace those three phase-to-neutral voltages inputs v_a, v_b, and v_c in the UPFC controller shown in Fig. 6-11(a).

On the other hand, if (6.1) is valid, the number of current measurements from the controlled transmission line can also be reduced. Only two measurements of line currents are necessary, provided that $i_c = -i_a - i_b$. In this case, the $\alpha\beta$-current components are given by

$$\begin{bmatrix} i_{S\alpha} \\ i_{S\beta} \end{bmatrix} = \sqrt{\frac{2}{3}} \begin{bmatrix} \frac{3}{2} & 0 \\ \frac{\sqrt{3}}{2} & \sqrt{3} \end{bmatrix} \begin{bmatrix} i_{Sa} \\ i_{Sb} \end{bmatrix} \quad (6.7)$$

The convenience of reducing measurements and computation efforts in the Clarke transformation should take into account performance requirements for the UPFC, principally when it is operating under fault conditions in the ac system. In this case, zero-sequence components may not be negligible.

In Fig. 6-11(a), the real and imaginary powers of the controlled transmission line are given by

$$\begin{bmatrix} p \\ q \end{bmatrix} = \begin{bmatrix} v_\alpha & v_\beta \\ v_\beta & -v_\alpha \end{bmatrix} \begin{bmatrix} i_{S\alpha} \\ i_{S\beta} \end{bmatrix} \quad (6.8)$$

These instantaneous powers are continuously compared with their references (real and imaginary power order, P_{REF} and Q_{REF}, respectively). The calculated error signals Δp and Δq serve as input variables to PI controllers that generate an "imaginary control power" signal q_C and a "real control power" signal p_C, respectively. Note that Δp and Δq are positive or negative values in accordance to those phasor diagrams shown in Fig. 6-7. For instance, if at any instant the real power order P_{REF} is equal to +1 pu and the calculated real power p flowing through the controlled transmission line is smaller than that reference value, an error signal Δp that is the input variable to the PI controller is positive, and its output control signal q_C assumes a positive value. As shown in Fig. 6-7, an orthogonal compensating voltage component, $\dot{V}_C(q_C > 0)$, lagging by 90° the controlled bus voltage $\dot{V}$ is generated, which forces the real power p of the controlled transmission line to increase, as desired, until the error signal Δp is reduced to zero. Similar analysis holds for $\dot{V}_C(p_C)$ that controls the imaginary power q of the controlled transmission line at the electrical point corresponding to the product of voltage v and current i_S (see Fig. 6-9).

The compensating voltage components on the $\alpha\beta$ axes are determined as

$$\begin{bmatrix} v^*_{C\alpha} \\ v^*_{C\beta} \end{bmatrix} = \frac{1}{v_\alpha^2 + v_\beta^2} \begin{bmatrix} v_\alpha & v_\beta \\ v_\beta & -v_\alpha \end{bmatrix} \begin{bmatrix} p_C \\ q_C \end{bmatrix} \quad (6.9)$$

From the above equation it is clear that p_C and q_C are auxiliary variables that were created based on the p-q Theory. However, they do not have dimension of power. Instead, their dimension is given by V^2. The inverse Clarke transformation of (6.9) is given by

$$\begin{bmatrix} v_{Ca}^* \\ v_{Cb}^* \\ v_{Cc}^* \end{bmatrix} = \sqrt{\frac{2}{3}} \begin{bmatrix} 1 & 0 \\ -\dfrac{1}{2} & \dfrac{\sqrt{3}}{2} \\ -\dfrac{1}{2} & -\dfrac{\sqrt{3}}{2} \end{bmatrix} \begin{bmatrix} v_{C\alpha}^* \\ v_{C\beta}^* \end{bmatrix} \qquad (6.10)$$

Comparing (6.9) with (3.21) one can see that the compensating voltage of the UPFC series converter is derived from the basic algorithm of shunt current compensation, instead of the dual p-q Theory for series voltage compensation, as given in (3.63). The above set of equations constitutes the entire control algorithm of the controller for the series converter of the UPFC. Next, the equation for the shunt converter is presented.

The controller for the UPFC shunt converter as shown in Fig. 6-11(b) realizes the voltage regulation as illustrated in Fig. 6-3 and Fig. 6-4. For convenience, the measured voltage of the controlled ac bus, already used and transformed into the $\alpha\beta$ axes in the series converter controller of Fig. 6-11(a), is provided to the shunt converter controller in Fig. 6-11(b).

In Chapter 2, the concept of instantaneous aggregate value was introduced [see (2.47)]. There is a correspondence between instantaneous aggregate values in phase mode and in $\alpha\beta0$ variables, that is,

$$v_\Sigma = \sqrt{v_a^2 + v_b^2 + v_c^2} = \sqrt{v_\alpha^2 + v_\beta^2 + v_0^2} \qquad (6.11)$$

In the present case, zero-sequence components are being neglected, and v_0 can be eliminated from (6.11). In an ideal case, the phase voltages are balanced, that is, there are no negative-sequence or zero-sequence components at the fundamental frequency, and are free from harmonics. As a result, the instantaneous aggregate voltage is a constant value, and equal to the rms value of the line-to-line voltage. However, in some cases, in order to avoid numerical instabilities or comply with the required performance specifications of the UPFC, some kind of filtering in v_Σ may be included in the corresponding control block of Fig. 6-11(b). Note that the error signal is given by $\Delta v = v_\Sigma - v_{REF}$. The error signal agrees with the control concept shown in Fig. 6-4 and the p-q Theory that gives negative values of imaginary power for capacitive currents, which tends to increase the magnitude of the controlled ac bus voltage.

A secondary function of the shunt converter of the UPFC is to provide energy balance inside the common dc link. In other words, the shunt converter must absorb (inject) an equal amount of average real power as that injected (absorbed) by the UPFC series converter. Instead of using direct calculation of power p_2 that is given by the product of v_C and i_S, a measurement of the dc link voltage v_{dc} is used to generate the compensating real power p_1 for the shunt converter, as shown in Fig. 6-11(b). By doing so, an extra real power needed to supply losses in the power circuit of the UPFC is conveniently included in control signal p_1.

The compensating current components on the $\alpha\beta$ axes are determined as

$$\begin{bmatrix} i_{C\alpha}^* \\ i_{C\beta}^* \end{bmatrix} = \frac{1}{v_\alpha^2 + v_\beta^2} \begin{bmatrix} v_\alpha & v_\beta \\ v_\beta & -v_\alpha \end{bmatrix} \begin{bmatrix} p_1 \\ q_1 \end{bmatrix} \quad (6.12)$$

The inverse Clarke transformation gives the instantaneous reference currents as

$$\begin{bmatrix} i_{Ca}^* \\ i_{Cb}^* \\ i_{Cc}^* \end{bmatrix} = \sqrt{\frac{2}{3}} \begin{bmatrix} 1 & 0 \\ -\frac{1}{2} & \frac{\sqrt{3}}{2} \\ -\frac{1}{2} & -\frac{\sqrt{3}}{2} \end{bmatrix} \begin{bmatrix} i_{C\alpha}^* \\ i_{C\beta}^* \end{bmatrix} \quad (6.13)$$

The above equations show that the control algorithm of the shunt converter provides instantaneous *current* reference values for the PWM control of the shunt converter, whereas the control algorithm of the series converter provides instantaneous *voltage* reference values.

It should be noted that the measured voltages of the controlled ac bus is used directly in the UPFC control algorithm. This works well if the system voltage can be considered almost balanced and free from harmonics. Otherwise, the compensating voltage in (6.9) will contain undesirable harmonic components. In this case, the current references in (6.12) for the shunt converter will also contain harmonic components as the product of the measured system voltage and the compensating real power p_1 and imaginary power q_1. Instead of using directly the measured system voltage in the controller of Fig. 6-11(a), the inclusion of a fundamental positive-sequence voltage detector like that shown in Fig. 4-27 and used in the controller of shunt active filters constitutes an alternative solution to avoid harmonic feedbacks in the UPFC controller.

The UPFC series converter generates compensating voltage components at the fundamental frequency, instead of harmonic voltage components, like active filters. However, the compensating voltage has an arbitrary phase angle and magnitude. Therefore, the series converter has to have a kind of PWM switching control that controls the ac output voltage independently of the magnitude of the dc voltage. In contrast, multipulse converters composed of voltage-source converters have switching patterns at the fundamental frequency. Hence, they have ac output voltage magnitudes that are proportional to their dc link voltages. This type of converter can replace only the shunt PWM converter in Fig. 6-10, and still perform the same compensation functionalities as established before. This is discussed in the next section.

6.1.3. UPFC Approach Using a Shunt Multipulse Converter

Multipulse converters are composed of voltage-source converters that can be conventional bridge converters or multilevel converters. Three-level converters are frequently used in high-power applications [27]. For the sake of simplicity, only multipulse converters based on conventional six-pulse converters will be analyzed.

Nowadays, gate-turn-off thyristors (GTOs), insulated-gate bipolar transistors (IGBTs), and integrated-gate commutated thyristors (IGCTs) are real options for designing self-commutated converters for high-power applications. An alternative is to build high-power converters with high efficiency, which is achieved by arranging several series and/or shunt connections of three-phase converter modules, each one formed with a reduced number of power semiconductor devices in series and/or parallel in each converter valve.

For high-power applications, converter losses also become an important issue. Switching losses must be reduced. This severely limits the maximum switching frequency of PWM converters, and constitutes the principal reason to consider, where possible, multipulse converters, because they switch at the fundamental frequency.

6.1.3.A. Six-Pulse Converter

Three-phase voltage-source converters can be understood as a bridges composed of six switches, as shown in Fig. 6-12. These switches operate with unidirectional voltage blocking capability and bidirectional current flow. The switches of Fig. 6-12 can be realized by associating power semiconductor devices, such as GTOs, IGBTs, or IGCTs and antiparallel diodes. For simplicity, the diodes are not represented in Fig. 6-12. All the time, in order to avoid short circuits on the dc bus, there may be only one conducting switch per leg. In Fig. 6-12, the legs are configured as:

phase *a* leg: S1 and S4
phase *b* leg: S3 and S6
phase *c* leg: S5 and S2

Therefore, for the *a*-phase leg, while switch S1 receives a firing pulse (positive gate signal), switch S4 must receive a blocking pulse (negative gate signal), and vice-versa. A similar idea holds for the phases *b* and *c*. For convenience, the dc bus capacitance in Fig. 6-12 was divided into two equal parts to create the dc midpoint

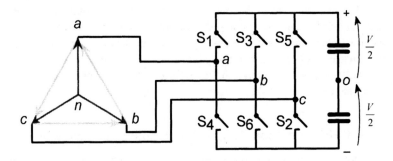

Figure 6-12. Idealized three-phase, voltage-source converter.

"o" that is used as the reference to the output voltages of the converter. They are v_{ao}, v_{bo}, and v_{co}. If each switch of a leg conducts 180° per cycle and the switching pulses of the legs are shifted by 120° from each other, the output voltages that appear at the electrical points a, b, and c referred to the dc midpoint o are as shown in Fig. 6-13. This figure indicates the switches that are conducting or not at each instant. If the current is in the opposite direction to the switch, which would correspond to a turned-on GTO in a real implementation, it will flow through its corresponding antiparallel diode, not represented in Fig. 6-12.

The output *line* voltages (v_{ab}, v_{bc}, and v_{ca}) and *phase* voltages (v_{an}, v_{bn}, and v_{cn}) can be easily derived from the voltages shown in Fig. 6-13. The line voltage outputs are:

$$\begin{cases} v_{ab} = v_{ao} - v_{bo} \\ v_{bc} = v_{bo} - v_{co} \\ v_{ca} = v_{co} - v_{ao} \end{cases} \quad (6.14)$$

and the resultant waveforms are given in Fig. 6-14. It is possible to see that the fundamental components of these line voltages lead the corresponding fundamental components of the output voltages of Fig. 6-13 by 30°. The phase voltages that would appear at a fictitious, balanced, Y-connected, three-phase load can be determined as

$$\begin{cases} v_{an} = (v_{ab} - v_{ca})/3 \\ v_{bn} = (v_{bc} - v_{ab})/3 \\ v_{cn} = (v_{ca} - v_{bc})/3 \end{cases} \quad (6.15)$$

The resultant waveforms are given in Fig. 6-15. Note that the fundamental components of the phase voltages v_{an}, v_{bn}, and v_{cn} are in phase with the output voltages v_{ao}, v_{bo}, and v_{co} (Fig. 6-13), respectively. These phase angle relations are important to create properly pulse time sequences in the switching control synchronized with the controlled ac bus voltage. Although it has no fundamental component, $3n$ ($n = 1, 3, 5, \ldots$) harmonic voltage components appear between the neutral point of the load and the midpoint of the dc side of the six-pulse converter. Fig. 6-16 shows the potential difference $v_{no} = v_{ao} - v_{an}$ between these points.

The output voltage waveform of the elementary six-pulse converter, as shown in Fig. 6-14 and Fig. 6-15, contains harmonic components with frequencies on the order of $[(6k \pm 1) \cdot f_0]$, where f_0 is the fundamental frequency and $k = 1, 2, 3, \ldots$. On the other hand, the current in the dc side has harmonic components at frequencies on the order of $[6k \cdot f_0]$. As is evident, the high harmonic content of the output voltage makes this simple converter impractical for direct use in high-power applications. Instead of using filters, several six-pulse converter modules can be arranged, using transformers as magnetic interfaces, to form a multipulse converter, which is a useful technique to perform harmonic neutralization. The higher the number of six-pulse converter modules, the lower the distortion of the resultant output voltage. For instance, eight six-pulse converters can be combined by means

284 COMBINED SERIES AND SHUNT POWER CONDITIONERS

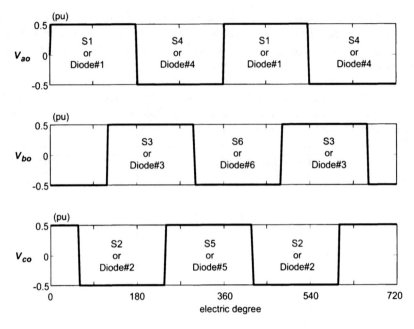

Figure 6-13. Output voltages of the converter referred to the dc midpoint "o."

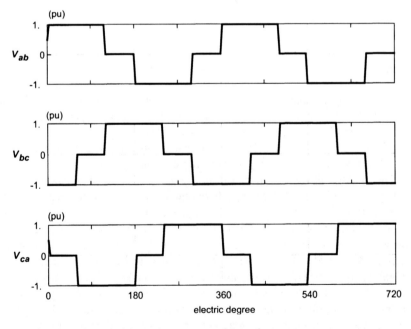

Figure 6-14. Line voltage output of the voltage-source converter.

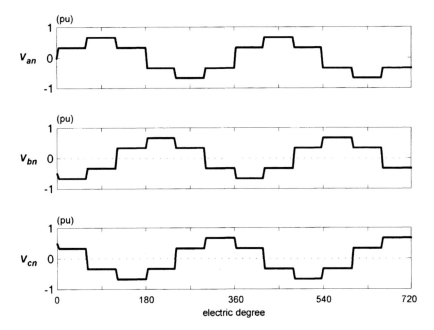

Figure 6-15. Phase voltages at a balanced three-phase load connected to the six-pulse converter.

of a magnetic interface (special transformers) to form an equivalent 48-pulse converter. In this case, the first harmonic order is the 47th in the ac voltage and the 48th in the dc current. The amplitude of the harmonics decreases as the harmonic order increases. The 48-pulse converter has a nearly sinusoidal output voltage. It can be directly applied in power systems without any additional filtering system. Although the resultant output voltage is nearly sinusoidal, each six-pulse converter still generates the quasisquare output voltages. A great advantage of this multi-

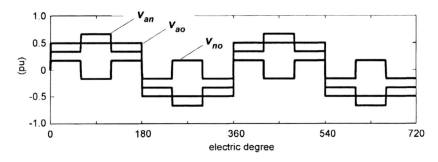

Figure 6-16. Voltage v_{no} between the load neutral and dc midpoint.

pulse converter is its low switching losses, which allow the use in equipment with very high power ratings.

As the switches of the converter have self-commutation capabilities, the six-pulse converter switching at the fundamental frequency can generate output voltages with arbitrary phase angle and frequency. However, the amplitude of the generated voltage is directly proportional to the dc bus voltage. Therefore, the switching control of a six-pulse converter can only vary the frequency and phase angle of the output voltage. The magnitude of the ac output voltage can vary only if the dc voltage changes accordingly. Note that the UPFC in Fig. 6-10 uses pulse-width-modulation (PWM) control techniques in the converters, so that they can control the amplitude, frequency, and phase angle of the ac output without the need of changing the dc bus voltage. This is the reason why the UPFC controller for the shunt converter as presented in Fig. 6-11(b) cannot be applied to such a multipulse converter. An adequate controller for the shunt multipulse converter of the UPFC is presented later.

6.1.3.B. Quasi 24-Pulse Converter

Figure 6-17 shows the basic configuration of a quasi 24-pulse converter that can be used as a shunt converter in a UPFC approach. There are four six-pulse converters connected in parallel to the dc capacitor. Four ordinary three-phase transformers are used to compose the magnetic interface. This approach is preferred due to its sim-

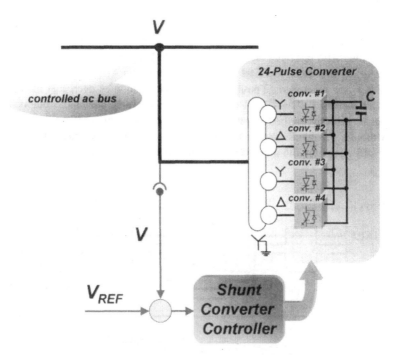

Figure 6-17. Basic configuration of a quasi 24-pulse shunt converter for the UPFC.

plicity compared to those magnetic interfaces (zig-zag transformers) used in [19] and [22]. If complex magnetic interfaces are used, it becomes theoretically possible to completely cancel low-order harmonics. For instance, in a 24-pulse arrangement, all low-order harmonics up to the 23rd order, of course, except for the fundamental one, are fully cancelled. However, the quasi 24-pulse arrangement, as given in Fig. 6-18, cannot cancel fully the 11th and 13th harmonics, although it reduces them significantly.

The quasi 24-pulse magnetic interface exploits the well-known technology of three-phase Y–Y and Y–Δ transformers and takes advantages of the different voltage waveforms, which excite the windings of those transformers. The voltages that are applied to the windings of a Y–Y transformer are the phase-to-neutral voltages v_{an}, v_{bn}, and v_{cn}, like those shown in Fig. 6-15, whereas the line-to-line voltages v_{ab}, v_{bc}, and v_{ca} of the six-pulse converter, like those shown in Fig. 6-14, are applied to the secondary windings of a Y–Δ transformer.

In Fig. 6-18, negative values for the switching sequence ($\phi_{CONV} < 0$) of the converters mean that the first pulse (turn-on) of the sequence for the switch S1 in Fig. 6-13 is delayed by the corresponding negative angle $\phi_{CONV} = \omega t_d$ with respect to the reference given by a synchronized phase angle reference of the primary side of the transformer, that is, the phase angle of the controlled ac bus in Fig. 6-17. For instance, if the system frequency is 50 Hz, the time delay given for the converter #1 is $t_d = (7.5\pi)/(180 \cdot 2\pi \cdot 50) = 416.67$ μs. Contrarily, the switching sequence of converter #3 is in advance by 416.67 μs. Converter #2 together with converter #1 form a 12-pulse converter, whereas converter #4 and converter #3 form another 12-pulse

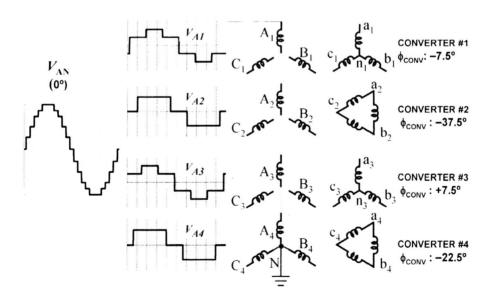

Figure 6-18. Transformer connections for the quasi 24-pulse converter.

converter. The voltage in primary windings A2 and A4 have the same waveforms of the secondary windings a_2b_2 and a_4b_4, multiplied by the turn ratio. The voltage on a_2b_2 winding is the line voltage v_{ab} shown in Fig. 6-14 that leads by 30° the voltage v_{ao}. Furthermore, v_{ao} has a phase angle determined by the switching sequence that is equal to ϕ_{CONV}. Therefore, if the switching sequence of an Y–Δ transformer is delayed by 30°, its line voltage v_{ab} gets in phase with the phase voltage v_{an} of the Y–Y transformer. Hence, if converter #1 is delayed by 7.5°, converter #2 must be delayed by 37.5° to form a 12-pulse converter. Similarly, converter #3 is advanced by 7.5° and converter #4 is delayed by 22.5°. This results in two 12-pulse voltage waveforms with phase angles of −7.5° and +7.5°, respectively. They are added up to form a quasi 24-pulse voltage with zero phase angle. Therefore, the resulting quasi 24-pulse voltage is synchronized with the controlled ac bus voltage. The switching sequences of the converters must keep the fixed phase shifts ϕ_{CONV} as given in Fig. 6-18, in order to preserve the quasi 24-pulse waveform. To shift the phase of the resultant quasi 24-pulse voltage, all converters must receive the same phase shifting, in addition to that fixed-phase shift for each converter. Note that the phase angles are fixed in radians, representing the product of the system frequency and the time delay, that is, $\phi_{CONV} = \omega t_d$. Thus, if the system frequency varies, the time delay of the switching sequences must vary accordingly.

If the turns ratio of the Y–Y transformers of converters #1 and #3 is given by

$$\frac{V_{A1}}{V_{a1}} = \frac{N_1}{N_2} \tag{6.16}$$

the turns ratio of the Y–Δ transformers of converters #2 and #4 is determined by

$$\frac{V_{A2}}{V_{a2b2}} = \frac{N_1}{\sqrt{3}\,N_2} \tag{6.17}$$

6.1.3.C. Control of Active and Reactive Power in Multipulse Converters

The voltage regulation principle of Fig. 6-3 and Fig. 6-4 is useful to understand the following discussion about the control of the active and reactive power generation by multipulse converters.

It is assumed that a multipulse converter is used as the shunt converter of a UPFC and is connected to the controlled ac bus, as shown in Fig. 6-19. The main elements in the power circuit involved with the variation of active and reactive power are represented in this figure. They are the dc capacitor of the common dc link, a single block representing the converters, and a single transformer representing the magnetic interface. As mentioned, the UPFC shunt converter should be able to control the active-power generation independently of the reactive power generation. In this way, it can perform energy balance in the common dc link, as well as voltage regulation on the controlled ac bus, as in the previous approach of UPFC that is composed of PWM converters.

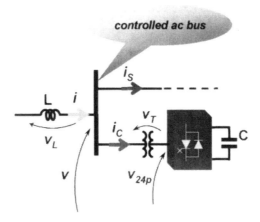

Figure 6-19. Shunt connected multipulse converter.

The 24-pulse voltage of Fig. 6-18 has a fundamental component with the amplitude directly proportional to the dc voltage on the common dc link. The phasor associated to this fundamental component, represented by v_{24p} in Fig. 6-19, is plotted together with the phasor of the controlled ac bus voltage v to form, in Fig. 6-20(a) and Fig. 6-20(b), four interesting situations of active and reactive power generation by the multipulse converter. If $\dot{V}_{24p}$ and $\dot{V}$ have the same frequency, the same amplitude and the same phase angle, no current flows through the converter transformer, and $\dot{V}_T$ is zero. If $\dot{V}_{24p}$ is in phase with $\dot{V}$, but they have different amplitudes, pure reactive current flows through the converter transformer, as shown in Fig. 6-20(a). If $\dot{V}_{24p}$ has a lower amplitude, $\dot{I}_C$ is inductive (lagging current). Contrarily, $\dot{I}_C$ is capacitive (leading current) if $\dot{V}_{24p}$ has an amplitude greater than that of $\dot{V}$. Therefore, the dc voltage must vary in order to control the generated reactive power of a multipulse converter. Hence, if the dc voltage increases, the fundamental component of the output voltage v_{24p} increases, and capacitive current i_C flows through the converter transformer whenever v_{24p} is greater than v.

The transmission line current i_S in Fig. 6-19 is not directly affected by i_C, because i_S is controlled by the series converter of the UPFC. Thus, the generated reactive current of the shunt multipulse converter flows to the system at the left-hand side of the UPFC and causes an additional voltage drop v_L on its equivalent series impedance. This additional voltage drop will cause an increment on the magnitude of the controlled ac bus voltage, as explained in Fig. 6-3 and Fig. 6-4, if the shunt converter is drawing capacitive current. In summary, the multipulse converter must vary accordingly the common dc-link voltage of the UPFC to regulate the controlled ac-bus voltage.

On the other hand, active power is controlled by phase shifting of the output voltage $\dot{V}_{24p}$ with respect to $\dot{V}$, as shown in Fig. 6-20(b). This will allow the realization of the second control function of the UPFC shunt converter, which is the energy balance in the dc link, as explained before. In fact, the phase shifting can be controlled in such

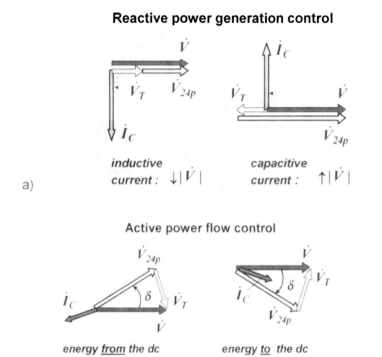

Figure 6-20. Principle of power control by multipulse converters: (a) reactive power control and (b) active power control.

a way that it injects or extracts temporarily an extra amount of energy beside that necessary for the energy balance, in order to adjust properly the dc-link voltage to control also the reactive power generation needed for the ac-bus voltage regulation.

6.1.3.D. Shunt Multipulse Converter Controller

As mentioned before, the multipulse converters to be analyzed here are composed of six-pulse, three-phase converters, which have gate-drive circuits that receive sequences of six turn-on/off pulses per cycle of the fundamental period, as illustrated in Fig. 6-13. Therefore, the controller in Fig. 6-11(b) for the UPFC shunt PWM converter must be changed, not in its compensation principles, but in its final output signal. Now, instead of producing instantaneous values of compensating current references, the switching logic of the shunt multipulse converter must receive the following information:

1. The phase angle $\theta = \omega t$, synchronized with the controlled ac bus voltage, which can be calculated by a phase-locked-loop (PLL) circuit

2. A shifting angle δ, as shown in Fig. 6-20(b), determined by the UPFC shunt converter controller

Figure 6-21 shows the principal control blocks of the controller of the multipulse shunt converter.

The switching logic circuit of the multipulse converter does not need any feedback from the generated output voltage or current, and the main shunt converter controller needs only a measurement from the controlled ac bus voltage. Surprisingly, no measurements from the common dc link are needed.

The PLL control circuit as introduced in Fig. 4-28 is an excellent alternative to generate the synchronized phase angle $\theta = \omega t$. As explained, it locks the phase angle of the fundamental positive-sequence component of the measured system voltage, instead of locking the phase angle of the phase-to-neutral voltage in a phase, as commonly calculated by other PLL circuits. Details of the PLL circuit can be found in Chapter 4, in the text dedicated to Fig. 4-28. For convenience, Fig. 6-22 repeats that PLL circuit, with some simplifications, as dictated by the needs here.

The power angle control block is basically composed of a PI controller that integrates the error between the instantaneous aggregate value of the controlled ac bus voltage, as calculated in (6.11), and its reference value. Fig. 6-23 shows the com-

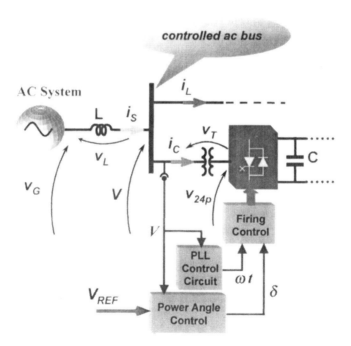

Figure 6-21. Controller for the UPFC multipulse shunt converter.

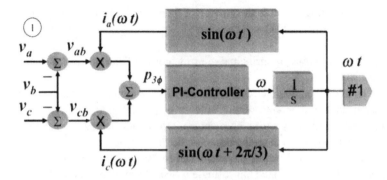

Figure 6-22. Synchronizing circuit—the PLL circuit.

plete power angle control block diagram. The single power angle control that determines the signal δ realizes simultaneously the two principal compensation functions of the shunt converter. The following simple example helps to clarify this point.

If the UPFC series converter, when controlling the reactive and active power flow through the controlled transmission line, needs to inject energy into the ac network (the active power determined by the product of v_C and i_S in Fig. 6-9), this energy leaves the dc capacitor, discharging it. This reduces the dc voltage and causes a reduction in the output voltage v_{24p} (see Fig. 6-21). From Fig. 6-20a, it is possible to say that the compensating current i_C becomes more inductive, which will cause less shunt (capacitive) reactive compensation, and the voltage magnitude of the controlled ac bus decreases. This produces an increasing negative voltage error (ΔV signal in Fig. 6-23), and the power angle δ becomes more negative (lagging v_{24p} in Fig. 6-20b), forcing the UPFC shunt converter to draw more energy from the network. Hence, the dc capacitor will be charged again, until the shunt reactive compensation (capacitive component of i_C) becomes effective enough to bring the controlled ac bus voltage to its reference value. Therefore, indirect changes in the shunt

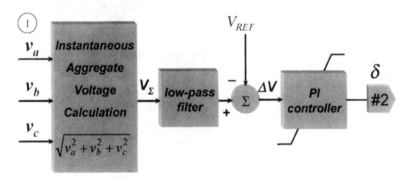

Figure 6-23. Power angle control circuit.

capacitive current compensation makes the PI controller of the power angle control circuit in Fig. 6-23 perform both compensation functions of the shunt converter: ac voltage regulation and energy balance inside the common dc link.

Fig. 6-24 shows the switching logic for six switches, represented by GTOs, that form one six-pulse converter of the quasi 24-pulse converter presented in Fig. 6-17. Therefore, a total of four switching circuits as shown in Fig. 6-24 are needed, and the difference among them is the argument ϕ_{CONV} in radians, corresponding to the phase shift given in degree for each converter number.

The phase delay of $-\pi/2$ included in all sine functions in Fig. 6-24 agrees with a particular characteristic of the PLL circuit. This means that it has a single point of stable operation given by $\sin(\omega t)$ leading by $\pi/2$ the fundamental component $\dot{V}_{+1}$ in the phase voltage v_a. The switching functions given by the signals GTO1 and GTO4, GTO3 and GTO6, and GTO5 and GTO2 in Fig. 6-24 force the six-pulse converters to generate v_{ao}, v_{bo}, and v_{co}, as shown in Fig. 6-13. The quasi 24-pulse voltage waveform is formed as shown in (Fig. 6-18). It is synchronized and in phase with voltage v of the controlled ac bus (Fig. 6-21), and has a phase shift controlled by δ. Note that the voltage v is the input voltage of the PLL circuit in Fig. 6-22.

6.2. THE UNIFIED POWER QUALITY CONDITIONER (UPQC)

Unified power quality conditioners (UPQCs) consist of combined series and shunt active power filters for simultaneous compensation of voltage and current [29,31,32,33]. They are applicable to power distribution systems, being connected close to loads that generate harmonic currents. The harmonic-producing loads may affect other harmonic sensitive loads connected at the same ac bus terminal. The

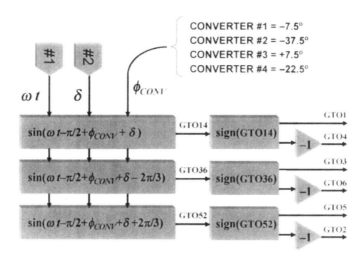

Figure 6-24. Firing logic of the quasi 24-pulse shunt converter.

UPQC constitutes one of the most flexible devices for harmonic compensation in the new concept of "custom power" [34]. The UPQC can compensate not only harmonic currents and imbalances of a nonlinear load, but also harmonic voltages and imbalances of the power supply. Hence, it improves the power quality delivered for other harmonic-sensitive loads. Thus, the UPQC joins both principles of shunt current compensation and series voltage compensation into a single device, as illustrated in Fig. 6-25.

In a practical implementation, the UPQC is realized by employing two PWM converters, coupled back-to-back through a common dc link, and an integrated controller that provides both voltage and current references for the series and shunt PWM converters, as shown in Fig. 6-26. An advantage of this integrated approach is that it produces a compact compensator with improved overall performance, since it is possible to coordinate the functionalities and dynamics of the series and the shunt active filters.

6.2.1. General Description of the UPQC

The unified power quality conditioner has two distinct parts:

1. Power circuit formed by the series and shunt PWM converters
2. UPQC controller

Fig. 6-26 shows the basic configuration of a unified power quality conditioner. If it is compared with the basic configuration of the unified power flow controller (UPFC), as shown in Fig. 6-2, they appear to be quite similar. Here, an additional current measurement of the nonlinear load is needed. Unlike the previous configu-

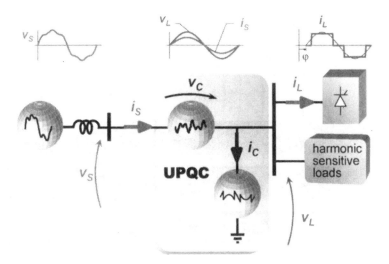

Figure 6-25. Combined series and shunt compensation (UPQC device).

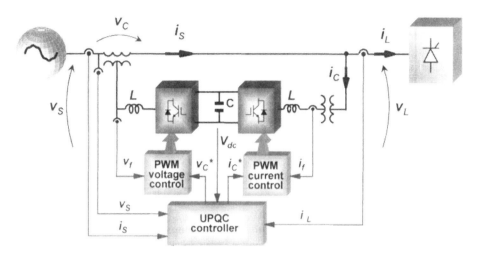

Figure 6-26. Basic configuration of the unified power quality conditioner, the UPQC.

ration of UPFC, the shunt converter of the UPQC must be connected as close as possible to the non-linear load, instead of the network side.

The series PWM converter of the UPQC behaves as a controlled voltage source, that is, it behaves as a series active filter, whereas the shunt PWM converter behaves as a controlled current source, as a shunt active filter. No power supply is connected at the dc link. It contains only a relatively small dc capacitor as a small energy storage element. The integrated controller of the series and shunt active filters of the UPQC realizes an "instantaneous" algorithm to provide the compensating voltage reference v_C^*, as well as the compensating current reference i_C^* to be synthesized by the PWM converters.

The UPQC approach is the most powerful compensator for a scenario as depicted in Fig. 6-25, where the supply voltage v_S is itself already unbalanced and/or distorted and is applied to a critical load that requires high power quality. On the other hand, part of the total load includes nonlinear loads that inject a large amount of harmonic current into the network, which should be filtered. Table 6.1 summarizes the tasks assigned to each active filter in the UPQC approach.

In Fig. 6-26, the current i_L represents all nonlinear loads that should be compensated. The shunt active filter of the UPQC can compensate all undesirable current components, including harmonics, imbalances due to negative- and zero-sequence components at the fundamental frequency, and the load reactive power as well. The same kind of compensation can be performed by the series active filter for the supply voltage, as listed in Table 6.1. Hence, the simultaneous compensation performed by the UPQC guarantees that both the compensated voltage v_L at load terminal and the compensated current i_S that is drawn from the power system become balanced, so that they contain no unbalance from negative- and zero-sequence components at the fundamental frequency. Moreover, they are sinusoidal and in phase, if the load reactive power is also compensated. Additionally, the shunt active filter

Table 6.1. Assignments for a combined series and shunt active power filter, the UPQC

Unified Power Quality Conditioner	
Series Active Filter	Shunt Active Filter
Compensates *source voltage* harmonics, including negative- and zero-sequence components at the fundamental frequency	Compensates *load current* harmonics, including negative- and zero-sequence components at the fundamental frequency
Blocks harmonic currents flowing to the source (harmonic isolation)	Compensates reactive power of the load
Improves the system stability (damping)	Regulates the capacitor voltage of the dc link

has to provide dc link voltage regulation, absorbing or injecting energy from or into the power distribution system, to cover losses in the PWM converters, and correct eventual transient compensation errors that lead to undesirable transient power flows into the UPQC. This point will be clarified better in the following sections.

It might be interesting to design UPQC controllers that allow different selections of the compensating functionalities listed in Table 6.1. For instance, if the UPQC is applied to three-phase systems without neutral conductor, the power circuit of the UPQC can be composed of three-phase, three-wire converters, since they do not need to compensate any zero-sequence voltage or current component. In this case, if the UPQC controller has a modular design, all control blocks involving compensation of zero-sequence components could be easily simplified, without degrading the other compensating functions listed in Table 6.1.

Further, in order to reduce the power rating of the shunt active filter, it might be interesting to exclude load reactive-power compensation from the UPQC approach, and to leave this function to other conventional or less expensive compensators. If the shunt active filter does not have to compensate the fundamental reactive power of the load, the power rating of the shunt active filter can be reduced significantly. Recalling that this power corresponds to one term in the general expression for $\bar{q}$, as given in (3.96), an easy solution would be to eliminate the average value of imaginary power $\bar{q}$ from the control strategy of the shunt active filter. In most cases, this fundamental reactive power component, $Q_{3\phi} = 3 \cdot V_{+1} \cdot I_{+1} \cdot \sin(\phi_{+1} - \delta_{+1})$, is much higher than the sum of all other harmonic power components, thus leading to an increased power rating for the UPQC shunt converter. This problem does not affect the UPQC series converter, because it compensates basically the harmonic voltage components of the power supply, which usually produce very small harmonic power. There are more cost-effective devices for fundamental reactive power compensation than active filters. The shunt active filter is expensive due to its necessity of high-frequency switching, generally ten times higher than the highest harmonic frequency in the load current that is to be compensated.

If no reactive power compensation is required, the UPQC of Fig. 6-26 can be combined with shunt passive filters (LC filters), to reduce the total costs of the installation and to extend its use in very high power applications. This combined approach is called here the hybrid UPQC, and is presented in Section 6.2.3.

6.2.2. A Three-Phase, Four-Wire UPQC

This section discusses the implementation of a three-phase, four-wire UPQC, including its modular controller. The principal difference between a three-phase, four-wire and a three-phase, three-wire UPQC is that the first one is able to deal with zero-sequence components in phase voltage and line current. This kind of imbalance and harmonic distortion is mostly caused by nonlinear loads in low-voltage distribution systems, where the neutral conductor is provided. In medium- or high-voltage systems, the absence of the neutral conductor naturally eliminates this problem, at least during normal operation. Although a three-phase, three-wire UPQC is a more realistic device and may find more application fields, this section presents a three-phase, four-wire UPQC to discuss all those compensating functions listed in Table 6.1. Moreover, since the following controller for the three-phase, four-wire UPQC realizes all compensating functions in a modular manner, simplifications on it are easily made to customize it for applications to three-phase, three-wire UPQCs. This feature is highlighted below.

6.2.2.A. Power Circuit of the UPQC

The power circuit of the combined series and shunt three-phase, four-wire active filter—the unified power quality conditioner (UPQC)—is presented in Fig. 6-27. A UPQC for application to three-phase systems *without* neutral conductors can be

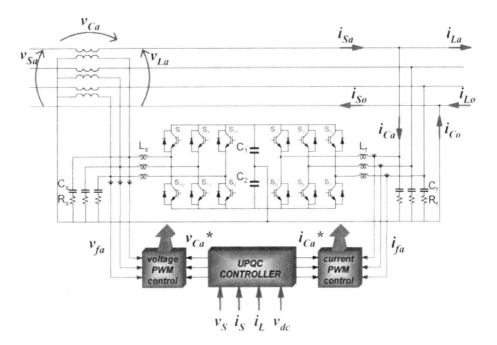

Figure 6-27. Combined series and shunt active filters (UPQC) to simultaneously compensate the voltages and currents in three-phase, four-wire systems.

easily derived from this power circuit configuration. The changes consist only of eliminating the neutral wires from the power circuit. In the UPQC controller, as will be explained later, the changes consist in canceling all variables related with zero-sequence components, like v_0, i_0, and p_0.

Two conventional three-leg converters are used in Fig. 6-27. The ac-system neutral conductor is directly connected to the electrical midpoint of the common dc bus. In the case of implementing a three-phase, three-wire UPQC, this connection should also be eliminated, and the center points of the Y-connected RC filters are left ungrounded, or grounded by high impedances. Otherwise, a low-impedance path for zero-sequence components would be formed through the ground, close to the ac outputs of the combined PWM converters.

A shunt active filter for three-phase, four-wire systems, described in Chapter 4, has the same converter topology and PWM *current* control as the shunt active filter in the UPQC in Fig. 6-27. This control forces the shunt converter to behave as a controlled current source. This converter topology has the advantage of using a conventional three-leg topology, instead of implementing an additional leg for the neutral wire. However, special care should be taken to avoid voltage deviations on the two dc capacitors. This problem was discussed in Section 4.3.2, where a special PWM switching pattern, the dynamic hysteresis-band control, was proposed to equalize the voltages on the dc capacitors using the split-capacitor converter topology.

Ideally, the shunt active filter can compensate the neutral current of the load i_{Lo}, so that no current flows through the neutral wire of the network, that is, $i_{So} = 0$. Of course, the shunt active filter for a three-phase, four-wire system is also able to perform all those compensation features of a three-phase, three-wire shunt active filter. In other words, it can also compensate undesirable portions of powers of a nonlinear load that are related with positive- and negative-sequence harmonics, including the negative-sequence component at the fundamental frequency. Thus, the three-phase, four-wire shunt active filter can compensate the neutral current, the fundamental reactive power, current imbalances, as well as all other current harmonics.

The series converter has a PWM *voltage* control. It should synthesize accurately the compensating voltages determined by the UPQC controller. This control forces the converter to behave as a controlled voltage source. The series active filter compensates the supply voltages v_{Sa}, v_{Sb}, and v_{Sc}, including zero-sequence components. Hence, the load voltages v_{La}, v_{Lb}, and v_{Lc} become balanced and free of harmonics. The combined action of the series and shunt active filters forces the compensated voltages v_{La}, v_{Lb}, and v_{Lc}, and the compensated currents i_{Sa}, i_{Sb}, and i_{Sc} to become sinusoidal and balanced. Moreover, the currents will be in phase with their corresponding voltages if the load reactive power is also compensated by the shunt active filter.

Other loads can be connected close to the nonlinear load (at the right side of the UPQC in Fig. 6-27), without being affected by harmonic currents of this non-linear load, while they are supplied with higher power quality than other loads at the left hand side of the UPQC.

The UPQC controller realizes almost instantaneous control algorithms to provide the current references i^*_{Ca}, i^*_{Cb}, and i^*_{Cc} to the shunt active filter, and the voltage references v_{Ca}^*, v_{Cb}^*, and v_{Cc}^* to the series active filter. The high switching frequency of the PWM converters produces the currents i_{fa}, i_{fb}, and i_{fc}, and the voltages v_{fa}, v_{fb}, and v_{fc} with some unwanted high-order harmonics that can be easily filtered by using small passive filters, represented by C_f, R_f, C_s, and R_s in Fig. 6-27. If the filtering is ideal, the compensating currents i_{Ca}, i_{Cb}, and i_{Cc} and voltages v_{Ca}, v_{Cb}, and v_{Cc} track strictly their references i_{Ca}^*, i_{Cb}^*, and i_{Cc}^* and v_{Ca}^*, v_{Cb}^*, and v_{Cc}^*, respectively.

6.2.2.B. The UPQC Controller

The control circuits of the series and shunt active filters can be merged into an integrated controller for the UPQC with the advantage of saving computation efforts. The functional control block diagram of the UPQC controller is illustrated in Fig. 6-28. Control signals, for example, the fundamental positive-sequence voltages v'_a, v'_b, and v'_c, are simultaneously used in both active filter controllers. The line currents

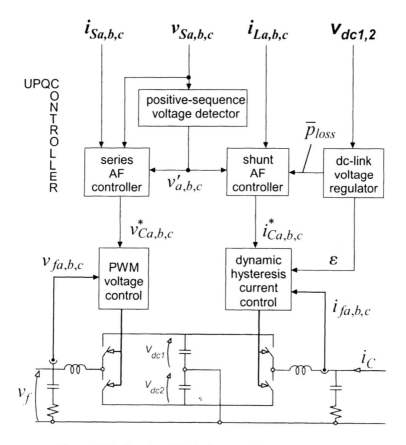

Figure 6-28. Functional block diagram of the UPQC controller.

i_{La}, i_{Lb}, and i_{Lc}, of the nonlinear load, and the compensated currents i_{Sa}, i_{Sb}, and i_{Sc}, as well as the phase voltages v_{Sa}, v_{Sb}, and v_{Sc}, constitute the inputs to the UPQC controller. The voltages v_{dc1} and v_{dc2} of the dc capacitors are used in the dc-link voltage regulator, whereas the actual values of the compensating voltages v_{fa}, v_{fb}, and v_{fc} and compensating currents i_{fa}, i_{fb}, and i_{fc} are fed back to the PWM controllers through the minor control loops, as indicated in Fig. 6-28.

In Fig. 6-28, six functional blocks are identified:

1. Positive-sequence voltage detector
2. Shunt active filter controller
3. dc-link voltage regulator
4. Dynamic hysteresis current control
5. Series active filter controller
6. PWM voltage control

In Table 6.1, the functionalities assigned to the shunt active filter of the UPQC agrees with the sinusoidal source current control strategy, presented in Chapter 4, Fig. 4-26. Moreover, the split-capacitor converter topology used in the UPQC can use the same dynamic hysteresis current control, as explained in Section 4.3.2. More details about the control part for the shunt active filter of the UPQC can be found in Chapter 4. In the following sections, the PWM voltage control and the control part for the series active filter, which corresponds to items (5) and (6) listed above, are described. Then, a general overview on the UPQC controller is presented.

6.2.2.B.1. PWM Voltage Control with Minor Feedback Control Loop.

The PWM voltage control should allow the series active filter to generate nonsinusoidal voltages according to their references v_{Ca}^*, v_{Cb}^*, and v_{Cc}^*, which can vary widely in frequency and amplitude. Therefore, conventional sine-PWM techniques (SPWM) may not fit due to their inherent amplitude attenuation [35]. Further, the LRC circuit at the ac output of the PWM converter, as shown in Fig. 6-27, causes phase displacements in the compensating voltages v_{fa}, v_{fb}, and v_{fc}. Therefore, three minor feedback control loops using the actual values of v_{fa}, v_{fb}, and v_{fc} are implemented to minimize possible deviations between the reference values v_{Ca}^*, v_{Cb}^*, and v_{Cc}^* and the compensating voltages v_{Ca}, v_{Cb}, and v_{Cc} generated at the primary sides of the series transformers. The proposed PWM voltage control is given in Fig. 6-29.

The gain K_V multiplies the errors between reference values and actual values of the compensating voltages. It is set as high as possible, considering that the new reference values given to the PWM control, equal to $v_C^* + K_v(v_{Ck}^* - v_{fk})$, $k = a, b, c$, should not exceed the amplitude of the triangular carrier.

Since the control loop is composed only of a proportional gain, the errors cannot be completely eliminated. Moreover, in a real implementation, the series transformer may increase the compensation errors, because measurements are made at

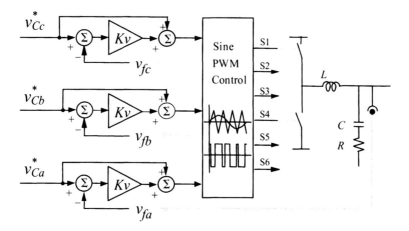

Figure 6-29. PWM voltage control with minor feedback control loops.

the secondary side, and a real transformer does not have a perfect linear response in a wide frequency range. The measurements of v_{fa}, v_{fb}, and v_{fc} are performed at the secondary side to avoid nonlinear effects of the transformers that may introduce instability to the control.

The use of PD (proportional-derivative) or PID (proportional-integral-derivative) controllers, replacing the gain K_V in Fig. 6-29, can improve system dynamics and almost eliminate steady-state errors. In this case, the feedforward loops are not necessary and the errors $(v_{Ck} - v_{fk})$, $k = a, b, c$, are just passed through PID controllers and their outputs are directly used in the sine-PWM control.

6.2.2.B.2. Series Active Filter Controller. The main goal of the series active filter of the UPQC is to compensate harmonics and imbalances in the supply voltage at the left side of the UPQC. In other words, the series active filter should compensate all voltage components in the supply voltage, which do not correspond to its fundamental positive-sequence component. It will be shown later that an auxiliary control algorithm can be included, which generates an auxiliary control signal that enhances the overall system stability, without degenerating the main compensation functions of the UPQC.

The voltage convention adopted in Fig. 6-27 leads to the following relation:

$$\begin{bmatrix} v_{La} \\ v_{Lb} \\ v_{Lc} \end{bmatrix} = \begin{bmatrix} v_{Sa} \\ v_{Sb} \\ v_{Sc} \end{bmatrix} + \begin{bmatrix} v_{Ca} \\ v_{Cb} \\ v_{Cc} \end{bmatrix} \qquad (6.18)$$

On the other hand, all portions of the supply voltages v_{Sa}, v_{Sb}, and v_{Cc} that do not correspond to the fundamental positive-sequence voltage, which should be compensated, can be calculated as

$$\begin{bmatrix} v'_{Ca} \\ v'_{Cb} \\ v'_{Cc} \end{bmatrix} = \begin{bmatrix} v'_a \\ v'_b \\ v'_c \end{bmatrix} - \begin{bmatrix} v_{Sa} \\ v_{Sb} \\ v_{Sc} \end{bmatrix} \quad (6.19)$$

where the voltages v'_a, v'_b and v'_c correspond to the instantaneous values of the fundamental positive-sequence component in v_{Sa}, v_{Sb}, and v_{Sc} on the right side of the UPQC, which are continuously determined by a positive-sequence voltage detector. This detector is also needed in the sinusoidal source current control strategy implemented in the shunt active filter controller. In this case, the main compensation function of the series active filter reduces in three subtractions, as described in (6.19). Note that all imbalances and harmonics, including the fundamental negative-sequence and zero-sequence components, which may be present in the supply voltage, are included in the signals v'_{Ca}, v'_{Cb}, and v'_{Cc}. If v'_{Ca}, v'_{Cb}, and v'_{Cc} given in (6.19) replace v_{Ca}, v_{Cb}, and v_{Cc} in (6.18), the compensated voltages at the right side of the UPQC become ideally

$$\begin{bmatrix} v_{La} \\ v_{Lb} \\ v_{Lc} \end{bmatrix} = \begin{bmatrix} v'_a \\ v'_b \\ v'_c \end{bmatrix} \quad (6.20)$$

which means that the compensated voltage will comprise only the fundamental positive-sequence component of the source voltage. This constitutes the main compensation function of the series active filter.

As mentioned before, an interesting auxiliary control algorithm can be implemented in the series active filter to provide damping in resonance phenomena and to enhance the overall system stability. The basic idea consists in offering an additional series resistance for *harmonic* currents flowing from the network to the load and to the RC passive filter of the shunt active filter [36]. It does not affect the fundamental current flowing through the series active filter. In other words, it behaves as a short circuit for the fundamental positive-sequence current ($\dot{I}_{+1}$).

An inherent characteristic of this control algorithm is that it drains energy from the network to the series active filter, since the series active filter behaves as a "series resistance" for harmonic currents [37]. Of course, in normal operation, this energy is so small that it is absorbed by the series active filter and is determined as the product between the harmonic compensating voltage and the harmonic currents flowing through the series transformer, because it is expected that the compensated currents i_{Sa}, i_{Sb}, and i_{Sc} become balanced and free of harmonics. Nevertheless, this drawback is particularly well overcome in a combined approach of series and shunt active filter (UPQC approach), where the shunt active filter regulates the voltage of the common dc link. This type of control algorithm that offers additional harmonic resistance will also be used in the hybrid UPQC approach, in the next section, where active and passive filters are combined.

To implement the damping control algorithm, the line currents i_{Sa}, i_{Sb}, and i_{Sc} flowing through the series transformers are measured. Their harmonic contents i_{ha}, i_{hb}, and i_{hc} are calculated from these currents. Figure 6-30 presents a way to calcu-

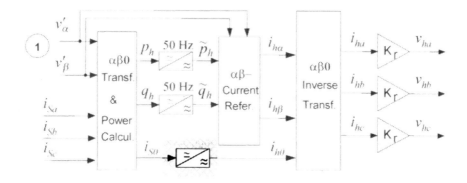

Figure 6-30. Block diagram of the control algorithm for damping of oscillations due to resonance phenomena.

late i_{ha}, i_{hb}, and i_{hc} using the instantaneous powers defined in the p-q Theory. The control signals v'_α and v'_β, marked by ①, come from the positive-sequence voltage detector, as will be better visualized later when the complete UPQC controller is presented. The currents i_{Sa}, i_{Sb}, and i_{Sc} are transformed into the $\alpha\beta$ reference frame as follows:

$$\begin{bmatrix} i_{S0} \\ i_{S\alpha} \\ i_{S\beta} \end{bmatrix} = \sqrt{\frac{2}{3}} \begin{bmatrix} \frac{1}{\sqrt{2}} & \frac{1}{\sqrt{2}} & \frac{1}{\sqrt{2}} \\ 1 & -\frac{1}{2} & -\frac{1}{2} \\ 0 & \frac{\sqrt{3}}{2} & -\frac{\sqrt{3}}{2} \end{bmatrix} \begin{bmatrix} i_{Sa} \\ i_{Sb} \\ i_{Sc} \end{bmatrix} \quad (6.21)$$

This control algorithm uses the fundamental positive-sequence component of voltages v_{Sa}, v_{Sb}, and v_{Sc}, represented by v'_a, v'_b, and v'_c in (6.19). They are already transformed to the $\alpha\beta$-reference frame (v'_α and v'_β), to calculate the instantaneous real and imaginary powers:

$$\begin{bmatrix} p_h \\ q_h \end{bmatrix} = \begin{bmatrix} v'_\alpha & v'_\beta \\ -v'_\beta & v'_\alpha \end{bmatrix} \begin{bmatrix} i_{S\alpha} \\ i_{S\beta} \end{bmatrix} \quad (6.22)$$

This approach of three-phase, four-wire UPQC was developed in 1995 [29,30]. The corresponding simulation and experimental results, which will be discussed in the following sections, were also obtained at that time. In 1995, only the original definition of imaginary power $q = v_\alpha i_\beta - v_\beta i_\alpha$, as introduced in [38], was in use, instead of $q = v_\beta i_\alpha - v_\alpha i_\beta$ that is adopted in this book. For convenience, this control method is kept unchanged here and all resulting consequences will be noted in the text.

High-pass filters are used in Fig. 6-30 to separate the oscillating real ($\tilde{p}_h$) and imaginary powers ($\tilde{q}_h$) from the calculated powers p_h and q_h, respectively. Ideally, the cutoff frequency should be set as low as possible, in order to well separate the dc levels $\bar{p}_h$ and $\bar{q}_h$ from p_h and q_h. In this case, $\tilde{p}_h$ and $\tilde{q}_h$ should contain all harmonics, as well as the current imbalance due to the fundamental negative-sequence component $\dot{I}_{-1}$ that also generates oscillating powers with $\dot{V}_{+1}$, at double frequency. However, high-pass filters with a low cutoff frequency have long settling times that degenerate the dynamic response of the control algorithm. Therefore, fifth-order Butterworth high-pass filters with a cutoff frequency of 50 Hz were used. Note that this choice stays close to "the maximal limit" of the possible cutoff frequency. For instance, from (3.98) and (3.99), it is possible to see that the product of a second harmonic from the positive-sequence component in the current ($\dot{I}_{+2}$) and the fundamental voltage component $\dot{V}_{+1}$ generates $\tilde{p}_h$ and $\tilde{q}_h$ at the fundamental frequency, which would not pass fully through the high-pass filters.

If necessary, a high-pass filter can also be inserted in the path of the calculated zero-sequence current i_{S0}. As mentioned, the shunt active filter compensates the neutral current of the nonlinear load. Hence, i_{S0} in Fig. 6-27 should be ideally zero. However, other loads eventually connected at the right side of the UPQC, which are not being compensated, can produce undesirable zero-sequence currents and cause i_{S0} to become nonzero.

Once the powers $\tilde{p}_h$ and $\tilde{q}_h$ are separated, the positive-sequence and negative-sequence harmonic components in the $\alpha\beta$-reference frames are determined in (6.23), in accordance with the old definition of imaginary power employed in (6.22). Then, together with the current signal i_{h0}, the inverse transformation is performed, that is,

$$\begin{bmatrix} i_{h\alpha} \\ i_{h\beta} \end{bmatrix} = \frac{1}{v_\alpha'^2 + v_\beta'^2} \begin{bmatrix} v_\alpha' & -v_\beta' \\ v_\beta' & v_\alpha' \end{bmatrix} \begin{bmatrix} -\tilde{p}_h \\ -\tilde{q}_h \end{bmatrix} \qquad (6.23)$$

$$\begin{bmatrix} i_{ha} \\ i_{hb} \\ i_{hc} \end{bmatrix} = \sqrt{\frac{2}{3}} \begin{bmatrix} \frac{1}{\sqrt{2}} & 1 & 0 \\ \frac{1}{\sqrt{2}} & -\frac{1}{2} & \frac{\sqrt{3}}{2} \\ \frac{1}{\sqrt{2}} & -\frac{1}{2} & -\frac{\sqrt{3}}{2} \end{bmatrix} \begin{bmatrix} -i_{h0} \\ i_{h\alpha} \\ i_{h\beta} \end{bmatrix} \qquad (6.24)$$

The harmonic currents are multiplied by a control gain K_r, representing a fictitious series "harmonic resistance," to produce the voltage references v_{ha}, v_{hb}, and v_{hc}:

$$\begin{bmatrix} v_{ha} \\ v_{hb} \\ v_{hc} \end{bmatrix} = K_r \begin{bmatrix} i_{ha} \\ i_{hb} \\ i_{hc} \end{bmatrix} \qquad (6.25)$$

The above control algorithm generates the voltage references v_{ha}, v_{hb}, and v_{hc} that can be added to the compensating voltage signals v'_{Ca}, v'_{Cb}, and v'_{Cc} in (6.19) to compose new compensating voltage references v^*_{Ca}, v^*_{Cb}, and v^*_{Cc}:

$$\begin{bmatrix} v^*_{Ca} \\ v^*_{Cb} \\ v^*_{Cc} \end{bmatrix} = \begin{bmatrix} v'_{Ca} \\ v'_{Cb} \\ v'_{Cc} \end{bmatrix} + \begin{bmatrix} v_{ha} \\ v_{hb} \\ v_{hc} \end{bmatrix} \qquad (6.26)$$

The new voltage references in (6.26) force the series active filter to compensate all harmonics and voltage imbalance, which may be present in the measured voltages v_{Sa}, v_{Sb}, and v_{Sc}, and additionally enhance the system stability, providing harmonic current isolation between source and load.

If a three-phase system without neutral wire is considered, the control circuit may be simplified by eliminating zero-sequence current variables as follows:

1. Eliminating i_{S0} in (6.21), and i_{h0} in (6.24)
2. Eliminating the high-pass filter that obtains i_{h0} in Fig. 6-30 (shaded area)

6.2.2.B.3. Integration of the Series and Shunt Active Filter Controllers.

Using the sinusoidal source current control strategy as presented in Chapter 4 and the control parts for the series converter as discussed above, the complete controller for a three-phase four-wire UPQC can be implemented, as shown in Fig. 6-31.

In Fig. 6-31, it is possible to see how the voltage references v^*_{Ca}, v^*_{Cb}, and v^*_{Cc} are composed, according to (6.19) and (6.26), comprising all voltage imbalances and harmonics to be compensated, as well as the auxiliary compensating voltages to improve stability and provide harmonic current isolation between the load and the network. All equations of each control box in this figure can be found in the previous sections, for the control part related to the series active filter, and in Chapter 4 for the rest of the UPQC controller.

The source currents i_{Sa}, i_{Sb}, and i_{Sc} should not contain any zero-sequence components, since the shunt active filter compensates the neutral current of the nonlinear load. Consequently, the calculated zero-sequence current i_{S0} should be zero. However, a dynamic hysteresis current control is used to control the dc voltage difference ($V_{dc2} - V_{dc1}$). It forces the shunt active filter to draw from the network the neutral current with the very low frequency equal to that of the signal in Fig. 6-31. This low-frequency component in i_{S0} may be cut out by applying a low-pass filter. In this way, only zero-sequence components at higher frequencies that eventually persist in the source currents are considered in the calculation of the harmonic currents i_{ha}, i_{hb}, and i_{hc}. Otherwise, the voltage references v_{ha}, v_{hb}, and v_{hc} would induce a very slow amplitude modulation on the phase voltages at the load side of the UPQC. A high-pass filter with cutoff frequency set at 5 Hz is applied to separate the signal i_{h0} from i_{S0}. Since the shunt active filter compensates the load harmonics, the values of i_{ha}, i_{hb}, and i_{hc} are expected to be very small in the steady state. The proportional gain K_r, applied to obtain v_{ha}, v_{hb}, and v_{hc} makes the

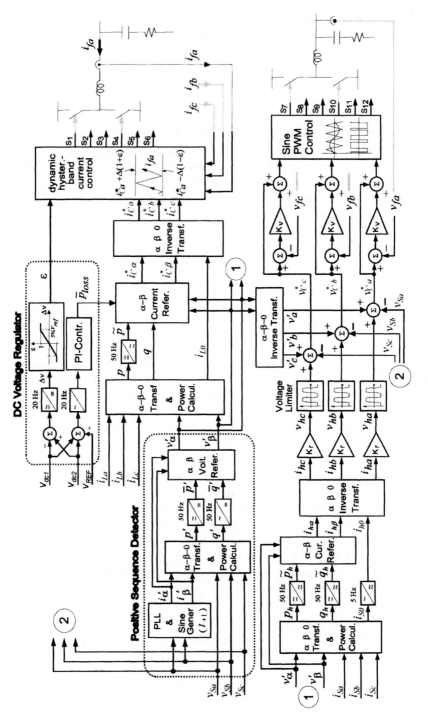

Figure 6-31. The controller of the unified power quality conditioner.

series active filter behave as an additional series resistance of K_r Ω, connected between source and load, but effective *only for harmonic currents,* including the harmonic zero-sequence current i_{h0}.

The series active filter does not generate any fundamental voltage component, except that required to compensate unbalances in supply voltage, caused by negative- and zero-sequence components at the fundamental frequency. This compensation function is included in the control signals v'_{Ca}, v'_{Cb}, and v'_{Cc}, as given by (6.19). The series active filter could also be controlled to regulate the fundamental positive-sequence voltage at the load bus; however, this point is not analyzed in this book.

Compensation errors may appear during transient responses in the UPQC. They are mostly caused by the dynamics of the filters (Butterworth) in the control circuits. While the source currents i_{Sa}, i_{Sb}, and i_{Sc} are changing, the filters cannot determine correctly the actual values of powers $\tilde{p}_h$ and $\tilde{q}_h$. Until a new operating point is reached, wrong values are determined for i_{ha}, i_{hb}, and i_{hc}, which can be significant in some cases. This problem is better explained in the next section. Therefore, in order to avoid hard voltage sags at load terminals (v_{La}, v_{Lb}, and v_{Lc}) during varying load conditions, the signals v_{ha}, v_{hb}, and v_{hc} are limited to 10% of the rated ac system voltage by using "voltage limiters" as indicated in Fig. 6-31.

6.2.2.B.4. General Aspects. At this point, some general aspects of series compensators should be commented on. In a real implementation, especially fast bypass switches, perhaps feasible only if employing power semiconductor devices, should be provided to bypass the load currents flowing through the series transformers, protecting them and the PWM converter against high short-circuit currents. Before the blocking command can be sent to the IGBTs of the series active filter, an alternative low-impedance path for the load current must be provided. Hence, at least for a short time, the switches of the series converter have to withstand short-circuit currents, since they may not be blocked before establishing a bypass on the series transformers. This is a problem that has to be dealt carefully.

The specification of the RLC filters at the output of the PWM series converter, from the switching frequency point of view, is relatively easy to accomplish [36]. However, depending on the equivalent impedance of the power generating system and the characteristics of the load, resonance phenomena may appear in the small capacitor C_S, which degenerate the compensation characteristics of the series active filter. An optimization of the passive RLC filters of the shunt PWM converter has also to be performed. In a preliminary analysis, some linearization in the power system can be performed, considering the series active filter as an ideal *voltage* source and the shunt active filter as an ideal *current* source, to derive transfer functions and use linear control techniques for this analysis. However, a final verification of the complete system in a digital simulator that allows more accurate representation of nonlinear behaviors is highly recommended, due to the high number of nonlinear devices involved in the problem, and to take into account the dynamic of the UPQC controller. As final results, nonprohibitive values of resistances have to be found to satisfactorily damp eventual resonance effects.

Alternatively, an interesting approach that includes one more minor feedback control loop in Fig. 6-29, as proposed in [39], can also be investigated. This approach is advantageous for minimizing losses, since the resistance R_S can be theoretically eliminated. However, additional current measurement circuits should be provided to sense the currents flowing through the filter capacitors C_S.

After the design of the RLC filters, the complete system must be optimized, in order to adjust all gains, time constants, and cutoff frequencies of filters in the controller. Again, the use of a digital simulator is suggested to perform this task.

6.2.2.C. Analysis of the UPQC Dynamic

The system shown in Fig. 6-32 is considered in the following performance analysis. The rated supply line-to-line voltage is 380 V. The average switching frequency of the PWM converters should lie between 10 kHz and 15 kHz. Hence, it is reasonable to assume that the UPQC can compensate harmonics up to about 1 kHz. Next, some issues are addressed to guide parameter optimization in the power circuit of the UPQC, as well as in its controller.

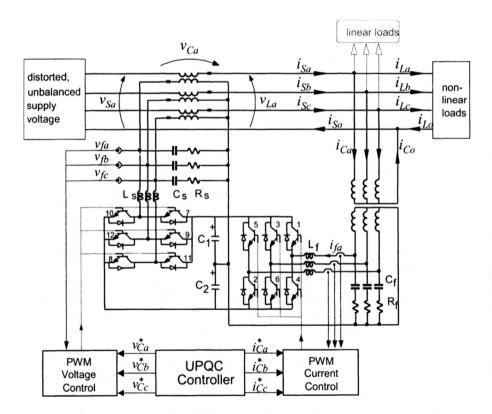

Figure 6-32. Combined series and shunt active filters (UPQC) to compensate voltages and currents in three-phase, four-wire systems.

6.2.2.C.1. Optimizing the Power System Parameters.
Assuming 600 V for the dc link ($V_{dc1} + V_{dc2}$) and a 380-V (50 Hz), three-phase power supply, the turns ratios of the transformers were fixed:

	Turns ratio Primary side[a] : Secondary side[b]
Series single-phase transformers	1 : 2
Shunt three-phase transformer	2 : 1

[a]primary side = network side (380 V).
[b]secondary side = PWM converter side.

Here, ideal transformers are considered. Once the turn ratios of the transformers are fixed and a value for the equivalent network impedance is estimated, iterative calculation can be used to optimize the commutation reactance (L_f and L_S) and the high-pass filters (C_f, R_f, C_S, and R_S). The equivalent circuit that has been considered is presented in Fig. 6-33. The series PWM converter is modeled as a *voltage* source, whereas the shunt PWM converter is modeled as a *current* source. The harmonic-generating load is modeled as generic current source. The equivalent impedance of the power supply is reduced to a discrete reactance L_n.

Functions in the frequency domain, $F(j\omega)$, relating the voltages $\dot{V}_1$ and $\dot{V}_2$ and currents $\dot{I}_1$ and $\dot{I}_2$, can be written, combining the impedances shown in Fig. 6-33. The turn ratios of the transformers have to be taken into account when reflecting impedances on the secondary side to the primary side, and vice-versa. Short-circuiting the voltage sources and opening the current sources, the following relations can be found:

- High-pass filter of the shunt PWM converter reflected to the primary side:

$$Z'_f = \left(\frac{2}{1}\right)^2 \left(R_f + \frac{1}{j\omega C_f}\right) \qquad (6.27)$$

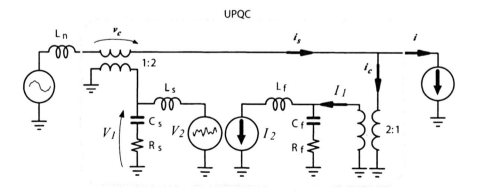

Figure 6-33. Equivalent circuit for optimizing main parameters of the UPQC.

- Z_f' and network impedance reflected to the secondary side of the series single-phase transformer:

$$Z'_{ns} = \left(\frac{2}{1}\right)^2 (Z'_f + j\omega L_n) \tag{6.28}$$

- Z'_{ns} in parallel with the high-pass filter of the series PWM converter:

$$Z_{1s} = \frac{Z'_{ns}\left(R_s + \dfrac{1}{j\omega C_s}\right)}{Z'_{ns} + \left(R_s + \dfrac{1}{j\omega C_s}\right)} \tag{6.29}$$

Using (6.29), the relation between $\dot{V}_1$ and $\dot{V}_2$ are found to be:

$$\frac{\dot{V}_1}{\dot{V}_{+2}} = \frac{Z_{1s}}{Z_{1s} + j\omega L_s} \tag{6.30}$$

From (6.30), Bode diagrams can be drawn to aid in the dimensioning of impedances of the UPQC power circuit. Since impedances of the series PWM converter (R_s, L_s, and C_s) influence the performance of the shunt PWM converter, and vice-versa, another function should be found for describing the current of the shunt PWM converter:

- Impedances of the series PWM converter reflected to the primary side:

$$Z'_s = \left(\frac{2}{1}\right)^2 \frac{j\omega L_s\left(R_s + \dfrac{1}{j\omega C_s}\right)}{j\omega L_s + \left(R_s + \dfrac{1}{j\omega C_s}\right)} \tag{6.31}$$

- Z'_s and the network impedance reflected to the secondary side of the shunt three-phase transformer:

$$Z'_{nf} = \left(\frac{1}{2}\right)^2 (Z'_s + j\omega L_n) \tag{6.32}$$

Using (6.32), the relation between $\dot{I}_1$ and $\dot{I}_2$ are found to be:

$$\frac{\dot{I}_1}{\dot{I}_2} = \frac{\left(R_f + \dfrac{1}{j\omega C_f}\right)}{Z'_{nf} + \left(R_f + \dfrac{1}{j\omega C_f}\right)} \tag{6.33}$$

The transfer function given by (6.33) should be analyzed together with (6.30), although the above equations alone are not sufficient for optimizing the impedances of the power circuit. The whole system, including the control of the UPQC, has to be checked as a unit, in an iterative way, to reach "optimal" values for the parameters in the power circuit simultaneously with those in the UPQC control. Hence, after this preliminary parameter optimization, changes in parameters are continuously verified in a digital simulator running the complete model of the UPQC prototype.

After some iterative analysis, the following set of parameters was chosen to be used in the UPQC performance analysis: $R_f = 1\,\Omega$, $L_f = 0.8$ mH, $C_f = 100\,\mu\text{F}$, $R_s = 3\,\Omega$, $L_s = 0.4$ mH, $C_s = 60\,\mu\text{F}$, and $L_n = 0.6$ mH.

6.2.2.C.2. Optimizing the Parameters in the Control Systems.
There are several parameters in the UPQC controller and in the PWM controllers that may be investigated for improving the performance of the UPQC. In the PWM controllers, two parameters are very important:

1. The hysteresis bandwidth, Δ, in the PWM current control (dynamic hysteresis-band current control) of the shunt active filter
2. The gain K_V in the sine PWM control of the series active filter

The switching frequency of the shunt PWM converter changes if the bandwidth Δ of the dynamic hysteresis current control varies. The hysteresis control must protect the IGBTs against very high switching frequencies, since it has an inherent variable switching frequency. Therefore, a supervision circuit for minimum interval between successive turn-on and turn-off pulses, set at 5 μs, has been implemented. Additionally, this circuit requires that a minimum interval between two successive turn-on pulses is preserved, which was set at 70 μs. The final value of Δ was set at $\Delta = 5$ A. The triangular carrier wave in the PWM voltage control in the series converter has amplitude of ± 300 V and frequency of 14.7 kHz. For the most cases, $K_V = 5$ was acceptable.

When the analysis was focused on the minimization of energy storage capacity in the common dc link, some changes had to be made in the dc-voltage regulator (see Fig. 6-31). The cut-off frequency of the low-pass filter—a fourth-order Bessel type—in the path that determines $\bar{p}_{loss}$ was increased to 25 Hz. The PI controller gains were made equal to $K_P = 60$ W/V (proportional) and $K_I = 400$ W/Vs (integral). The cut-off frequency of the other fourth-order Bessel-type filter in the path that determines ε has been decreased to 15 Hz, and the limit function that generates the signal ε was changed to 10% of V_{ref}, where $V_{ref} = 600$ V. Each dc capacitor (C_1 and C_2) was made equal to 2000 μF.

In the control circuit that generates the voltage references v^*_{Ca}, v^*_{Cb}, and v^*_{Cc}, special attention has been paid to the following parameters:

1. Cut-off frequency of the high-pass filters
2. Proportional gain K_r
3. Limit value of the voltage references v_{ha}, v_{hb}, and v_{hc}

Three high-pass filters are used in the control circuit of the series active filter. Two of them separate higher frequencies of powers ($\tilde{p}_h$ and $\tilde{q}_h$) and the third one separates harmonic currents (i_{h0}), but all these filters are employed for the same purpose: to determine the harmonic currents flowing into the network (i_{ha}, i_{hb}, and i_{hc}). Ideally, no harmonic currents should flow into the network, since the shunt active filter compensates harmonic currents of the nonlinear load. Furthermore, the voltage references v'_{Ca}, v'_{Cb}, and v'_{Cc} given by (6.19) make the series active filter compensate all harmonics and imbalances present in the supply voltages v_{Sa}, v_{Sb}, and v_{Sc}. Therefore, the compensated voltages v_{La}, v_{Lb}, and v_{Lc} should not excite any harmonic current to flow between network and load. However, there is at least one kind of harmonic current that is not compensated, neither by the shunt active filter, nor by v'_{Ca}, v'_{Cb}, and v'_{Cc}. It consists of harmonic current excited by the active filters themselves. This analysis should investigate the total impedance Z_{res} connected to the power supply, determined by the sum of the impedances given in (6.27), (6.31), and $j\omega L_n$ when there is no load connected to it. This total impedance ($Z_{res} = j\omega L_n + Z'_s + Z'_f$) may amplify resonance effects at frequencies at which it assumes low values. At a specified operating range of frequency, the series active filter should be able to generate compensating voltage v_{ha}, v_{hb}, and v_{hc}, as given in (6.25), in order to attenuate harmonic currents through the series active filter, enhancing the overall system stability.

Most simulation cases omit the high-pass filter for separating i_{h0} from i_{So}. For these cases, i_{So} was directly passed to the next control box: the $\alpha\beta 0$ inverse transformation. This solution should be always adopted when i_{Lo} is well compensated and the signal ε does not induce high zero-sequence currents in the network. At the beginning, the cut-off frequency for the other two high-pass filters that separate $\tilde{p}_h$ and $\tilde{q}_h$ was set at 50 Hz. Then, the final results presented in the next section have successfully used 150 Hz as the cutoff frequency.

The voltage references v_{ha}, v_{hb}, and v_{hc} are added to v^*_{Ca}, v^*_{Cb}, and v^*_{Cc} to form v'_{Ca}, v'_{Cb}, and v'_{Cc} in Fig. 6-31, provide harmonic isolation between load and network by inserting a fictitious "harmonic" resistance in series with the system. The proportional gain K_r corresponds to the value of this constant, fictitious resistance, being effective only in the presence of harmonic currents i_{ha}, i_{hb}, and i_{hc}. Most simulation cases used K_r between 8 and 10 Ω. The voltages signals v_{ha}, v_{hb}, and v_{hc} were limited to ± 30 V.

6.2.2.C.3. Simulation Results.
To compose an unbalanced, nonlinear load, power electronics devices listed in Table 6.2 are used.

Table 6.2. Specification of the composed nonlinear load

loads	One three-phase thyristor converter; firing angle = 30°; I_{dc} = 8 A; commutation inductance = 3 mH
	One single-phase thyristor converter connected between the *a* phase and neutral wire; firing angle = 15°; I_{dc} = 10 A; commutation inductance = 3 mH
	One single-phase diode bridge connected between the *b* phase and neutral; L_{dc} = 300 mH ; R_{dc} = 20 Ω; commutation inductance = 3 mH

Several simulation cases with different compositions of harmonic pollution and voltage imbalances in the supply voltage were investigated. The connection of the series and shunt active filters of the UPQC does not disturb the system. The active filters act almost instantaneously after unblocking the PWM converters if the dc capacitors are precharged at their reference value (300 V in each capacitor) and the UPQC controller is already in operation and has reached the steady state. As an example, Fig. 6-34 shows the compensated voltages and currents with a successive connection of the shunt at $t = 80$ ms, and series active filter at $t = 140$ ms. It is possible to see spikes in the currents, even after the connection of the series active filter. The principal reason is that, in this simulation case, the value of L_f was made equal to 2 mH, which disagrees with that "optimal" set of parameters listed on page 311. The commutation inductance L_f is too large, which imposes limited ratios of di/dt on the shunt PWM converter, so that the compensating currents i_{Ca}, i_{Cb}, and i_{Cc}, as shown in Fig. 6-32, cannot fully cancel the high di/dt present in the load cur-

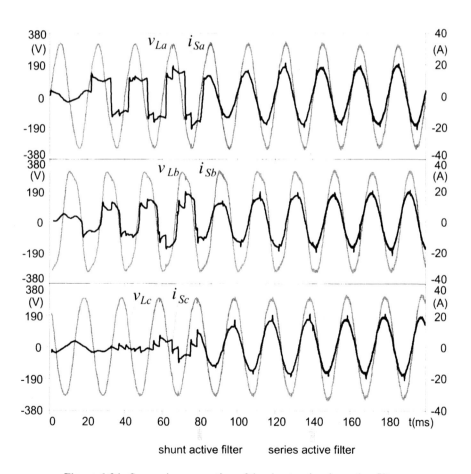

Figure 6-34. Successive connection of the shunt and series active filters.

rents.

Another simulation, now using the optimized set of parameters, is presented in Fig. 6-35. It presents better compensated currents than those in Fig. 6-34. A short period that involves the connection of the series active filter is focused on. At this instant, $t = 140$ ms, the load and the shunt active filter are already connected, but the system is not in steady state.

The connection of the series active filter is made by opening the bypass switches at the primary side of the single-phase transformers and starting the firing of the IGBTs of the series PWM converter. When the series active filter is inserted, the total impedance Z_{res} determined by the sum of impedances given in (6.27), (6.31) and $j\omega L_n$, changes, and a new resonant frequency appears. This can be noted in the currents i_{Sa}, i_{Sb}, and i_{Sc} in Fig. 6-35. The series active filter is able to produce compensating voltages at this new frequency, providing damping to the system, as well as harmonic current isolation between source and load. Fig. 6-35 and subsequent figures refer to a case involving the following events:

Event #1: starting the shunt PWM converter at 40 ms

Event #2: connecting a single-phase diode bridge rectifier at 60 ms, as a load between *b phase* and neutral conductor

Event #3: connecting a single-phase thyristor bridge rectifier at 61.7 ms ($\cong 30°$ firing angle), as a load between *a phase* and neutral conductor

Event #4: connecting a six-pulse thyristor bridge rectifier at 63.3 ms ($\cong 30°$ firing angle), as a three-phase load

Event #5: connecting the series active filter at 140 ms

Event #6: disconnecting the six-pulse thyristor bridge rectifier at 185 ms

The total simulation time interval was 260 ms. All graphics in Fig. 6-36 to Fig. 6-42 refer to this simulation case.

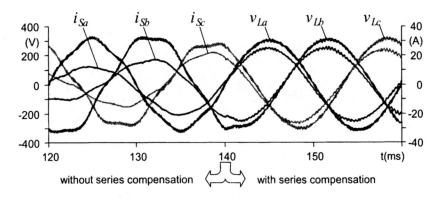

Figure 6-35. Connection of the series active filter.

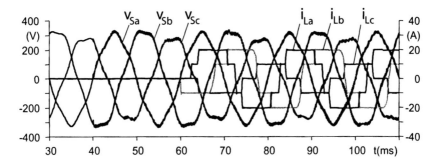

Figure 6-36. Supply voltages and load currents.

At first, the connection of the load is presented. This is perhaps the worst transient period for the UPQC, because the dc capacitors begin to discharge, supplying the whole power of the load at the beginning of the transient period. Figure 6-36 shows the load currents and the supply voltages at the left side (source side) of the UPQC. Figure 6-37 shows the compensated currents drawn from the network and the compensated voltages at the load terminal (right-hand side of the UPQC). The supply voltages and the compensated voltages have similar waveshapes, since the series active filter was still disconnected during this period. Figure 6-36 shows that the load current is inductive (lagging currents). Figure 6-37 shows that the compensated currents drawn from the network are a little capacitive (they are leading currents). This capacitive-reactive power corresponds to the reactive power of the shunt high-pass filter, represented by C_f in Fig. 6-33. Current ripples due to the high-frequency switching of the shunt PWM converter are also flowing to this filter. Moreover, harmonic currents excited by the harmonic voltages in the source are also flowing to this shunt passive filter. This can be seen in Fig. 6-37, for $t < 60$ ms, which corresponds to the period when no load and active filters are connected to the system. It should be noted that this kind of harmonic current is not compensated by the shunt active filter. Only the damping algorithm, represented by signals v_{ha}, v_{hb},

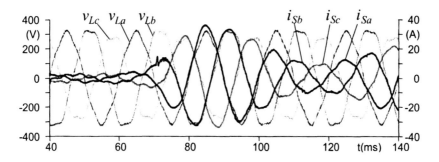

Figure 6-37. Compensated voltages and currents.

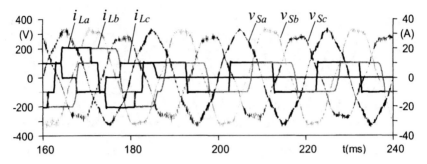

Figure 6-38. Supply voltages and load currents.

and v_{hc} of the series active filter in Fig. 6-31 attenuates this type of harmonic current. High transient currents appear after the connection of the load, and they are caused by the control signal $\bar{p}_{loss}$ of the dc voltage regulator in Fig. 6-31. Higher capacitance at the dc link should be provided if a smooth transient for the source currents is required.

After the transient period presented in Fig. 6-37, another event occurs at $t = 140$ ms: the connection of the series active filter. This event was already presented in Fig. 6-35. Then, the six-pulse thyristor bridge rectifier is disconnected at $t = 185$ ms, as shown in Fig. 6-38. At the disconnecting of this load, the dc capacitors tend to charge, increasing the voltage at the dc link around $t = 190$ ms, as shown in Fig. 6-40, due to an error in calculation of the shunt active filter controller. This is caused by the dynamics of the fifth-order Butterworth high-pass filter (see Fig. 6-31). This decreases relatively slowly the source currents, as shown in Fig. 6-39. While the source currents are decreasing, the series active filter controller is also inducing compensation errors. The dynamics of the fifth-order Butterworth high-pass filters, which separate the powers $\tilde{p}_h$ and $\tilde{q}_h$ in the auxiliary control circuit for the harmonic isolation that generates v_{ha}, v_{hb}, and v_{hc}. This leads to an increase in the amplitude of the compensated voltages v_{La}, v_{Lb}, and v_{Lc}. The slower feedback

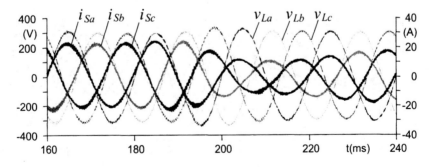

Figure 6-39. Compensated voltages and currents.

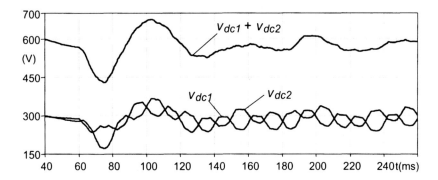

Figure 6-40. The dc capacitor voltages.

control loops formed by the control signals $\bar{p}_{loss}$ and ε from the dc voltage regulator proved to be effective in correcting the compensation errors caused by the UPQC controller, as can be seen in Fig. 6-40, where the dc capacitor voltages v_{dc1} and v_{dc2} are well regulated.

From Fig. 6-40, it is possible to conclude that the connection of load is a critical phenomenon that has to be taken into account during dimensioning of the dc capacitors. Another steady-state critical phenomenon that should be taken into account when specifying the dc capacitors is the power oscillation resulting from caused by the compensation of the neutral current of the load i_{Lo}. This is the lowest frequency (50 Hz) of voltage fluctuation at the dc link. The compensation of $\tilde{p}$ of the load does not cause significant power oscillations, if compared with these two critical phenomena.

A survey of the instantaneous powers involved in this simulation case is shown in Fig. 6-41 and Fig. 6-42. The real powers and imaginary powers in these figures are calculated as follows.

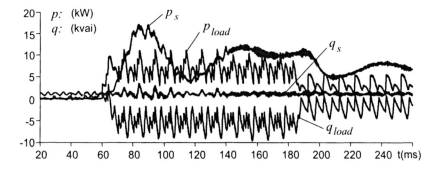

Figure 6-41. Real and imaginary power of the load and the compensated voltage and current.

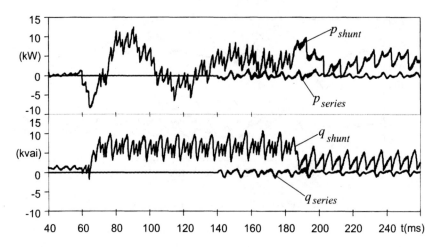

Figure 6-42. Real and imaginary powers of the series and shunt active filters.

Voltages and currents* transformed into the $\alpha\beta$-reference frame are:

$$\begin{bmatrix} v_{L\alpha} \\ v_{L\beta} \end{bmatrix} = \sqrt{\frac{2}{3}} \begin{bmatrix} 1 & -\frac{1}{2} & -\frac{1}{2} \\ 0 & \frac{\sqrt{3}}{2} & -\frac{\sqrt{3}}{2} \end{bmatrix} \begin{bmatrix} v_{La} \\ v_{Lb} \\ v_{Lc} \end{bmatrix}$$

$$\begin{bmatrix} i_{S\alpha} \\ i_{S\beta} \end{bmatrix} = \sqrt{\frac{2}{3}} \begin{bmatrix} 1 & -\frac{1}{2} & -\frac{1}{2} \\ 0 & \frac{\sqrt{3}}{2} & -\frac{\sqrt{3}}{2} \end{bmatrix} \begin{bmatrix} i_{Sa} \\ i_{Sb} \\ i_{Sc} \end{bmatrix}$$

$$\begin{bmatrix} v_{C\alpha} \\ v_{C\beta} \end{bmatrix} = \sqrt{\frac{2}{3}} \begin{bmatrix} 1 & -\frac{1}{2} & -\frac{1}{2} \\ 0 & \frac{\sqrt{3}}{2} & -\frac{\sqrt{3}}{2} \end{bmatrix} \begin{bmatrix} v_{Ca} \\ v_{Cb} \\ v_{Cc} \end{bmatrix}$$

$$\begin{bmatrix} i_{C\alpha} \\ i_{C\beta} \end{bmatrix} = \sqrt{\frac{2}{3}} \begin{bmatrix} 1 & -\frac{1}{2} & -\frac{1}{2} \\ 0 & \frac{\sqrt{3}}{2} & -\frac{\sqrt{3}}{2} \end{bmatrix} \begin{bmatrix} i_{Ca} \\ i_{Cb} \\ i_{Cc} \end{bmatrix}$$

*Note: the sumbols of voltages and current used here agree with those used in Fig. 6-32.

$$\begin{bmatrix} i_{L\alpha} \\ i_{L\beta} \end{bmatrix} = \sqrt{\frac{2}{3}} \begin{bmatrix} 1 & -\frac{1}{2} & -\frac{1}{2} \\ 0 & \frac{\sqrt{3}}{2} & -\frac{\sqrt{3}}{2} \end{bmatrix} \begin{bmatrix} i_{La} \\ i_{Lb} \\ i_{Lc} \end{bmatrix}$$

Powers of the load are:

$$p_{load} = v_{L\alpha} i_{L\alpha} + v_{L\beta} i_{L\beta}$$

$$q_{load} = v_{L\alpha} i_{L\beta} - v_{L\beta} i_{L\alpha}$$

Powers of the shunt active filter are:

$$p_{shunt} = v_{L\alpha} i_{C\alpha} + v_{L\beta} i_{C\beta}$$

$$q_{shunt} = v_{L\alpha} i_{C\beta} - v_{L\beta} i_{C\alpha}$$

Powers of the series active filter are:

$$p_{series} = v_{C\alpha} i_{S\alpha} + v_{C\beta} i_{S\beta}$$

$$q_{series} = v_{C\alpha} i_{S\beta} - v_{C\beta} i_{S\alpha}$$

Compensated powers are:

$$p_s = v_{L\alpha} i_{S\alpha} + v_{L\beta} i_{S\beta}$$

$$q_s = v_{L\alpha} i_{S\beta} - v_{L\beta} i_{S\alpha}$$

Fig. 6-42, shows that the power rating of the series active filter is much smaller than that of the shunt active filter, although the source currents flow through the series active filter. The shunt active filter compensates almost instantaneously the changes in the imaginary power of the load, as can be seen by comparing the curves of q_{load} in Fig. 6-41 with that of q_{shunt} in Fig. 6-42. The reason is that the total imaginary power ($q = \bar{q} + \tilde{q}$) of the load is being compensated, so that no high-pass filter is needed in this control path, as can be seen in Fig. 6-31. However, the compensation of the oscillating real power ($\tilde{p}$) of the load is troublesome. The real power of the shunt active filter (p_{shunt}), as well as the real power drained from the network (p_s), has presented high oscillations during transients, which perhaps could be unacceptable in a real case. Moreover, the digital model of the UPQC showed poor performance concerning losses in the UPQC. Figure 6-41 shows that the real power drained from the network is far greater than the average real power of the load, even during steady-state conditions, as can be observed in the period of 160 ms $< t <$ 180 ms. Although this drawback in the UPQC digital model does not invalidate the present analysis, more accurate digital models of IGBTs and diodes should be used in the PWM converters.

The above simulation results were compared with experimental ones, obtained from a UPQC prototype that was tested in the laboratory. The UPQC prototype was tested in similar conditions of loading and voltage ratings as to above simulation results. A short discussion about the principal experimental results is presented in the next section.

6.2.2.C.4. Experimental Results.

The power circuit of the UPQC prototype was tested with the same nonlinear loads and power supply ratings as those used in the tests of the prototype of shunt active filter presented in Chapter 4. Here, the UPQC prototype replaced the shunt active filter in Fig. 4-80. The following experimental results confirm the dynamic performance shown above and obtained in the digital simulator.

First, a successive connection of the shunt and series active filters is presented. The results of this connection can be compared with the simulation results presented in Fig. 6-34. The recorded currents are shown in Figure 6-43. Figure 6-44 shows the compensated voltages and Fig. 6-45 shows the dc capacitor voltages. The imbalance of the phase voltages consists of 33 V (rms) of negative-sequence component at fundamental frequency.

The unbalanced and distorted supply voltage causes saturation in the three-phase transformer of the shunt active filter, which excites relatively high harmonic currents flowing from the network to this transformer, as can be seen in curve i_{Ca} in Fig. 6-43, for $t < 0$ ms. These currents are not compensated by the shunt active filter, and the compensated currents remain distorted, even after the connection of the shunt active filter, as can be seen in the period of $0 < t < 80$ ms. After the connection of the series active filter, the source currents i_{Sa}, i_{Sb}, and i_{Sc} in Fig. 6-43, and the

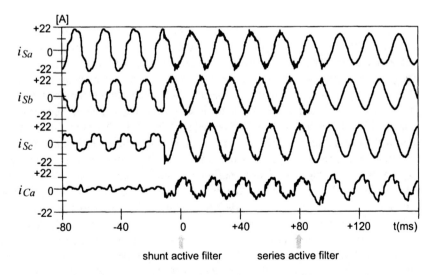

Figure 6-43. Currents from a successive connection of the shunt and series active filters.

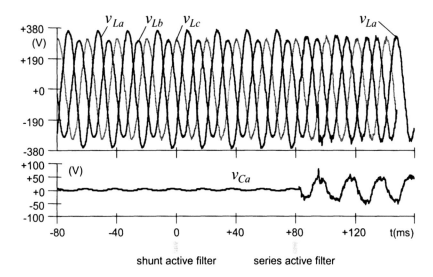

Figure 6-44. Voltages from a successive connection of the shunt and series active filters.

voltages v_{La}, v_{Lb}, and v_{Lc} at load terminal in Fig. 6-44, become almost sinusoidal and balanced.

The recorded measurements were decomposed in Fourier series. Then, the symmetrical components were calculated. A typical result that has been found is a reduction in voltage imbalance from 14% to 3%. It is expected that this relation can be improved further, if more care is taken in using adequate transformers and commutation reactances, as well as making improvements in the control circuits, especially to avoid noises in the measurements used in the controller of the laboratory prototype of the UPQC.

The dc voltage regulator implemented in the UPQC prototype does not correspond to that given in Fig. 6-31. Only a proportional gain was used, instead of a PI controller. Fig. 6-45 shows that the dc voltage regulator stabilizes this voltage around 270 V, which corresponds to a steady-state error of 30 V to its reference value. Four capacitors of 3300 μF (350 Vdc), connected in parallel, for each capaci-

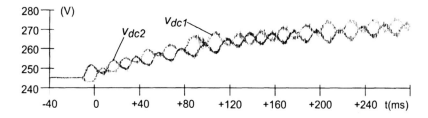

Figure 6-45. Experimental results. dc capacitor voltages.

tance C_1 and C_2 were used, although simulation results indicated that a single 3300 µF capacitor could be enough. Consequently, the dc voltage variations in the experimental results were lower than those in the simulation results. In the worst case, the connection of the load was realized successfully.

The loading conditions and the power circuit that was mounted to test the UPQC prototype have the same characteristics as those illustrated in Fig. 4-80, which were used to test a three-phase, four-wire, shunt active filter. The voltage imbalance that was imposed on the test system is shown in Fig. 6-44. A case of load connection was realized, having the UPQC (series *and* shunt active filters) already in operation. Figure 6-46 shows the load currents and Fig. 6-47 the compensated voltages at load terminals. After the connection of the load at $t = 0$ ms, the currents drained from the network (compensated currents) begin to increase slowly, and do not present any overshoots, as can be seen in Fig. 6-48. As mentioned before, the connection of the load causes transitory errors in calculations of $\tilde{p}_h$ and $\tilde{q}_h$ in the series active filter controller. This excites excessive "harmonic damping" produced by the compensating signals v_{ha}, v_{hb}, and v_{hc}. However, these signals are passed through voltage limiters, which allow only noncritical voltage sags at the load terminal, as can be seen during $0 < t < +40$ ms in Fig. 6-47.

The phase voltages at the load terminal were well balanced, as were the compensated currents. The expressive load imbalance that draws the neutral current i_{Lo}, as shown in Fig. 6-49, was well compensated by the shunt active filter (i_{Co}), so that, for the power supply, it seems like "a balanced linear load." The line currents of the shunt active filter, which are also compensating the imaginary power of the load, are given in Fig. 6-50. The series active filter generates the voltages v_{Ca}, v_{Cb}, and v_{Cc}, as shown in Fig. 6-51, to compensate the fundamental negative-sequence component and harmonics in the supply voltage, as well as to provide harmonic isolation between source and load.

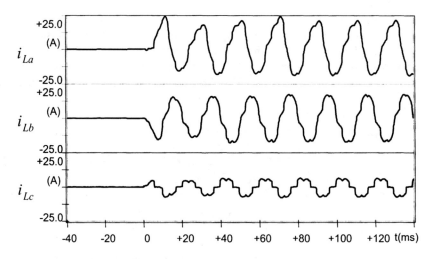

Figure 6-46. Experimental results during connection of the load. load currents.

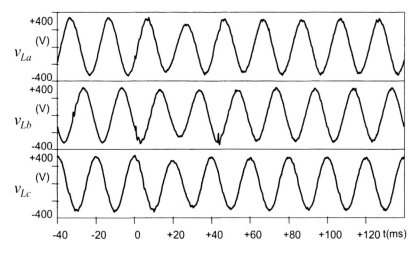

Figure 6-47. Experimental results during connection of the load. voltages at load terminal.

Some calculations of powers have been carried out from the measured voltages and currents. Figure 6-52 shows the instantaneous active three-phase power (sum of the real and zero-sequence powers), the imaginary power of the load, and that generated by the compensated voltages and currents for the above experimental case (the connecting of the load).

Finally, an interesting experimental case involving zero-sequence voltages is described. Again, the same conditions of loading were used. With the load already in

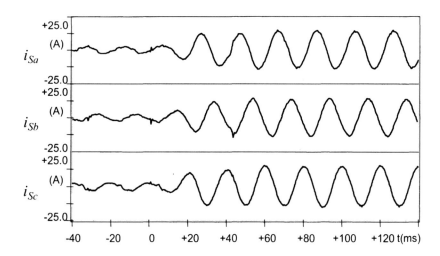

Figure 6-48. Experimental results during connection of the load. compensated currents drained from the network.

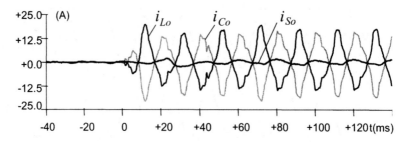

Figure 6-49. Experimental results during connection of the load. neutral currents.

operation, a successive connection of both shunt active filter and series active filter was recorded. The shunt three-phase transformer with a turn ratio of 2:1 in Fig. 6-33 has a very low zero-sequence impedance that allows high magnetization currents if excited by zero-sequence voltages. For the following experiment, the connection of the variable, series transformers in Fig. 4-80 was changed to generate an imbalance from zero-sequence components.

In the next figures, the imbalance of zero-sequence components corresponds to half, that is, 15 V (rms) ≅ 7% of imbalance, of that used in the previous case regarding imbalance of the negative-sequence component. The unbalanced voltages that appear at the load terminal are shown in Fig. 6-53. The zero-sequence components, $V_0 = (v_{Sa} + v_{Sb} + v_{Sc})/3$, begin to be compensated after $t = 120$ ms, when the series active filter is connected. Note that the scale used for the graphic of V_0 is only ±28 V, whereas the scale for v_{Sa}, v_{Sb}, v_{Sc} is ±400 V. Fig. 6-54 shows better the instant of connection of the series active filter. The fundamental zero-sequence voltage ($\dot{V}_{01}$) is in phase with the c-phase voltage (v_{Sc}).

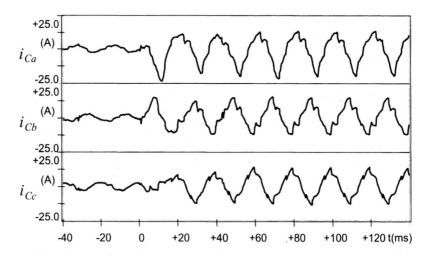

Figure 6-50. Experimental results during connection of the load: currents of the shunt active filter.

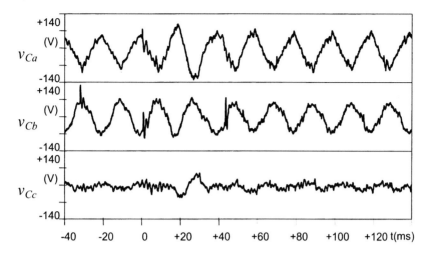

Figure 6-51. Experimental results during connection of the load. voltages of the series active filter.

With the load already in operation, the shunt active filter was connected at $t = 0$ ms. Fig. 6-55 shows the compensated currents i_{Sa}, i_{Sb}, and i_{Sc} drawn from the network, and the current i_{So} in the neutral conductor of the power supply. This neutral current has a peak value greater than those of the line currents, even during the period of $0 < t < 120$ ms, when only the shunt active filter of the UPQC is connected. This means that the zero-sequence currents of the load, which have the same order of magnitude as presented in Fig. 6-49, are not the principal components in the neutral current i_{So} of the source. A significant portion of i_{So} consists of the magnetization current of the three-phase shunt transformer, which is still present during $0 < t < 120$ ms. Hence, the phase of i_{So} lags $\dot{V}_0$ by 90° that is in phase with v_{Sc}. Thus, the phase displacement between i_{So} and the *active* portion of the load current i_{Lb} that is in phase with v_{Sb} is equal to 150°. Since $i_{So}/3$ is flowing through b-phase conductor,

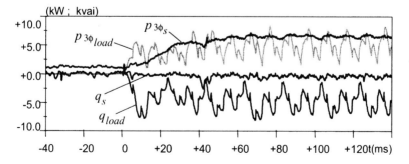

Figure 6-52. Experimental results. instantaneous active three-phase power and imaginary power.

326 COMBINED SERIES AND SHUNT POWER CONDITIONERS

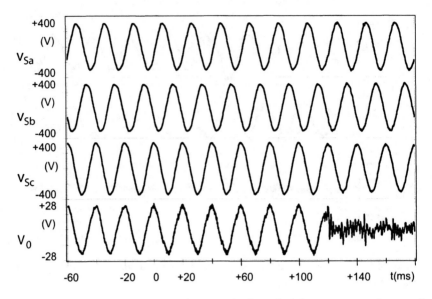

Figure 6-53. Experimental results. voltages at load terminal for a compensation case involving zero-sequence components.

this is the reason why the compensated current i_{Sb} is smaller than i_{Sa} and i_{Sc} during $0 < t < 120$ ms. Then, the series active filter is connected, and for $t > 120$ ms the zero-sequence voltage imbalance of the power supply is being compensated. The neutral current i_{So} is almost eliminated and the line currents become balanced, as shown in Fig. 6-55.

6.2.3. The UPQC Combined with Passive Filters (the Hybrid UPQC)

If the shunt converter of the UPQC does not have to generate high reactive current at the fundamental frequency, a more cost-effective arrangement that combines ac-

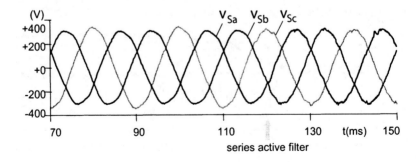

Figure 6-54. The connection of the series active filter.

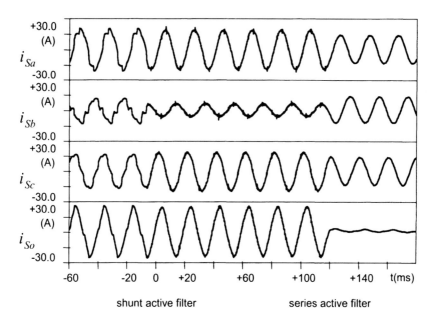

Figure 6-55. Experimental results: currents of the source.

tive and passive filters, as introduced in [31], may be considered. It uses a small-power-rating PWM converter as a shunt active filter, connected in series with passive filters, in a shunt branch, as shown in Fig. 6-56. To distinguish this approach from that presented previously, this compensator will be called the hybrid UPQC. The control strategy for the hybrid UPQC realizes compensation functions that are different from those listed in Table 6.1. This point will be clarified in the following discussion.

The network supplies the current i_S to the load through the series active filter. This current is the input for the controller of the series active filter and the voltage v_{ZF} across the shunt passive filters is the input for the controller of the shunt active filter. The fundamental component $\bar{i}_S$ of i_S is "instantaneously" calculated in the UPQC controller and excluded from the measured line current. This operation results in the current $i_{Sh} = i_S - \bar{i}_S$ that represents all harmonics present in the network current. If i_S is not purely sinusoidal, the series active filter inserts a compensating voltage calculated as

$$v_{C1} = K_1 \cdot i_{Sh} \tag{6.34}$$

Note that in case of imbalances due to negative-sequence as well as zero-sequence components at the fundamental frequency, these components should be excluded from i_{Sh}. Otherwise, the controller would force the series active filter to generate compensating voltage components at the fundamental frequency. This could increase significantly the power rating of the series PWM converter and also induce high-volt-

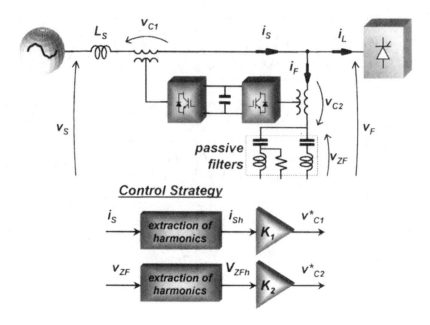

Figure 6-56. The hybrid UPQC approach.

age oscillations at double the system frequency in the dc link. If the fundamental negative-sequence component ($\dot{I}_{-1}$) is not excluded from (6.34), a compensating voltage $\dot{V}_{C1} = \dot{V}_{-1} = K_1 \dot{I}_{-1}$ is generated, which would produce $\tilde{p}$ at 2ω with the fundamental positive-sequence component ($\dot{I}_{+1}$) in i_S flowing to the load, as given in (3.98). Thus, the series active filter represents a high resistance, as high as the control gain K_1 can be, but effective only for harmonic currents. Ideally, the series active filter represents a short circuit (zero impedance) for currents at the fundamental frequency.

Similarly, the fundamental component $\bar{v}_{ZF}$ of the voltage v_{ZF} that appears across the passive filter terminals is "instantaneously" calculated and excluded from the measured voltage v_{ZF}. The resulting voltage, $v_{ZFh} = v_{ZF} - \bar{v}_{ZF}$, represents all harmonics in the voltage across the passive filter. The high resistance represented by the series active filter ($v_{C1} = K_1 \cdot i_{Sh}$) forces all current harmonics of the load, including noncharacteristic ones, for which the passive filters are not tuned, to flow through the shunt passive filters. However, the shunt passive filters represent relatively high impedances for current harmonics at different frequencies than those for which they are tuned. This can produce relatively high harmonic voltage, v_{ZFh}, across the passive filters, which would be reflected to the terminal load voltage v_F. To mitigate harmonic voltages at load terminal, the shunt active filter generates compensating voltage as

$$v_{C2} = K_2 \cdot v_{ZFh} \tag{6.35}$$

Therefore, in an ideal case, the simultaneous compensation of the series and shunt active filters results in almost sinusoidal voltage v_F at the load terminal, as well as almost sinusoidal current i_S drawn from the network. The following simplified discussion will help to clarify this point.

Using the principle of superposition and the control strategies for the series ($v_{C1} = K_1 \cdot i_{Sh}$) and shunt ($v_{C2} = K_2 \cdot v_{ZFh}$) active filters, under some restrictions, a single-phase circuit diagram can be used to represent the system shown in Fig. 6-56. Further, it can be divided as the sum of two other circuits, one for the fundamental frequency and other for the harmonics, as shown in Fig. 6-57. For the fundamental frequency, both series and shunt active filters act as short circuits. Contrarily, both active filters are effective in providing harmonic current isolation, as well as damping of harmonic-voltage propagation.

In Fig. 6-57, the equivalent circuit for the harmonics shows that the voltage at load terminal is given, in the frequency domain, by

$$\dot{V}_{Fh} = \dot{V}_{ZFh} - \dot{V}_{C2} = \mathbf{Z_F} \dot{I}_{Fh} - K_2 \mathbf{Z_F} \dot{I}_{Fh} = (1 - K_2) \mathbf{Z_F} \dot{I}_{Fh} \qquad (6.36)$$

The equivalent impedance for the shunt branch composed of the passive filter together with the active filter is found to be

$$\mathbf{Z} = \frac{\dot{V}_{Fh}}{\dot{I}_{Fh}} = (1 - K_2) \mathbf{Z_F} \qquad (6.37)$$

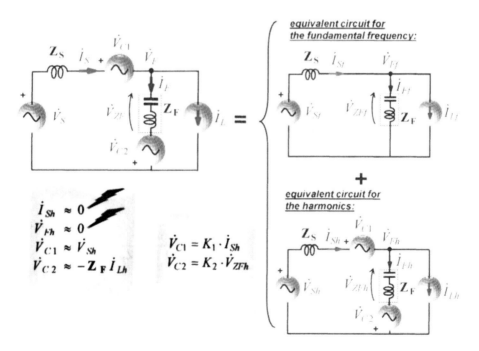

Figure 6-57. Single-phase equivalent circuits of the power system represented in Fig. 6-56.

From (6.37), an optimal value for the control gain K_2 is found to be $K_2 = 1$, in order to obtain the ideally zero harmonic voltage ($v_{Fh} = 0$) at load terminal. This corresponds to one of the two principal objectives of a hybrid UPQC, that is, to provide high voltage quality to critical loads, attenuating-harmonic voltage propagations from the network to the load terminal.

The second main objective is to prevent harmonic currents from flowing into the network. In an ideal case, the harmonic currents injected by the nonlinear load should be confined as close as possible to it. The shunt branch consisting of the passive filter and the active filter forms a path close to the nonlinear load to confine harmonic currents. The shunt active filter of the hybrid UPQC imposes a "short circuit" for harmonic currents in the shunt branch if $K_2 = 1$, whereas the series active filter imposes an "open circuit" on the network path if $K_1 \approx \infty$.

The principle of superposition is useful for finding the harmonic current flowing into the network, $\dot{I}_{Sh}$, and the harmonic voltage appearing at load terminal, $\dot{V}_{Fh}$. The harmonic current $\dot{I}_{Sh}$ is the sum of the contributions of a term excited by the harmonic voltage $\dot{V}_{Sh}$ and another excited by the harmonic load current $\dot{I}_{Lh}$. The component excited by $\dot{V}_{Sh}$ is calculated with the current source $\dot{I}_{Lh}$ opened. The second component, now with $\dot{V}_{Sh}$ short circuited, corresponds to the split of $\dot{I}_{Lh}$ into the shunt branch (opposite direction to the filter current convention, $-\dot{I}_{Fh}$) and into the network (in the same direction as $\dot{I}_{Sh}$):

$$\dot{I}_{Sh} = \frac{1}{\mathbf{Z_S} + K_1 + (1 - K_2)\mathbf{Z_F}} \dot{V}_{Sh} + \frac{(1 - K_2)\mathbf{Z_F}}{\mathbf{Z_S} + K_1 + (1 - K_2)\mathbf{Z_F}} \dot{I}_{Lh} \quad (6.38)$$

Similarly, the harmonic voltage that appears at load terminal can be calculated as the sum of contributions from $\dot{V}_{Sh}$ and $\dot{I}_{Lh}$. The two terms that excite $\dot{I}_{Fh}$ in (6.36) are found to be

$$\dot{V}_{Fh} = \frac{(1 - K_2)\mathbf{Z_F}}{\mathbf{Z_S} + K_1 + (1 - K_2)\mathbf{Z_F}} \dot{V}_{Sh} - \frac{(\mathbf{Z_S} + K_1)(1 - K_2)\mathbf{Z_F}}{\mathbf{Z_S} + K_1 + (1 - K_2)\mathbf{Z_F}} \dot{I}_{Lh} \quad (6.39)$$

From (6.38) and (6.39), if the control gains tend to their optimal values, $K_1 \to \infty$ and $K_2 \to 1$, the two principal compensation objectives are simultaneously reached, that is,

$$\lim_{K_1 \to \infty}\left[\lim_{K_2 \to 1} \dot{I}_{Sh}\right] = \\ \lim_{K_1 \to \infty}\left[\lim_{K_2 \to 1}\left(\frac{1}{\mathbf{Z_S} + K_1 + (1 - K_2)\mathbf{Z_F}} \dot{V}_{Sh} + \frac{(1 - K_2)\mathbf{Z_F}}{\mathbf{Z_S} + K_1 + (1 - K_2)\mathbf{Z_F}} \dot{I}_{Lh}\right)\right] \approx 0 \quad (6.40)$$

$$\lim_{K_1 \to \infty}\left[\lim_{K_2 \to 1} \dot{V}_{Fh}\right] = \\ \lim_{K_1 \to \infty}\left[\lim_{K_2 \to 1}\left(\frac{(1 - K_2)\mathbf{Z_F}}{\mathbf{Z_S} + K_1 + (1 - K_2)\mathbf{Z_F}} \dot{V}_{Sh} - \frac{(\mathbf{Z_S} + K_1)(1 - K_2)\mathbf{Z_F}}{\mathbf{Z_S} + K_1 + (1 - K_2)\mathbf{Z_F}} \dot{I}_{Lh}\right)\right] \approx 0 \quad (6.41)$$

The compensating voltage of the series active filter, given in (6.34), is determined by multiplying (6.38) by K_1, as follows:

$$\dot{V}_{C1} = \frac{K_1}{Z_S + K_1 + (1 - K_2)Z_F} \dot{V}_{Sh} + \frac{K_1(1 - K_2)Z_F}{Z_S + K_1 + (1 - K_2)Z_F} \dot{I}_{Lh} \quad (6.42)$$

The compensating voltage of the shunt active filter, dictated by (6.35), can be written by calculating $\dot{V}_{ZFh}$ as a function of Z_S, Z_F, K_1, K_2, $\dot{V}_{Sh}$, and $\dot{I}_{Sh}$. From (6.37), it is seen that

$$\dot{I}_{Fh} = \frac{\dot{V}_{Fh}}{(1 - K_2)Z_F} \quad (6.43)$$

Replacing $\dot{V}_{Fh}$ from (6.39) in (6.43) and making $\dot{V}_{ZFh} = Z_F \dot{I}_{Fh}$, the compensating voltage $\dot{V}_{C2}$ is given by

$$\dot{V}_{C2} = K_2 \dot{V}_{ZFh} = K_2 Z_F \dot{I}_{Fh}$$

$$\dot{V}_{C2} = \frac{K_2 Z_F}{Z_S + K_1 + (1 - K_2)Z_F} \dot{V}_{Sh} + \frac{K_2(Z_S + K_1)Z_F}{Z_S + K_1 + (1 - K_2)Z_F} \dot{I}_{Lh} \quad (6.44)$$

In the ideal case, when $K_1 \to \infty$ and $K_2 \to 1$, the compensating voltages as determined in (6.42) and (6.44) become

$$\lim_{K_1 \to \infty} \left[\lim_{K_2 \to 1} \dot{V}_{C1} \right] =$$
$$\lim_{K_1 \to \infty} \left[\lim_{K_2 \to 1} \left(\frac{K_1}{Z_S + K_1 + (1 - K_2)Z_F} \dot{V}_{Sh} + \frac{K_1(1 - K_2)Z_F}{Z_S + K_1 + (1 - K_2)Z_F} \dot{I}_{Lh} \right) \right] \approx \dot{V}_{Sh} \quad (6.45)$$

$$\lim_{K_1 \to \infty} \left[\lim_{K_2 \to 1} \dot{V}_{C2} \right] =$$
$$\lim_{K_1 \to \infty} \left[\lim_{K_2 \to 1} \left(\frac{K_2 Z_F}{Z_S + K_1 + (1 - K_2)Z_F} \dot{V}_{Sh} - \frac{K_2(Z_S + K_1)Z_F}{Z_S + K_1 + (1 - K_2)Z_F} \dot{I}_{Lh} \right) \right] \approx -Z_F \dot{I}_{Lh} \quad (6.46)$$

as summarized in Fig. 6-57.

6.2.3.A. Controller of the Hybrid UPQC

As mentioned before, the source current and the voltage on the passive filter in the shunt branch (i_S and v_{ZF} in Fig. 6-56) are the inputs of the hybrid UPQC controller. The main goal is to mitigate harmonic components in the source current and in the load bus voltage. Therefore, they must be "instantaneously" and continuously determined and multiplied by the gains K_1 and K_2 to form the compensating voltages $v_{C1} = K_1 \cdot i_{Sh}$, for the series active filter and $v_{C2} = K_2 \cdot v_{ZFh}$, for the shunt active filter.

A simple way to extract the fundamental component from a generic three-phase voltage or current is the use of an algorithm that combines the Park transformation with the Clarke transformation. This approach was previously applied to a shunt active filter controller, discussed in Chapter 4. Its relation with the original control algorithm based on the *p-q* Theory is summarized in (4.27). The Park transformation, a synchronizing circuit. This circuit is generally called a PLL circuit (phase-locked-loop circuit) that determines the synchronous angular frequency $\theta = \omega t$. An example of robust PLL circuit that can also be used here is shown in Fig. 6-22.

Once the fundamental components have been determined, the differences between these components and their corresponding measured voltages and currents are the harmonic contents that are needed to form the compensating voltage references for the active filters of the hybrid UPQC. Thus, the hybrid UPQC controller is divided in three parts:

1. Circuit for extraction of the fundamental component ($\bar{v}_{ZF}$) of the voltage (v_{ZF}) on the shunt passive filter (Fig. 6-58)
2. Circuit for extraction of the fundamental component ($\bar{i}_S$) of the current (i_S) flowing through the series active filter (Fig. 6-59)
3. dc voltage regulator and composition of the compensating voltages (v_{C1} and v_{C2}) of series and shunt active filters (Fig. 6-60)

In Fig. 6-58, the measured voltages on the shunt passive filters (v_{ZFa}, v_{ZFb}, and v_{ZFc}) are the inputs to the PLL circuit (Fig. 6-22) used to determine the angular frequency $\theta = \omega t$ of the fundamental positive-sequence component ($\dot{V}_{+1}$) contained in those voltages. In parallel, the Clarke transformation is used to transform them into the $\alpha\beta$ axes. Then, the Park transformation is applied and the average values on the dq axes ($\bar{v}_{ZFd}$ and $\bar{v}_{ZFq}$) are calculated. Finally, the inverse Park transformation and the inverse Clarke transformation are applied to determine the fundamental components ($\bar{v}_{ZFa}$, $\bar{v}_{ZFb}$, and $\bar{v}_{ZFc}$) of the measured voltages on the shunt passive filter.

In a similar way, the fundamental components ($\bar{i}_{Sa}$, $\bar{i}_{Sb}$, and $\bar{i}_{Sc}$) of the measured line currents (i_{Sa}, i_{Sb}, and i_{Sc}) flowing through the series active filter can be determined, as shown in Fig. 6-59. The same angular frequency, numbered with ② in Fig. 6-58, can be used here in the Park transformation. Thus, the same dynamic behavior of the PLL circuit that affects the extraction of the fundamental component of the measured voltages affects also the extraction of the fundamental current component. Moreover, the dynamics of the low-pass filters in both circuits also introduce errors during transients. Unfortunately, this kind of error is unavoidable in all controllers of active power line conditioners presented hitherto. They lead to dc voltage variations, and a slower feedback loop provided by a dc voltage regulator has been used to maintain the dc voltage around its reference value in the steady state.

As mentioned, by subtracting the calculated fundamental components from the measured voltages and currents, the harmonic contents are determined. Multiplying by gain K_1, the compensating voltages of the series active filter are achieved:

$$\begin{bmatrix} v_{C1a} \\ v_{C1b} \\ v_{C1c} \end{bmatrix} = K_1 \begin{bmatrix} i_{Sa} - \bar{i}_{Sa} \\ i_{Sb} - \bar{i}_{Sb} \\ i_{Sc} - \bar{i}_{Sc} \end{bmatrix} \quad (6.47)$$

For the shunt active filter, an additional control signal from a dc voltage regulator should be added to correct the dc voltage variations caused by losses and transient compensation errors. In order to force energy to flow into or from the dc capacitor, the shunt active filter should behave as a "resistor" or as a "generator" in the network. For the voltages and currents directions adopted in Fig. 6-57, the shunt PWM converter is in "generator convention," that is, for voltages in phase with the currents, only a positive active power is produced. This means that energy is flowing from the dc link to the network, discharging the dc capacitor. Contrarily, to charge the dc capacitor, the shunt PWM converter should synthesize a fundamental ac voltage component in counterphase with the current flowing through the shunt branch. The control signal from the dc voltage regulator, labeled $\bar{p}_{loss}$ in Fig. 6-60, together with the fundamental components $\bar{v}_{ZF\alpha}$ and $\bar{v}_{ZF\beta}$, already determined in Fig. 6-58, realizes dc voltage regulation. The shunt active filter controller determines a compensating volt-

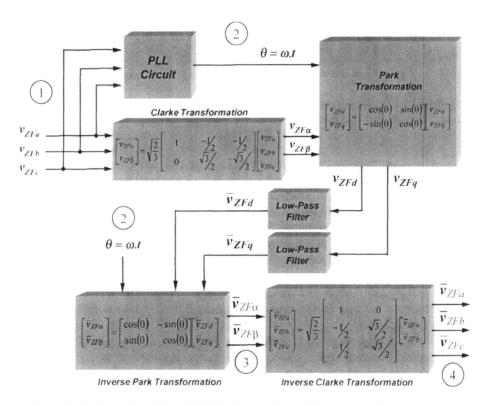

Figure 6-58. Controller of the hybrid UPQC: extraction of the fundamental voltage component on the shunt passive filter.

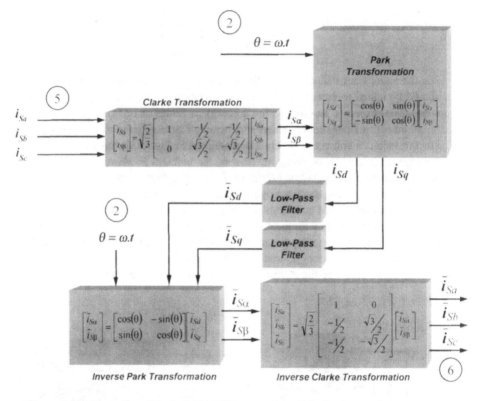

Figure 6-59. Controller of the hybrid UPQC: extraction of the fundamental current component flowing through the series active filter.

age component at the fundamental frequency, which is in phase or in counterphase with the fundamental current flowing through the shunt branch composed of the passive and the active filters. The following discussion helps to clarify this point.

In order to avoid additional measurements of the current flowing through the shunt branch, that is, the current I_F in Fig. 6-57, it is reasonable to assume that the equivalent impedance of the shunt passive filter, at the fundamental frequency, is almost purely capacitive. Disregarding ohmic losses in the passive filter, ideally, the fundamental component of the voltages on this filter that is determined in the control algorithm illustrated in Fig. 6-58 ($\bar{v}_{ZFa}$, $\bar{v}_{ZFb}$, and $\bar{v}_{ZFc}$) lag by 90° the fundamental currents and can be used as references in the dc voltage regulator of the UPQC controller.

For the voltages and currents directions adopted in Fig. 6-57 and the signal $\bar{p}_{loss}$ generated as shown in Fig. 6-60, Fig. 6-61 illustrates the phasor diagram for the fundamental components that should be used as references, to generate compensating voltages that act against dc voltage variations. As seen in Fig. 6-60, the dc voltage error ($V_{REF} - V_{dc}$) is positive for $V_{REF} > V_{dc}$, which makes $\bar{p}_{loss}$ increase and become positive. Under this situation, in order to reduce the error, energy should flow

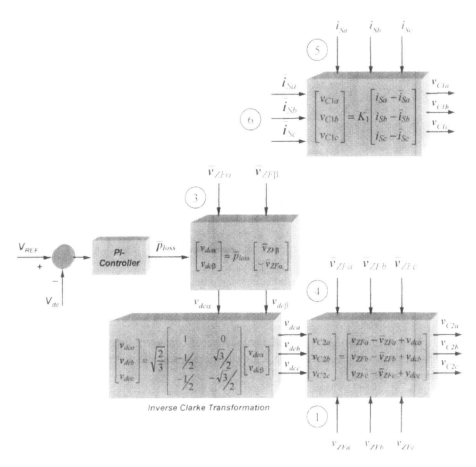

Figure 6-60. Controller of the hybrid UPQC: dc voltage regulator and composition of the compensating voltages references.

into the dc capacitor, to charge it. For this purpose, the fundamental voltage component, $\dot{V}_{C2dc}$, generated by the shunt active filter, must be in counterphase with the fundamental current $\dot{I}_{Ff}$. Considering $\dot{I}_{Ff}$ to be orthogonal and leading by 90° the fundamental voltage $\dot{V}_{ZFf}$ on the shunt passive filter, the shunt active filter must generate compensating voltages lagging by 90° [$\dot{V}_{C2dc}(\bar{p}_{loss} > 0)$] with respect to that fundamental voltage component.

Now, a question is how to generate compensating voltages that are proportional to the amplitude of the control signal $\bar{p}_{loss}$ and orthogonal to the calculated voltages $\bar{v}_{ZFa}$, $\bar{v}_{ZFb}$, and $\bar{v}_{ZFc}$; lagging for $\bar{p}_{loss} > 0$, or leading for $\bar{p}_{loss} < 0$. The fundamental component of the measured voltages on the shunt passive filters, transformed to the $\alpha\beta$ axes, are given as $\bar{v}_{ZF\alpha}$ and $\bar{v}_{ZF\beta}$ in Fig. 6-58. From these calculated fundamental voltages, it is possible to generate the compensating voltages references, defined as v_{dca}, v_{dcb}, and v_{dcc} in Fig. 6-60. The correctness of the phase angle of v_{dca}, v_{dcb}, and

336 COMBINED SERIES AND SHUNT POWER CONDITIONERS

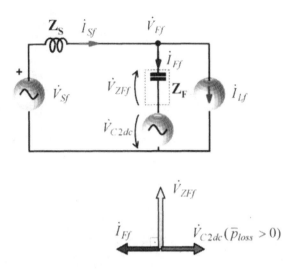

Figure 6-61. Single-phase equivalent circuit and phasor diagram for the fundamental frequency, for dc voltage regulation.

v_{dcc} can be understood by recalling the concept of current decomposition given in Chapter 3. For convenience, the basic equation for current decomposition is repeated below:

$$\begin{bmatrix} i_\alpha \\ i_\beta \end{bmatrix} = \frac{1}{v_\alpha^2 + v_\beta^2} \begin{bmatrix} v_\alpha & v_\beta \\ v_\beta & -v_\alpha \end{bmatrix} \begin{bmatrix} p \\ 0 \end{bmatrix} + \frac{1}{v_\alpha^2 + v_\beta^2} \begin{bmatrix} v_\alpha & v_\beta \\ v_\beta & -v_\alpha \end{bmatrix} \begin{bmatrix} 0 \\ q \end{bmatrix} \qquad (6.48)$$

From (6.48), the following conclusions can be summarized. The real currents are given by the first term and the imaginary currents by the second term in the right-hand side of (6.48). They are calculated as functions of the voltage and the real power, or imaginary power, respectively. A fundamental characteristic of this current decomposition is that if average values of real and imaginary powers are considered and the voltages are purely sinusoidal and balanced (they contain only $\dot{V}_{+1}$), then the calculated real and imaginary currents will be also sinusoidal and balanced, that is, they contain only $\dot{I}_{+1}$. Moreover, $\dot{I}_{+1}$ is in phase with $\dot{V}_{+1}$, for a constant positive value of p. On the other hand, the imaginary current would also contain only $\dot{I}_{+1}$. However, it is orthogonal to $\dot{V}_{+1}$ for a constant value of q (lagging for $q > 0$ or leading for $q < 0$). If the voltages v_α and v_β in (6.48) are replaced by $\bar{v}_{ZF\alpha}$ and $\bar{v}_{ZF\beta}$, and q by making $p = 0$, it follows that

$$\begin{bmatrix} v_{dc\alpha} \\ v_{dc\beta} \end{bmatrix} = \frac{1}{\bar{v}_{ZF\alpha}^2 + \bar{v}_{ZF\beta}^2} \begin{bmatrix} \bar{v}_{ZF\alpha} & \bar{v}_{ZF\beta} \\ \bar{v}_{ZF\beta} & -\bar{v}_{ZF\alpha} \end{bmatrix} \begin{bmatrix} 0 \\ \bar{p}_{loss} \end{bmatrix} \qquad (6.49)$$

Since $\bar{v}_{ZF\alpha}$ and $\bar{v}_{ZF\beta}$ contain only $\dot{V}_{+1}$, the squared sum corresponding to the denominator in (6.49) is a constant value in steady-state conditions, and could be replaced by this constant value. Moreover, since $\bar{p}_{loss}$ replaces q, it is possible to affirm that the resulting voltages $v_{dc\alpha}$ and $v_{dc\beta}$ are orthogonal to $\bar{v}_{ZF\alpha}$ and $\bar{v}_{ZF\beta}$, and $v_{dc\alpha}$, $v_{dc\beta}$, and v_{dcc} lag by 90° the voltages $\bar{v}_{ZFa}$, $\bar{v}_{ZFb}$, and $\bar{v}_{ZFc}$, respectively, for $\bar{p}_{loss} > 0$. Under different steady-state operation points, the squared sum of $\bar{v}_{ZFa}$ and $\bar{v}_{ZF\beta}$ has different values. However, this is accommodated by the integral part of the PI controller used in the feedback control loop that generates the signal $\bar{p}_{loss}$. Hence, (6.49) can be simplified further, resulting in the expression used in Fig. 6-60, that is,

$$\begin{bmatrix} v_{dc\alpha} \\ v_{dc\beta} \end{bmatrix} = \bar{p}_{loss} \begin{bmatrix} \bar{v}_{ZF\beta} \\ -\bar{v}_{ZF\alpha} \end{bmatrix} \quad (6.50)$$

The inverse Clarke transformation is applied to the compensating voltages of (6.50). Since the optimal value of K_2 in (6.35) is unity, and the harmonic voltage v_{ZFh} is given by the difference ($v_{ZF} - \bar{v}_{ZF}$), the new compensating voltages of the shunt active filter that avoids harmonic-voltage propagation to the load terminal (right side of the hybrid UPQC) and also provides voltage regulation in the common dc link of the hybrid UPQC are found to be

$$\begin{bmatrix} v_{C2a} \\ v_{C2b} \\ v_{C2c} \end{bmatrix} = \begin{bmatrix} v_{ZFa} - \bar{v}_{ZFa} + v_{dca} \\ v_{ZFb} - \bar{v}_{ZFb} + v_{dcb} \\ v_{ZFc} - \bar{v}_{ZFc} + v_{dcc} \end{bmatrix} \quad (6.51)$$

6.2.3.B. Experimental Results

This section shows some experimental results from a laboratory prototype of hybrid UPQC [31].

Figure 6-62 shows the experimental power system for testing the hybrid UPQC. Each active filter of 0.5 kVA (AF1 is the series active filter and AF2 is the shunt active filter) consists of three single-phase voltage-source PWM converters using

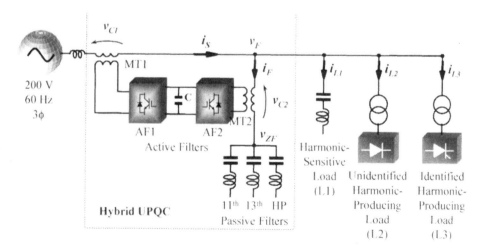

Figure 6-62. System configuration for testing the Hybrid UPQC.

power MOSFETs. The dc terminals of the single-phase converters are connected in parallel to a dc capacitor of 2200 μF. The matching transformers, MT1 and MT2, have a 1:20 turns ratio. The passive filter (PF) of 8 kVA consists of 11th and 13th tuned LC filters and a high-pass filter. Table 6.3 shows the circuit constants of the passive filters.

A harmonic-sensitive load L1 and two harmonic-producing loads L2 and L3 are connected on a common bus, where the bus voltage v_F is 200 V. A three-phase, twelve-pulse thyristor rectifier of 20 kVA represents the identified harmonic-producing load L3. This load dominantly generates the 11th- and 13th-order current harmonics. A three-phase diode rectifier of 3 kVA represents the unidentified load L2 that dominantly generates the fifth and seventh current harmonics. Note that neither a fifth nor a seventh tuned LC filter is included in the passive filter for this experiment, although it should be included in a practical application. This mistake is intentionally introduced in order to experimentally prove that the hybrid UPQC shown in Fig. 6-56 is useful for harmonic compensation in a power system feeding harmonic-sensitive loads and unidentified harmonic-producing loads. Note that the previous approach of UPQC (Fig. 6-26) does not compensate unidentified harmonic-producing loads, if their currents are not included in the measured current i_L of the load.

A power capacitor of 3 kvar in series with a reactor of 5% of the capacitor rating is considered as hypothetical harmonic-sensitive load. Relatively small voltage distortion at the common bus causes large harmonic currents flowing into this load, so that i_{L1} becomes very distorted. The equivalent network reactance in Fig. 6-62 is 3% on a 20 kVA, 200 V, 60 Hz base. The nominal dc voltage on the capacitor in the common dc link is 120 V. The switching frequency of both PWM converters is fixed at 15 kHz. At so high a switching frequency, the leakage inductances of the matching transformers MT1 and MT2 play an important role in reducing ripples at the switching frequency. Thus, no additional filtering is needed.

As discussed earlier, the ideal gain of the series active filter (AF1) is $K_1 = \infty$ and that of shunt active filter (AF2) is $K_2 = 1$. In this experiment, however, K_1 is set to 2.2 Ω (= 100% on a 200 V, 60 A, 20 kVA base), and K_2 to 0.8, in order to avoid system instability that may be caused by time delays in detecting and calculating the "instantaneous" references of compensating voltages in the UPQC controller. The previous analysis for choosing optimal gains considers the system at steady state. Certainly, dynamic interactions between the control circuits for the shunt and series active filters can induce instability that was not carefully investigated. Moreover, depending on the parameters in the power system, nonlinear loads, shunt capacitor banks, and the passive filters in the hybrid UPQC, such dynamic interactions can be

Table 6.3. Circuit constants of the passive filters

	Inductance	Capacitance	Q
11th	380 μH	150 μF	20
13th	300 μH	140 μF	20
HP	40 μH	260 μF	20

amplified further. All those issues are still to be addressed. Although the gains in the controller have been experimentally found in the present stage, the relationship between these gains and impedances has to be made clearer in future studies.

Figures 6-63, 6-64, and 6-65 show experimental waveforms in the case of disconnection of the harmonic-producing loads L2 and L3 from the common bus in Fig. 6-62. Table 6.4 shows the experimental values of harmonics of voltage and current with respect to their fundamental components. Due to the existence of fifth and seventh background harmonic voltages in the supply, fifth- and seventh-harmonic currents of 3.3% and 9.1% are present in the supply current, respectively, as shown in Fig. 6-63. Since the active filters of the hybrid UPQC are out of operation, an amount of harmonic voltage appears on the common bus, so that a fifth-harmonic current of 12% is flowing into the harmonic-sensitive load L1. After the series active filter AF1 is switched on, the fifth-harmonic current is reduced to 0.7%, as shown in Fig. 6-64, because this active filter acts as a blocking high resistor at the fifth-harmonic frequency. As a result, the harmonics in v_F and i_{L1} are reduced by two-thirds. Further improvements were not possible, even if both active filters of the hybrid UPQC are in operation, as can be seen in Fig. 6-65. One of the reasons was the relatively low harmonic contents in the voltage on the common bus, when the AF1 was operating, so that the harmonic voltage was not accurately detected by the control circuit of the shunt active filter (AF2).

The next experimental results are taken with the harmonic-sensitive load L1 and the harmonic-producing loads L2 and L3 connected to the common bus.

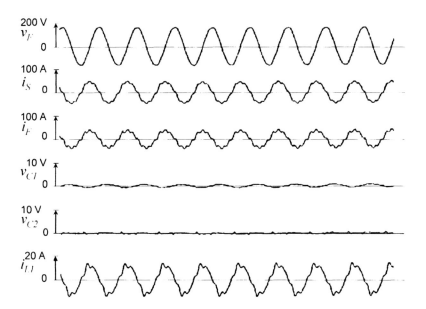

Figure 6-63. Experimental results when only the harmonic-sensitive load L1 is connected and the active filters AF1 and AF2 are out of operation.

340 COMBINED SERIES AND SHUNT POWER CONDITIONERS

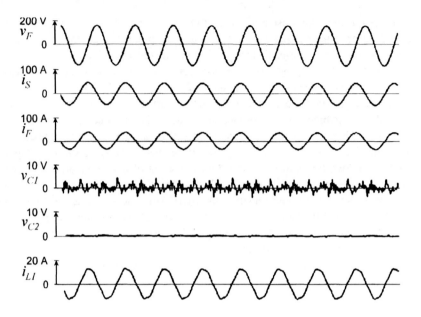

Figure 6-64. Experimental results when only the harmonic-sensitive load L1 is connected and only the series active filter (AF1) is operating.

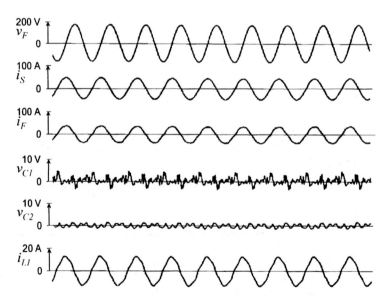

Figure 6-65. Experimental results when only the harmonic-sensitive load L1 is connected and with the hybrid UPQC in operation.

6.2. THE UNIFIED POWER QUALITY CONDITIONER (UPQC)

Table 6.4. Experimental values in case of disconnection of L2 and L3

Operating condition	i_S [%]		i_{L1} [%]		v_B [%]	
	5th	7th	5th	7th	5th	7th
PF	3.3	9.1	12.	8.3	0.8	1.5
PF + AF1	0.7	0.7	4.4	0.8	0.3	0.0
PF + AF1 + AF2	1.5	0.7	5.2	0.8	0.3	0.2

Figure 6-66, 6-67, and 6-68 show the waveforms, and Table 6.5 shows the experimental values of harmonics. The waveforms in Fig. 6-66 correspond to the case in which both active filters of the hybrid UPQC are out of operation. As in the previous experimental results, during this state, all upper IGBTs (two IGBT in each single-phase PWM converter) of the series and shunt active filters are permanently kept conducting to provide short circuits at the secondary windings of the transformers. The harmonic spectrum changes considerably with the connection of the harmonic-producing loads L2 and L3, as can be seen by comparing

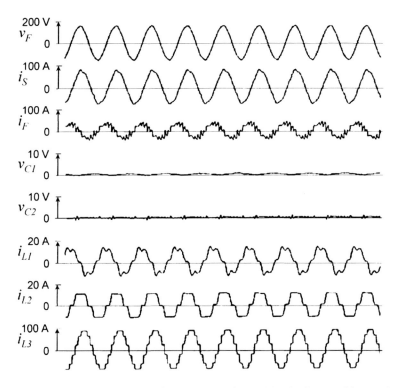

Figure 6-66. Experimental results with the harmonic-sensitive load L1 and harmonic-producing loads L2 and L3 connected and the active filters AF1 and AF2 out of operation.

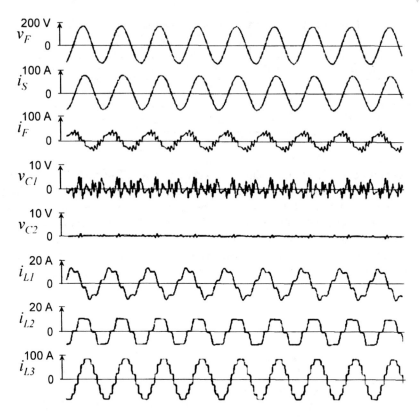

Figure 6-67. Experimental results with the harmonic-sensitive load L1, the harmonic-producing loads L2 and L3, and the series active filter AF1 connected.

Fig. 6-63 with Fig. 6-66, and the line "PF" in Table 6.4 and Table 6.5. Since the hybrid UPQC has no passive filters tuned to the fifth- and seventh-order harmonics, the amount of these harmonic currents flowing into the network are very large in both situations. In both cases, this hardly distorts the voltage on the common bus that, in turn, affects the current i_{L1} flowing into the harmonic-sensitive load.

After the connection of the series active filter, the fifth-order harmonic flowing to the network is reduced from 3.8% to 1.1%, and the seventh from 1.8% to 0.2%, so that the current i_S becomes almost sinusoidal, as shown in Fig. 6-67. However, the fifth- and seventh-order harmonics produced by the unidentified load L2 are forced to flow into the passive filter of the hybrid UPQC that is not tuned to them, thus causing high voltage distortions on the voltage at the common bus. Hence, a high amount of harmonic current is still flowing into the harmonic-sensitive load L1. Although still not optimal, now it is possible to verify that the shunt active filter of the hybrid UPQC can improve the performance of the whole system by reducing further the voltage distortion on the common bus. This also reduces a little more the

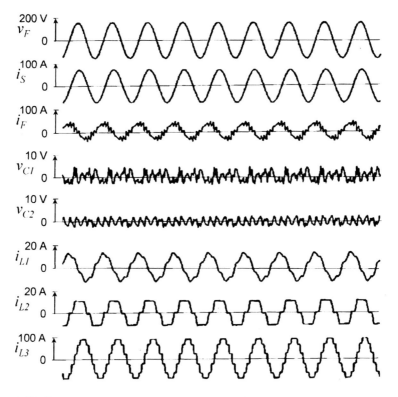

Figure 6-68. Experimental results with the harmonic-sensitive load L1, the harmonic-producing loads L2 and L3, and the hybrid UPQC connected.

fifth-harmonic flowing into the harmonic-sensitive load L1, as can be seen in the waveform for i_{L1} in Fig. 6-68 and the values in Table 6.5.

6.3. THE UNIVERSAL ACTIVE POWER LINE CONDITIONER (UPLC)

An interesting approach to FACTS devices is the unified power flow controller (UPFC) [14,17,23,24,25,26,28]. It allows the possibility of controlling together all three principal parameters (magnitude of the terminal voltage, series impedance of the line, and phase angle of the terminal voltage) that determine the power flow through a transmission line. This section proposes a new control strategy for the combined series and shunt converter topology shown in Fig. 6-1. It joins harmoniously into a single compensator all those compensation functions of the UPFC (Fig. 6-2), which involves compensating voltages and currents at the fundamental frequency, with those active filtering capabilities of the unified power quality conditioner (UPQC), as presented in Table 6.1. Due to this reason, this device is being called Universal Active Power Line Conditioner (UPLC). The next sections discuss

Table 6.5. Experimental values in case of connection of L1, L2, and L3

Operating condition	i_S [%]		i_{L1} [%]		v_B [%]	
	5th	7th	5th	7th	5th	7th
PF	3.8	1.8	21.	5.5	1.4	1.0
PF + AF1	1.1	0.2	20.	2.1	1.4	0.8
PF + AF1 + AF2	1.1	0.2	13.	2.1	1.0	0.8

two control strategies of the UPLC for two different necessities of series active filtering of harmonic voltages are detailed. Then, the integrated controllers for the UPLC are presented and the dynamic performances are analyzed by simulation using complete digital models of the UPLC.

6.3.1. General Description of the UPLC

Two scenarios are described, in which there are nonlinear loads connected to power supplies containing distorted voltages. In a scenario #1, as shown in Fig. 6-69, the distorted supply voltage and the nonlinear load are on the same side, forming subsystem A. This subsystem is connected to subsystem B through a transmission line, in which a UPLC is installed, close to subsystem A, to control the voltage at the line end terminal, as well as to control the active and reactive power flow through this transmission line. Further, harmonic voltages and currents of subsys-

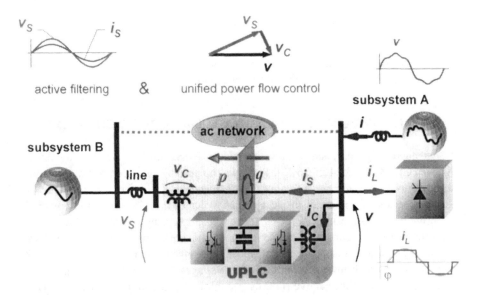

Figure 6-69. Combined series and shunt compensators with active filtering and power flow control capabilities. Configuration #1 of the UPLC.

tem A should not affect subsystem B. The UPLC for this purpose will be called configuration #1.

As shown in Fig. 6-69, configuration #1 of the UPLC is suitable when the active filtering of both voltage and current are needed on the same side (right side of the UPLC). In this case, the voltage v of the controlled bus is already distorted by power subsystem A and the nonlinear load (i_L) is also connected at the same side. The magnitude of voltage v should be controlled by the shunt converter of the UPLC injects a controlled reactive current, based on the same principle as a STATCOM [19] and the shunt converter of a UPFC, as summarized in Fig. 6-3. Additionally, the shunt converter of the UPLC should also act as a shunt active filter to compensate the current i_L of the nonlinear load.

On the other hand, the harmonic voltages and eventual imbalances from negative-sequence as well as zero-sequence components at the fundamental frequency, which can be present in voltage v, should be compensated by the active filtering capability of the series converter of the UPLC. Therefore, the harmonics and imbalances present in the voltage v of the controlled bus should not propagate further to the left side of the UPLC (v_S) and through the controlled transmission line. The current i_S flows through the controlled transmission line and should be adjusted by the series converter of the UPLC, using the same principle of power flow control of the UPFC as summarized in Fig. 6-7.

Figure 6-70 shows another scenario that was investigated, for which the configuration #2 of UPLC was developed. If compared with configuration #1, the difference is to require series active filtering of voltage v_S on the left side of the UPLC, whereas the shunt active filtering of current stays on the right side. As will be shown later, the controller of configuration #2 utilizes an additional measurement of voltage—the voltage v_S must be also measured—and an additional positive-sequence detector for this voltage is included.

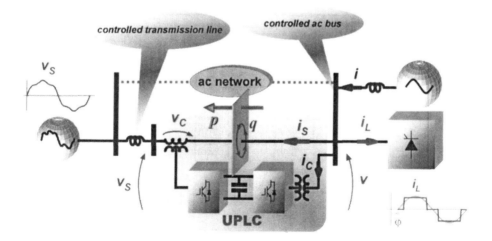

Figure 6-70. Configuration #2 of the universal active power line conditioner (UPLC).

346 COMBINED SERIES AND SHUNT POWER CONDITIONERS

For comparison, the real and imaginary powers produced by the voltage v (right side of the UPLC) and the current i_S are considered as the powers of the controlled transmission line in both configurations. They must track their power orders, using the same principle of power flow control implemented in the previous approach of the UPFC (see Fig. 6-7).

Because of the active filtering capability of the UPLC, the shunt and series converters should accurately compensate harmonic voltages and currents up to 1 kHz, if the same range of harmonics compensation as that considered in the previous chapters regarding active filters is adopted. This forces the use of PWM converters with a switching frequency of 10 kHz or even higher, which would make impracticable a real implementation of the UPLC for very high power applications, using the presently available self-commutated power semiconductors. Some considerations about the powers involved in the UPLC will be given later, through simulation results. Figure 6-71 shows the principal parts of the UPLC and the voltage and current measurements that are necessary in the UPLC controller of configuration #1. Note that this configuration of the UPLC uses the same number of voltage and current measurements as that of the UPQC shown in Fig. 6-26. However, some symbols and conventions of currents and voltages have been changed here.

A three-phase, four-wire system will be considered in the following, although it might be more realistic to think about a UPLC for a three-phase system *without* neutral conductor. The principal reason of using three-phase, four-wire systems is

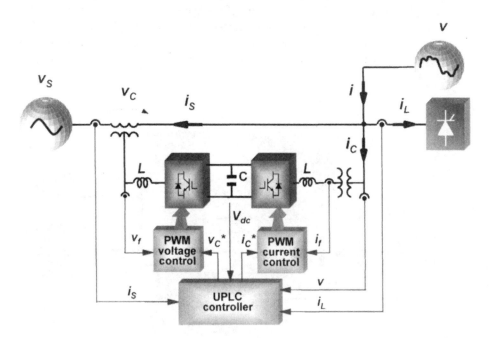

Figure 6-71. Needed measurements for the controllers of the UPLC operating according to configuration #1.

to develop a complete control algorithm, where it is possible to analyze the performance of the UPLC in the presence of all kinds of imbalances and harmonics, including zero-sequence components. The developed UPLC controller can be realized in a modular form, in order to allow the elimination of some compensation functions without degrading the other functionalities of the controller.

Later, it will be verified that the controller developed for the UPLC of Fig. 6-71 includes almost all concepts of active power line conditioning based on the p-q Theory that were described in the previous sections of this chapter.

It is well known that only positive- and negative-sequence components are present in the currents if a three-phase system *without* a neutral conductor is considered. General equations relating the symmetrical components theory [40], which is based on phasors (frequency domain) for a system in the steady state, and the powers in the $\alpha\beta0$-reference frame (the p-q Theory), are given in Chapter 3. From (3.91) and (3.92), it is possible to conclude that α and β components are not affected by zero-sequence components. Therefore, the controller of the UPLC that will be presented here can be simplified by canceling all control signals that are related to zero-sequence components, for example, v_0, i_0, and p_0, making this controller appropriate for applications to three-phase systems *without* a neutral conductor.

The power circuit topology of the UPLC is the same as that of the unified power quality conditioner (UPQC), shown in Fig. 6-27, whereas the measured voltages and currents and the positive directions of these measurements are those as shown in Fig. 6-71. For convenience, the same PWM *current* control (dynamic hysteresis-band current controller) for the shunt converter as described in Section 4.3.2 is used. Moreover, the same PWM *voltage* control as used in the series converter of the previous approach of the UPQC (see Fig. 6-29) is considered here. In fact, the UPLC controller that will be presented is developed by adapting the functionalities of the UPFC controller shown in Fig. 6-11 to the UPQC controller presented in Fig. 6-31.

Hence, the UPLC joins all active filtering capabilities of the UPQC and voltage regulation, as well as power flow control capabilities of the UPFC, into a single device. All these compensation functionalities are summarized in Table 6.6. Note that the shunt converter of the UPLC compensates the neutral current of the nonlinear load and also the power factor of this load. Table 6.6 shows that the UPLC is able to solve almost all problems related to power quality. Only the functionality for compensating voltage sags and swells are omitted. In fact, this functionality could also be aggregated in the UPLC controller, since the power converter topology used can also perform this kind of compensation. However, this improvement in the UPLC controller is left to future work. Next, the controller for configuration #1 of UPLC in Fig. 6-71 is detailed. Then, the needed changes for configuration #2, illustrated in Fig. 6-70, are indicated.

6.3.2. The Controller of the UPLC

All principles of active filtering, power factor compensation, voltage regulation, as well as power flow control included in the UPLC controller are detailed in the pre-

348 COMBINED SERIES AND SHUNT POWER CONDITIONERS

Table 6.6. Assignments for a universal active power line conditioner (UPLC)

UPLC *Series* Converter	UPLC *Shunt* Converter
Unifed Power Quality Conditioner (UPQC)	
• Compensation of *voltage* harmonics, including negative- and zero-sequence components at the fundamental frequency	• Compensation of *current* harmonics, including negative- and zero-sequence components at the fundamental frequency
• Suppression of harmonic currents through the power line (harmonic isolation)	• Compensation of the reactive power of the load
• Improvement of the system stability (damping)	• Regulation of the capacitor voltage of the common dc link
Unified Power Flow Controller (UPFC)	
• Control of the active and reactive power flow through the line	• Regulation of the terminal voltage of the line (controlled bus)

vious sections of this book. Here, they will be only briefly addressed and the particularities to adapt the controller to the scenario #1 (Fig. 6-71) indicated. Figure 6-72 presents the part of the UPLC controller that determines the compensating *current* references for the shunt converter, whereas Fig. 6-73 shows the other part that determines the compensating *voltage* references for the series converter.

It can be recognized that all active filtering functions of the shunt converter of the UPLC correspond to a control strategy for three-phase, four-wire, shunt active filter, called the sinusoidal current control strategy, as illustrated in Fig. 4-70. As a consequence, the UPLC controller for the shunt converter in Fig. 6-72 contains the whole control circuit shown in that figure without changes. To perform the extra function of voltage regulation on the controlled ac bus, the signal $\bar{q}_v$ is added in Fig. 6-72, to force the shunt converter to generate a reactive current component at the fundamental frequency; it is capacitive ($\bar{q}_v < 0$) to increase the voltage, or inductive ($\bar{q}_v > 0$) to reduce the voltage. Again, this feature was detailed in the previous section dealing with the UPFC, and a control circuit for this purpose is shown in Fig. 6-11b. The same compensation principle is realized here, through the signal $\bar{q}_v$. Note that the concept of instantaneous aggregate voltage as introduced in Chapter 2 [see (2.47)] is used to determine "instantaneously" the rms value of the line voltage in Fig. 6-72, comprising only the fundamental positive-sequence component of the measured phase voltages on the right side of the UPLC (Fig. 6-71), that is,

$$v'_\Sigma = \sqrt{v'^2_a + v'^2_b + v'^2_c} \tag{6.52}$$

The total imaginary power ($q = \bar{q} + \tilde{q}$) of the nonlinear load currents i_{La}, i_{Lb}, and i_{Lc}, is compensated. Hence, the fundamental power factor is compensated, since $\bar{q}$ comprises the fundamental reactive current of this load. Since the sinusoidal current control strategy is implemented in the shunt converter of the UPLC, the compensated powers $\tilde{p}$ and $\tilde{q}$, calculated from i_{La}, i_{Lb}, and i_{Lc} and the fundamental positive-sequence voltages v'_α and v'_β comprise all harmonics, as well as the imbalances from

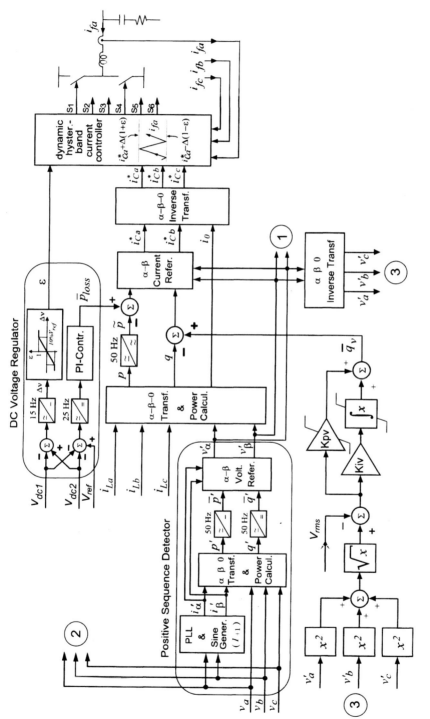

Figure 6-72. Controller of the shunt converter of the UPLC for configuration #1.

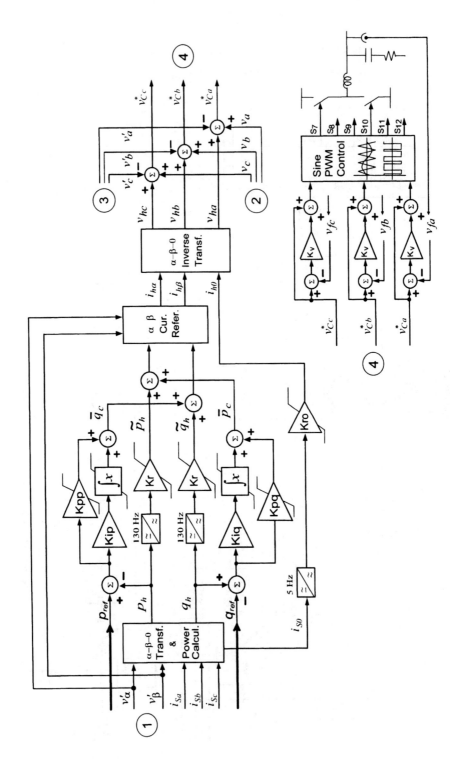

Figure 6-73. Controller of the series converter of the UPLC for configuration #1.

the fundamental negative-sequence component, which may be present in the load currents. Hence, the nonlinear load does not pollute the currents i_{Sa}, i_{Sb}, and i_{Sc} flowing through the controlled transmission line, and does not increase the harmonic content in current i_a, i_b, and i_c drawn from the subsystem that has distorted and/or unbalanced voltages (right side of the UPLC in Fig. 6-69). Only these already distorted/unbalanced voltages could still excite harmonic currents to flow through the controlled transmission line. However, they are blocked by the series converter of the UPLC. Two compensation features contribute to blocking current harmonics excited by the distorted voltages. They are (1) the differences $(v_a - v'_a)$, $(v_b - v'_b)$, and $(v_c - v'_c)$, and (2) the compensating voltages v_{ha}, v_{hb}, and v_{hc} in the controller for the series converter (see Fig. 6-73). The first one inserts the same degree of voltage harmonic/imbalance in the opposite direction, whereas the second one imposes an additional high resistance in series with the line, which is effective only for harmonics in i_{Sa}, i_{Sb}, and i_{Sc}.

The dc voltage regulator in Fig. 6-72 is the same as that used in Fig. 4-70, which provides total $(V_{dc1} + V_{dc2})$ dc link voltage regulation through the signal $\bar{p}_{loss}$ and voltage equalization through the signal ε, which is necessary in the "split-capacitor" converter topology. The symbol v_{rms} in Fig. 6-72 represents the reference value of the line voltage of the controlled ac bus (fundamental positive-sequence component). It can be set dynamically by a remote master control, such as a power delivery center, or just fixed locally to a predefined value. Although it is theoretically possible to impose fast dynamic response to the signal $\bar{q}_v$, here it was not designed to be fast enough for compensating voltage sags or swells. It was, rather, used only for voltage regulation in long-term dynamics. For voltage-sag compensation, more effective functionalities could be implemented in the series converter.

Only one positive-sequence voltage detector is needed in configuration #1 of UPLC controller (Fig. 6-72). Although it was included as part of the shunt converter controller, it is also a fundamental part for determining the compensating voltage references of the series converter. The interconnections between controllers, such as the measured voltages and their calculated positive-sequence voltages signals in the $\alpha\beta$-reference frame, and its inverse transformation into the abc-phase mode, are indicated with ②, ①, and ③, respectively in Fig. 6-72. The positive-sequence voltage detector in Fig. 6-72 is the same as that used in the shunt active filter controller for the sinusoidal current control strategy. All equations for this control circuit are shown in Fig. 4-27 and the phase-locked-loop circuit (PLL circuit) is shown in Fig. 4-28.

The difference between the measured voltage on the right side of the UPLC and its fundamental positive-sequence component comprises all harmonics and imbalances from negative- and zero-sequence components at the fundamental frequency. This constitutes the principle of series voltage compensation (active filtering) of the series PWM converter. It is determined as

$$\begin{bmatrix} v'_{Ca} \\ v'_{Cb} \\ v'_{Cc} \end{bmatrix} = \begin{bmatrix} v_a \\ v_b \\ v_c \end{bmatrix} - \begin{bmatrix} v'_a \\ v'_b \\ v'_c \end{bmatrix} \qquad (6.53)$$

In the steady state, the current of the controlled transmission line is ideally sinusoidal and balanced. However, during transients or due to inaccuracies in the UPLC shunt converter controller, some harmonic pollution or imbalances can appear in i_{Sa}, i_{Sb}, and i_{Sc}. A control circuit for the extraction of harmonic contents and imbalances at the fundamental frequency is used in the controller of the series active filter of the UPQC, as shown in Fig. 6-30. When the UPQC presented previously was developed, the old convention of imaginary power, $q = v_\alpha i_\beta - v_\beta i_\alpha$, was used, instead of the new one ($q = v_\beta i_\alpha - v_\alpha i_\beta$) adopted in this book, was used. As mentioned, the UPLC series converter controller integrates all active filtering functions of that UPQC together with the real and imaginary power flow control functions of the UPFC (Fig. 6-11). To conform both approaches of the UPQC and UPFC to the UPLC, some equations in the control blocks of Fig. 6-30 have to be changed before they can be used in Fig. 6-73. To summarize these changes, all equations used in the UPLC series converter controller are listed below.

The currents i_{Sa}, i_{Sb}, and i_{Sc} are transformed into the $\alpha\beta 0$-reference frame as

$$\begin{bmatrix} i_{S0} \\ i_{S\alpha} \\ i_{S\beta} \end{bmatrix} = \sqrt{\frac{2}{3}} \begin{bmatrix} \frac{1}{\sqrt{2}} & \frac{1}{\sqrt{2}} & \frac{1}{\sqrt{2}} \\ 1 & -\frac{1}{2} & -\frac{1}{2} \\ 0 & \frac{\sqrt{3}}{2} & -\frac{\sqrt{3}}{2} \end{bmatrix} \begin{bmatrix} i_{Sa} \\ i_{Sb} \\ i_{Sc} \end{bmatrix} \qquad (6.54)$$

Then, the control signals v'_α, and v'_β, marked by ① that come from the positive-sequence voltage detector of Fig. 6-72 are used together with these currents to calculate instantaneous real and imaginary powers as

$$\begin{bmatrix} p_h \\ q_h \end{bmatrix} = \begin{bmatrix} v'_\alpha & v'_\beta \\ v'_\beta & -v'_\alpha \end{bmatrix} \begin{bmatrix} i_{S\alpha} \\ i_{S\beta} \end{bmatrix} \qquad (6.55)$$

In the controller of the UPQC (Fig. 6-30), the powers p_h and q_h are filtered and the oscillating portions $\tilde{p}_h$ and $\tilde{q}_h$ are used to determine the harmonic currents $i_{h\alpha}$ and $i_{h\beta}$. Only after the inverse transformation of these currents into the abc-phase mode, three control gains K_r are applied to determine the compensating voltage references v_{ha}, v_{hb}, and v_{hc} in Fig. 6-30. Here, these gains must be anticipated in the control block diagram of Fig. 6-73, in order not to affect the power flow control functionalities performed by the signals $\bar{p}_c$ and $\bar{q}_c$. Thus, in Fig. 6-73, $\tilde{p}_h$ and $\tilde{q}_h$ are amplified by the gains K_r. Note that the oscillating portion of the zero-sequence current i_{S0} now is amplified by a gain K_{ro}. The gains K_r and K_{ro} may now be different. The reason is that K_r, $\tilde{p}_h$, $\tilde{q}_h$, and i_{S0} in the UPQC approach generate compensating voltages that makes the series active filter act as a series resistance ($v_h = K_r \cdot i_h$) that is effective only in harmonic frequencies other than the fundamental one. In other words, it represents a short circuit at the fundamental frequency. It is useful to suppress harmonic currents through the power line. Now, K_r, $\tilde{p}_h$, and $\tilde{q}_h$ realize the same active filtering functions, and they are also useful to improve stability, and play an important role during transients, mainly during changes in the power orders p_{ref} and q_{ref} of the controlled transmission line. Changes in the power orders will cause changes in $\bar{p}_c$ and $\bar{q}_c$, which will force changes in i_{Sa}, i_{Sb}, and i_{Sc} that transitorily will be interpreted as "current harmonics," and $\tilde{p}_h$ and $\tilde{q}_h$ tend to act against these changes. Therefore, the tuning of gain K_r depends on the gains K_{pp}, K_{ip}, K_{pq}, and K_{iq} in the PI controllers that determine $\bar{p}_c$ and $\bar{q}_c$, which realize power flow control functions as in the UPFC. There are several dynamic constraints and system parameters, which are involved with this problem, which can make the tuning of the control gains a tedious task.

Figure 6-69 mirrors Fig. 6-9, but the symbols and current directions keep the same relations, which results in identical circuit equations for the power flow control and voltage regulation functionalities. Hence, the phasor diagrams in Fig. 6-7 are also valid to verify the correctness of signs of $\bar{p}_c$ and $\bar{q}_c$ in Fig. 6-73. For instance, if the calculated real power, p_h, through the controlled transmission line is smaller than the power order, p_{ref}, a positive error appears, which produces a positive signal $\bar{q}_c$. This generates a compensating voltage component at the fundamental frequency that is orthogonal to and lag 90° the voltage v of the controlled ac bus on the right side of the UPLC. This situation is illustrated in the phasor diagram of Fig. 6-7(b). In this case, the voltage v_G should be interpreted as the terminal voltage of the controlled transmission line that is the voltage on the bus that connects the line with subsystem B in Fig. 6-69. Hence, the phase angle of voltage v_S at the sending end of the controlled transmission line will be shifted, and will lead the voltage v. This causes an increment in the power angle between the terminal voltages of the controlled transmission line, increasing the transmitted active power through this line, until reaching the set point dictated by the power order, p_{ref}.

On the other hand, if the reactive power at the sending end of the controlled transmission line, determined as q_h in Fig. 6-73, is more inductive than its reactive power order q_{ref}, both q_h and q_{ref} are positive values and $q_h > q_{ref}$. The resulting positive error will produce a positive control signal $\bar{p}_c$, which forces the series converter to insert a compensating voltage component in phase with voltage v. This situation corresponds to the phasor diagram of Fig. 6-7(c), in which the current of the controlled transmission line is forced to become more capacitive, reducing q_h until its set point given by q_{ref} is reached.

The "$\alpha\beta$-current references" in Fig. 6-73, which in fact are used as *voltage* references, are calculated as

$$\begin{bmatrix} i_{h\alpha} \\ i_{h\beta} \end{bmatrix} = \frac{1}{v_\alpha'^2 + v_\beta'^2} \begin{bmatrix} v_\alpha' & v_\beta' \\ v_\beta' & -v_\alpha' \end{bmatrix} \begin{bmatrix} \tilde{p}_h + \bar{p}_c \\ \tilde{q}_h + \bar{q}_c \end{bmatrix} \quad (6.56)$$

Finally, the compensating voltage components that provide harmonic isolation, damping of oscillations, and also active and reactive power flow control in the controlled transmission line, are determined as

$$\begin{bmatrix} v_{ha} \\ v_{hb} \\ v_{hc} \end{bmatrix} = \sqrt{\frac{2}{3}} \begin{bmatrix} \frac{1}{\sqrt{2}} & 1 & 0 \\ \frac{1}{\sqrt{2}} & -\frac{1}{2} & \frac{\sqrt{3}}{2} \\ \frac{1}{\sqrt{2}} & -\frac{1}{2} & -\frac{\sqrt{3}}{2} \end{bmatrix} \begin{bmatrix} i_{h0} \\ i_{h\alpha} \\ i_{h\beta} \end{bmatrix} \qquad (6.57)$$

The above control algorithm that generates the voltage references v_{ha}, v_{hb}, and v_{hc} can be added with the compensating voltage components given by (6.53), which provides active voltage filtering, including compensation of unbalances from negative- and zero-sequence components at the fundamental frequency. Thus, the compensating voltage references v^*_{Ca}, v^*_{Cb}, and v^*_{Cc} to the series PWM converter is given by

$$\begin{bmatrix} v^*_{Ca} \\ v^*_{Cb} \\ v^*_{Cc} \end{bmatrix} = \begin{bmatrix} v_a \\ v_b \\ v_c \end{bmatrix} - \begin{bmatrix} v'_a \\ v'_b \\ v'_c \end{bmatrix} + \begin{bmatrix} v_{ha} \\ v_{hb} \\ v_{hc} \end{bmatrix} \qquad (6.58)$$

Figure 6-73 suggests the use of a PWM voltage control with minor feedback control loops composed by a feedforward loop and a proportional gain, as employed in the UPQC approach of Fig. 6-29. Although the following simulation results show that this type of PWM voltage control can reduce satisfactorily the errors between the compensating voltage references (v^*_{Ca}, v^*_{Cb}, and v^*_{Cc}) and the secondary voltages (v_{fa}, v_{fb}, and v_{fc}), more recent studies have proved that the use of a PID (proportional-integral-derivative) or a PD (proportional-derivative) controller, as shown in Fig. 6-74, can be more effective. A limiter in the output of the PID controller may be necessary to confine the output values inside the amplitude range of the triangular (carrier) wave in the sine-PWM Control. In fact, all PI-controllers in the UPLC control, as shown in Fig. 6-72 and Fig. 6-73, are properly limited to pro-

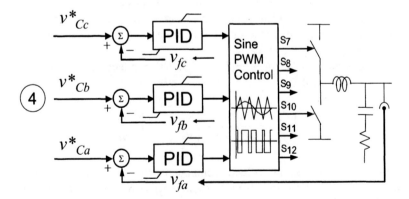

Figure 6-74. PWM voltage control with minor feedback control loops.

tect the PWM converters of the UPLC against overloads during abnormal situations, for instance, during operation under a faulty power system.

6.3.2.A. Controller for Configuration #2 of the UPLC

Configuration #2 of the UPLC is shown in Fig. 6-70. As previously mentioned, the UPLC can be installed at the end of a transmission line that has harmonic *currents* coming from the load side (right side of the UPLC) and harmonic *voltages* coming from the source side (left side of the UPLC). In terms of symbols used in Fig. 6-70, this means that the active filtering characteristic of the UPLC should be used to compensate v_S and i_L, instead of v and i_L, as performed in the previous approach in configuration #1. In this new situation, the voltage imbalance and harmonics that are present in v_S should be compensated to keep the voltage v balanced and sinusoidal. All other previous compensation functions in configuration #1 of UPLC should not be changed.

To implement this new approach, the voltage v_S has to be measured and used as input to the UPLC controller, besides all those shown in Fig. 6-71. An additional positive-sequence voltage detector must be included in the controller for the series converter (Fig. 6-73). It has the same structure as that used in the controller for the shunt converter (Fig. 6-72), but has v_{Sa}, v_{Sb}, and v_{Sc} as inputs. Only this additional circuit is sufficient to implement the UPLC controller that satisfies the new application constraints described in Fig. 6-70. If the outputs of this additional voltage detector is named as v'_{Sa}, v'_{Sb}, and v'_{Sc}, which would correspond to the instantaneous values of the fundamental positive-sequence component of voltage v_S, the following changes should be made:

1. The signals marked with ② ↔ (v_a, v_b, v_c) in Fig. 6-73 should be replaced by the measured voltages on the left side of the UPLC, that is, v_{Sa}, v_{Sb}, and v_{Sc};
2. The signals marked with ③ ↔ (v'_a, v'_b, v'_c) should be replaced by the outputs of an additional positive-sequence voltage detector that calculates v'_{Sa}, v'_{Sb}, and v'_{Sc}.

Digital simulations using both approaches have been carried out, and the principal results are presented in the next section.

6.3.3. Performance of the UPLC

In this section, the most important simulation results of a complete model of the UPLC are presented and discussed. Generally, the dynamic response of the UPLC has been very fast, which might be unachievable if traditional power theories described in Chapter 2 were used in the controller. A UPFC based on traditional concepts of powers can be found in [17].

6.3.3.A. Normalized System Parameters

The normalization of system parameters involves ac and dc quantities, as well as three- and single-phase circuits. There is no universal system basis for power systems involving ac and dc quantities, especially when PWM converters are included. The authors have tried to find a set of definitions to use, so that dc and ac impedances can have some sense of correlation when they are given as per unity quantities (p.u.). The following nonusual definition of the system basis including dc and ac quantities is adopted to normalize impedances and other system parameters.

Usual definitions are preserved on the following extended basis, that is, the fundamental values are:

Three-phase apparent power basis: $\quad S_{3\phi@} \quad [\text{VA}] \quad$ (6.59)

Line-voltage basis: $\quad V_{3\phi@} \quad [\text{V}] \quad$ (6.60)

which are arbitrarily chosen. If necessary, a single-phase basis can be derived from the above three-phase basis, that is,

Single-phase apparent power basis: $\quad S_{1\phi@} = \dfrac{S_{3\phi@}}{3} \quad [\text{VA}] \quad$ (6.61)

Phase-voltage basis: $\quad V_{1\phi@} = \dfrac{V_{3\phi@}}{\sqrt{3}} \quad [\text{V}] \quad$ (6.62)

From the single-phase or three-phase basis of voltage and power, all other usual quantities can be defined:

Current basis: $\quad I_@ = \dfrac{S_{3\phi@}}{\sqrt{3}V_{3\phi@}} = \dfrac{S_{1\phi@}}{V_{1\phi@}} \quad [\text{A}] \quad$ (6.63)

Impedance basis: $\quad Z_@ = \dfrac{V_{3\phi@}^2}{S_{3\phi@}} = \dfrac{V_{1\phi@}^2}{S_{1\phi@}} \quad [\Omega] \quad$ (6.64)

The fundamental frequency of the system complements the set of definitions,

Angular frequency basis: $\quad \omega_@ = \omega_1 = 2\pi f_1 \quad [\text{rad/s}] \quad$ (6.65)

and a capacitance and inductance basis for the ac side are found to be

Capacitance basis: $\quad C_@ = \dfrac{1}{2\pi f_1 Z_@} \quad [\text{F}] \quad$ (6.66)

Inductance basis: $\quad L_@ = \dfrac{Z_@}{2\pi f_1} \quad [\text{H}] \quad$ (6.67)

Additionally, for the quantities on the dc side of the PWM converters a set of bases is necessary. The power basis at the dc side has a direct correspondence with that of the ac side, that is,

dc power basis: $\qquad P_{dc@} \equiv S_{3\phi@}$ [W] $\qquad$ (6.68)

However, another arbitrarily chosen value has to be established for the dc voltage basis. It is worthwhile to adopt the nominal voltage at the dc side of the active power line conditioner as the dc voltage basis:

dc voltage basis: $\qquad V_{dc@}$ [V] $\qquad$ (6.69)

From (6.68) and (6.69), a dc current basis is defined:

dc current basis: $\qquad I_{dc@} = \dfrac{P_{dc@}}{V_{dc@}}$ [A] $\qquad$ (6.70)

and a dc resistance basis is found to be

dc resistance basis: $\qquad R_{dc@} = \dfrac{V_{dc@}}{I_{dc@}}$ [Ω] $\qquad$ (6.71)

Finally, a criterion has to be established to determine a dc capacitance and inductance basis. Generally, the energy stored in the dc link of a PWM converter plays an important role. Therefore, a way to correlate the energy stored in the ac and dc sides of the converter is adopted. The idea consists of finding a dc capacitance basis that stores the same energy as that stored in a 1-p.u. three-phase capacitor bank at the ac side. In an attempt to define this constraint, the following relation can be written:

Energy basis: $\qquad W_@ = \dfrac{3}{2} C_@ V^2_{1\phi@} = \dfrac{1}{2} C_{dc@} V^2_{dc@}$ [J] $\qquad$ (6.72)

or

Energy basis: $\qquad W_@ = \dfrac{3}{2} L_@ I^2_@ = \dfrac{1}{2} L_{dc@} I^2_{dc@}$ [J] $\qquad$ (6.73)

From (6.72) and (6.73), it is possible to extract the value of the dc capacitance basis and dc inductance basis, since all other quantities in these equations were previously determined.

A perunity quantity is achieved by dividing the electrical quantity by its corresponding basis value. For instance, the p.u. current of a load is determined by

$$i_a[\text{p.u.}] = \dfrac{i_a}{I_@} \,;\, i_b[\text{p.u.}] = \dfrac{i_b}{I_@} \,;\, i_c[\text{p.u.}] = \dfrac{i_c}{I_@} \,;\, i_o[\text{p.u.}] = \dfrac{i_o}{I_@} \qquad (6.74)$$

Similar equations hold for all other quantities.

The basic configuration of the simulated system is that presented in Fig. 6-69 and Fig. 6-70. The principal components of the power circuit of the UPLC are shown in Fig. 6-75. The only interconnection between ac subsystems "A" and "B" is through the controlled transmission line, since no other connection, like that suggested by dotted lines in Fig. 6-69 and Fig. 6-70, is implemented.

For a given basis, that is, for arbitrarily chosen values of $S_{3\phi(a)}$, $V_{3\phi(a)}$, $\omega_{(a)}$, and $V_{dc(a)}$, all other system basis parameters were calculated as shown before. Then, the normalized parameters for the power system as summarized in Table 6.7 were used to implement the system in a digital simulator. Table 6.7 shows that an unbalanced, distorted voltage source is employed in the subsystem called as G1. It contains fundamental zero- and negative-sequence components, as well as a third- and a fifth-order harmonic component. The voltage source of subsystem G2 is balanced and sinusoidal, but it has greater equivalent impedance. These subsystems are placed on the left and right sides of the UPLC, accordingly, to implement configuration #1 (Fig. 6-69) or configuration #2 (Fig. 6-70) of the UPLC. A nonlinear load consisting of the same kind of line-commutated converters as those used in the UPQC approach (see Table 6.2) is connected to the right side of the UPLC.

The set of control parameters that has been used in the following analysis of dynamic response of the UPLC is shown in Table 6.8. All parameters are given p.u. and information is given for each parameter on how to obtain the value for an implementation using actual (no p.u.) quantities. For example, $K_p = 1.5$ should be multiplied by the ratio between the three-phase power basis and the dc voltage basis $(S_{3\phi(a)}/V_{dc(a)})$.

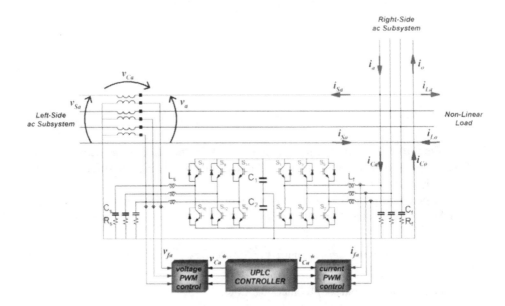

Figure 6-75. Three-phase circuit of the UPLC system for digital simulation purposes.

6.3. THE UNIVERSAL ACTIVE POWER LINE CONDITIONER (UPLC)

Table 6.7. Principal parameters of the power circuit

Parameter description	Value [p.u.]
Power generating subsystem G1:	
Voltage source	$V_{-1} = 1 \angle 0°$; $V_{-1} = 0.0456 \angle 0°$; $V_{01} = 0.0228 \angle 0°$; $V_{+3} = 0.0228 \angle 90°$; $V_3 = 0.0228 \angle 180°$; $V_{03} = 0.0228 \angle 270°$; $V_{-5} = 0.0456 \angle 0°$
Equivalent impedance ($R_{L1} + j\omega L_1$)	$Z_{L1} = 0.01039 + j0.06527$
Power generating subsystem G2:	
Voltage source	$V_{+1} = 1 \angle 0°$
Equivalent impedance ($R_{L2} + j\omega L_2$)	$Z_{L2} = 0.01039 + j0.1044$
Shunt converter:	
High-pass filter ($R_F + 1/j\omega C_F$)	$Z_F = 2.494 - j44.10$
Commutation reactor ($R_{LF} + j\omega L_F$)	$Z_{LF} = 0.1663 + j0.3917$
Series converter:	
High-pass filter ($R_S + 1/j\omega C_S$)	$Z_S = 1.039 - j11.02$
Commutation reactor ($R_{LS} + j\omega L_S$)	$Z_{LS} = 0.04155 + j0.05222$
dc capacitors:	$C_1 = C_2 = 5.654$
Nonlinear loads:	
One three-phase thyristor converter	Commutation reactor = $+j0.3263$
Two single-phase thyristor converters	Commutation reactor = $+j0.3263$

Connections and disconnections of loads, as well as of the complete digital model of the UPLC, have been simulated. The complete model of the UPLC requires a long time of simulation in a normal personal computer (PC). Nevertheless, several simulations have been carried out in order to adjust the system parameters, and to analyze the dynamic response of the UPLC. Finally, two final simulations show successively all important transient phenomena that may occur in a power system.

Table 6.8. Principal parameters of the control circuit

Parameter description	Value (pu)	
dc voltage regulator:		
PI-Controller	$K_p = 1.5$; $K_i = 4.0$;	$(S_{3\phi(a)}/V_{dc(a)})$
rms voltage control:		
PI controller	$K_{pv} = 2.03$; $K_{iv} = 405$;	$(S_{3\phi(a)}/V_{3\phi(a)})$
Limits of output	$\bar{q}_v = \pm 0.667$	$(S_{3\phi(a)})$
Power flow control:		
PI controller	$K_{pp} = K_{pq} = 3.2$; $K_{ip} = K_{iq} = 144$;	
Limits of output	$\bar{p}_c = \bar{q}_c = \pm 0.667$	$(S_{3\phi(a)})$
Damping circuit:		
α and β components	$K_r = 2$	
Zero-sequence components	$K_{ro} = 4$	
Limits of output	unlimited	

All values of parameters, voltage source distortion, and nonlinear load were the same in both simulations. Only the UPLC controller as shown in Fig. 6-72 and Fig. 6-73, or with the additional positive-sequence detector as explained on page 355, was exchanged and the power subsystems G1 and G2 were permuted from the left-hand to the right-hand sides, in order to show simulation results from both configurations of UPLC.

6.3.3.B. Simulation Results of Configuration #1 of the UPLC

The total simulation time interval is 300 ms, and involves the following events:

Event #1: starting of the shunt converter of the UPLC at 55 ms

Event #2: starting of the series converter of the UPLC at 60 ms

Event #3: connecting of a single-phase thyristor bridge rectifier as a load between a phase and neutral wire, and another between b phase and neutral wire, at 101.7 ms ($\cong 30°$ firing angle)

Event #4: connecting of a six-pulse thyristor bridge rectifier, as a three-phase load, at 103.3 ms ($\cong 30°$ firing angle)

Event #5: change in the imaginary power order q_{ref} from –0.33 pu to +0.33 pu, at 150 ms

Event #6: change in the real power order p_{ref} from –1 pu to +1 pu, at 180 ms

Event #7: disconnecting of the six-pulse thyristor bridge rectifier at 245 ms

The remaining figures refer to the above sequence of events. The phase voltages were divided by the rms value of the phase-voltage basis, and are given in p.u. quantities. Currents were divided by the rms current basis, and the real and imaginary powers were divided by the three-phase power basis.

The first simulation case refers to configuration #1 of the UPLC, in which the distorted power subsystem G1 is located at the right side of the UPLC, together with the nonlinear loads. Figure 6-76 shows the phase voltages that appear on the right side of the UPLC. This corresponds to voltage v in Fig. 6-69. They are distorted and unbalanced. The fundamental zero-sequence component in G1 excites a high neutral current that flows through the controlled transmission line, before the connection of the series converter of the UPLC ($t < 60$ ms). The neutral currents are shown in Fig. 6-77. During this period, the neutral current i_o of subsystem G1 is equal to the current i_{S_o} of the controlled transmission line, although the shunt converter of the UPLC was already connected at $t = 55$ ms. The shunt converter compensates neither the current i nor the current i_S, but only the current i_L (see Fig. 6-69). A sub-harmonic at a low frequency remains in i_o. It is related to the signal ε in the dc voltage regulator that equalizes the voltages on the dc capacitors in the employed "split-capacitor" converter topology, as explained in Chapter 4. This unwanted phenomenon disappears in a UPLC approach for three-phase systems without a neutral conductor.

The compensating voltages of the series converter are presented in Fig. 6-78, and

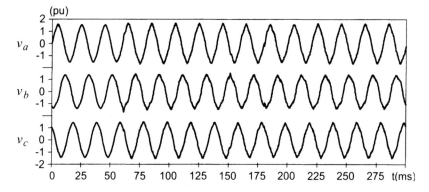

Figure 6-76. Phase voltages on the right side of the UPLC.

the compensated voltage v_S on the left side of the UPLC is shown in Fig. 6-79. Although the voltages of the series converter are not sinusoidal, it is possible to identify a fundamental component that is needed to control the power flow in the transmission line. In the steady state, this component is smaller than ±0.4 p.u. The active filtering characteristics of the unified power quality conditioner (UPQC) as presented in previous sections have been successfully preserved in the UPLC approach. The voltage v_S is almost sinusoidal and balanced. Fig. 6-80 shows better the waveshape of voltages v_{Sa}, v_{Sb}, and v_{Sc} before and after connecting the series converter. In a real implementation, high-frequency components in v_S would not be as high as shown in Fig. 6-80 because the real series transformer itself acts as "low-pass filter." This would cause damping in high-frequency components on the primary side connected in series with the controlled transmission line. This feature was not well modeled in the present simulation case, since ideal transformer models were used in the simulation.

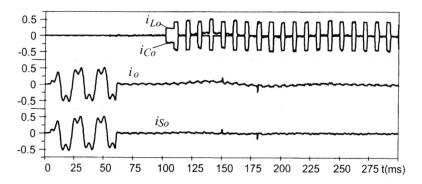

Figure 6-77. Neutral currents.

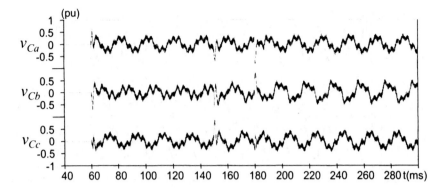

Figure 6-78. Compensating voltages of the series converter of the UPLC.

The nonlinear loads are connected after $t = 100$ ms, successively, according to the unblocking time of their firing angles, as shown on page 360. Fig. 6-81 presents the totalized line currents of these loads. They are distorted and unbalanced. Moreover, they produce also significant average imaginary power $\bar{q}$, due to the phase shift between fundamental voltage and current components, which is being compensated by the shunt converter. The neutral current i_{Lo} of the load is also compensated, as shown in Fig. 6-77. The total real and imaginary powers of these loads are illustrated in Fig. 6-82.

To cover losses in the UPLC and regulate the voltages on the dc capacitors of the common dc link of the UPLC, the shunt converter must draw an active current component at the fundamental frequency, as explained in previous sections dealing with active filters. Here, besides this current component, the shunt converter of the UPLC must draw an additional, fundamental, imaginary, current component for

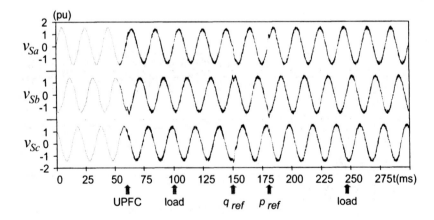

Figure 6-79. Compensated voltages on the left side of the UPLC.

6.3. THE UNIVERSAL ACTIVE POWER LINE CONDITIONER (UPLC) 363

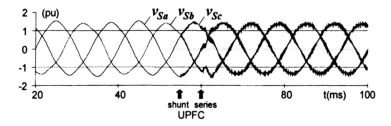

Figure 6-80. Compensated voltages at the left side of the UPLC.

regulating the magnitude of the fundamental positive-sequence voltage v (right side of the UPLC) of the controlled ac bus, as well as for compensating the reactive current of the nonlinear load. This leads to a relatively high fundamental current component drawn by the UPLC shunt converter that, in the present simulation case, has become greater than 1 p.u. Of course, this indicates a poor solution, at least from an economical point of view.

Although successfully implemented, it has been verified that the voltage control realized by the control signal $\bar{q}_v$ in Fig. 6-72 may lead to a very high need for imaginary power generation in the shunt converter, in order to regulate the voltage v. This functionality is similar to that of a STATCOM [19] and is useful if the UPLC is located in a weak power system. Figure 6-83 shows the line currents of the shunt converter. They increase significantly after $t = 200$ ms, the reason for which will be clarified later.

The UPLC is connected to the system with real power order p_{ref} for the controlled transmission line set to -1 p.u., and the imaginary power order q_{ref} set to -0.33 p.u. This means that, after the connection of the series converter ($t = 60$ ms), the current i_S and voltage v should be adjusted to produce a real power p_r and an imaginary power q_r equal to the reference values p_{ref} and q_{ref}, respectively. This occurs quickly, as can be seem in Fig. 6-84. This figure also shows the powers p_s and

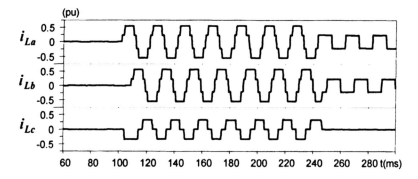

Figure 6-81. Totalized currents of the nonlinear loads.

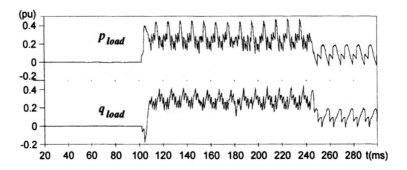

Figure 6-82. Instantaneous real and imaginary powers of the nonlinear loads.

q_s that are generated by voltage v_S and current i_S (left side of the UPLC). These powers are smoother than those on the right side of the UPLC, because the compensated voltage v_S is almost balanced and sinusoidal. However, they differ slightly from the power orders p_{ref} and q_{ref}. The differences are equal to the instantaneous powers that the series converter has to generate to control the real power p_r and imaginary power q_r of the controlled transmission line. The fast dynamic response verified in Fig. 6-84 evidences, once again, that the use of the p-q Theory in the control of active power line conditioners constitutes a powerful tool to obtain fast dynamic responses. A full power reversion from −1 p.u. to +1 p.u. is realized, and the steady state is achieved within 30 ms. It is so fast that, in case of real systems, the gains in the controller should be lowered to avoid system instability.

Unfortunately, Fig. 6-84 shows that the control of the real and imaginary powers of the controlled transmission line are not perfectly decoupled. Changes in the transmitted real power p_r affect transitorily the generated imaginary power q_r, and vice-versa. There are some ways to improve the UPLC controller for better decoupling of real and imaginary power flow control. A method to do so is presented in [28].

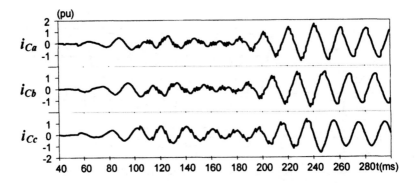

Figure 6-83. Compensating currents of the shunt converter of the UPLC.

6.3. THE UNIVERSAL ACTIVE POWER LINE CONDITIONER (UPLC)

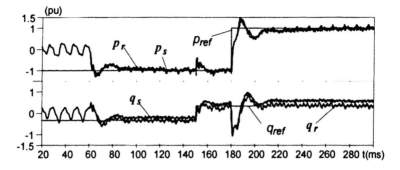

Figure 6-84. Instantaneous powers of the controlled transmission line.

The line currents i_a, i_b, and i_c of power subsystem G1 (distorted system voltages on the right side of the UPLC), and i_{Sa}, i_{Sb}, and i_{Sc} of the controlled transmission line are presented in Fig. 6-85 and Fig. 6-86, respectively. Since the current of the controlled transmission line is "fixed" by the series converter of the UPLC to produce p_r and q_r according to their reference values p_{ref} and q_{ref}, changes in the nonlinear loading conditions affect almost only the current of subsystem G1. This can be verified around $t = 100$ ms and $t = 250$ ms, when the three-phase thyristor bridge is connected to and disconnected from the system.

An interesting characteristic of the proposed control is that no errors occur in the neutral current compensation by the shunt converter of the UPLC, even during transient responses, as can be seen in Fig. 6-77. Before $t = 150$ ms, the current i_S leads v_S ($q_{ref} = -0.33$ p.u.), and lags v_S after this instant ($q_{ref} = +0.33$ p.u.). This can be better seen in Fig. 6-87. Before $t = 180$ ms, the phase displacement between i_S and v_S is greater than 90° ($p_{ref} = -1$ pu), and smaller than 90° after this instant ($p_{ref} = +1$ pu). Since q_{ref} remains equal to +0.33 p.u. after $t = 150$ ms, the current i_S is lagging v_S, before and after $t = 180$ ms.

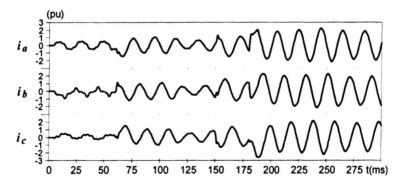

Figure 6-85. Currents of the power subsystem G1.

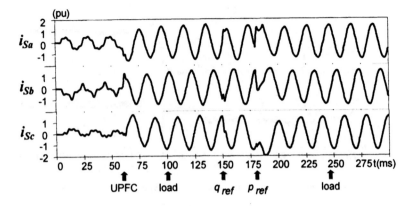

Figure 6-86. Currents of the controlled transmission line.

The required power rating of the series converter is relatively small, even when it should reflect severe power reversions as indicated above. In the steady state, the real and imaginary power p_{series} and q_{series}, given by the product of v_C (Fig. 6-78) and i_S (Fig. 6-86), is smaller than ±0.25 p.u., as can be seen in Fig. 6-88.

In contrast, the power rating of the shunt converter reaches almost the same order of magnitude as the rated power of the system (≈ 1 p.u.). The function of voltage regulation of the controlled ac bus (right side of the UPLC) is the principal cause of this increment of power rating. After the power reversion at $t = 180$ ms, the fundamental positive-sequence component of voltage v tends to decrease, and the shunt converter starts to increase the generation of capacitive reactive power ($q < 0$) to neutralize this voltage drop tendency. This feature is confirmed in Fig. 6-83 (compensating current of the shunt converter) and Fig. 6-89 (real and imaginary powers of the shunt converter). A high limit of reactive power generation given by signal $\bar{q}_v$ in the UPLC controller of Fig. 6-72, which performs voltage regulation on the controlled ac bus, has been established: $-0.67 \leq \bar{q}_v \leq +0.67$ (p.u.). However, this high generation of reactive power, which reaches -0.67 p.u. around $t = 240$ ms,

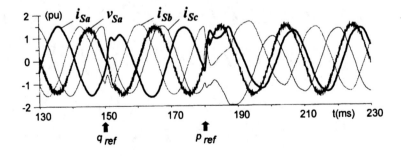

Figure 6-87. Currents of the controlled transmission line.

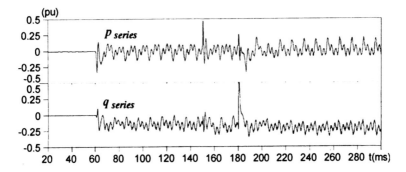

Figure 6-88. Instantaneous powers of the series converter of the UPLC.

could not raise the rms value of the fundamental positive-sequence voltage up to its reference value, that is, $V_{rms} = 1$ p.u.

Figure 6-90 shows the instantaneous aggregate values $\|v'_\Sigma\|$ and $\|v'_{s\Sigma}\|$ of the fundamental positive-sequence components of voltages v and v_S, respectively. Again, a very fast response is verified. However, this happens under unpractical conditions, since it is difficult to impose such changes in a real power system. Additionally, the shunt converter is also compensating the imaginary power of the nonlinear loads (Fig. 6-82). At $t = 245$ ms, the three-phase thyristor rectifier is disconnected, which alleviates this difficult condition of imaginary power generation in the shunt converter. After $t = 245$ ms, it is still generating imaginary power at the negative limit, but now it is able to better regulate the voltage of the controlled ac bus, as can be seen in the curve of the aggregate voltage $\|v'_\Sigma\|$ in Fig. 6-90, which almost reaches 1 p.u.

From the dc capacitors dimensioning point of view, which corresponds to C_1 and C_2 in Fig. 6-75, power reversion in the controlled transmission line is not as critical as connection of loads. Although a high capacitance has been employed, as indicated in Table 6.7, this item could surely be better optimized if greater dc voltage variations could be allowed or other nonlinear loads that draw smaller neutral

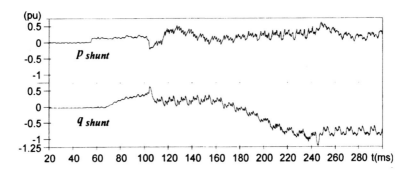

Figure 6-89. Instantaneous powers of the shunt converter of the UPLC.

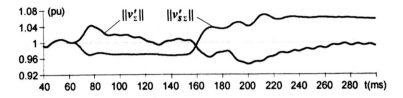

Figure 6-90. Instantaneous aggregate values of the fundamental positive-sequence voltages v and v_S.

current be considered. Fig. 6-91 shows that the highest dc voltage variation occurs after the connection and disconnection of loads, instead of during step changes in the power order of the controlled transmission line. During $200 < t < 240$ ms, the total dc voltage $V_{dc} = V_{dc1} + V_{dc2}$ is constantly decaying because of the increasing reactive power generation in the shunt converter. Thus, increasing losses in the PWM converter. In fact, the analysis of losses in the digital model of the UPLC is not trustworthy due to the poor digital models of IGBTs and diodes employed in the PWM converters.

6.3.3.C. Simulation Results of Configuration #2 of the UPLC

Next, some graphics from a second simulation case are presented. Now, the power subsystem G1 that contain unbalanced and distorted voltage source is located on the left side of the UPLC, and G2 replaces G1 on the right side. Therefore, the voltage v_S (left side in Fig. 6-70) should now be compensated in order to keep the voltage v balanced and sinusoidal. The changes implemented in the UPLC controller are explained on page 365. The nonlinear loads were the same as those in Fig. 6-81 and the sequence of events is equal to that given on page 360.

The compensated voltages on the right side of the UPLC are given in Fig. 6-92. After connecting the UPLC ($t = 60$ ms), the voltages v_a, v_b, and v_c become balanced and sinusoidal. These phase voltages do not have considerable high-frequency components, as do the voltages v_{Sa}, v_{Sb}, and v_{Sc} (compare with the previous simulation case, Fig. 6-80) because the high-pass filter (C_f and R_f) of the shunt converter attenuates high-frequency voltage harmonics on the right side of the UPLC.

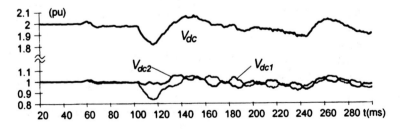

Figure 6-91. Voltages at the common dc link of the UPLC.

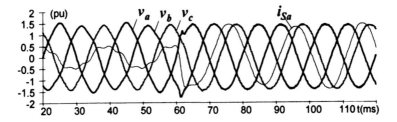

Figure 6-92. Compensated voltages at the right side of the UPLC.

Since the currents of the nonlinear loads are well filtered, the currents of the controlled transmission line (i_{Sa}, i_{Sb}, and i_{Sc}) are almost balanced and sinusoidal, and the real power p_r and imaginary power q_r generated by voltage v (right side) and current i_S are smoother than those presented in Fig. 6-84. The powers p_r and q_r for configuration #2 of UPLC are illustrated in Fig. 6-93. It shows that the power reversions have been successfully realized also in the new approach. The active filtering capabilities of the unified power quality conditioner (UPQC), as presented in Table 6.1, have been satisfactorily preserved from a voltage and current compensation point of view. This is confirmed by Fig. 6-92 and Fig. 6-94.

The voltage regulation provided by the shunt converter yielded better results than those given by Fig. 6-90. This can be explained by the fact of now having greater equivalent impedance (L_2 of the power subsystem G2) at the right side of the UPLC (see Table 6.7). In other words, shunt compensators for voltage regulation are more effective when connected to a weak ac bus of the power system. In this case, smaller reactive currents should be injected into the controlled ac bus for producing the necessary voltage drop across L_2, which regulates the voltage v. This is the reason why static var compensators can be more profitable if connected in weak areas of the power system. Figure 6-95 shows the instantaneous aggregate

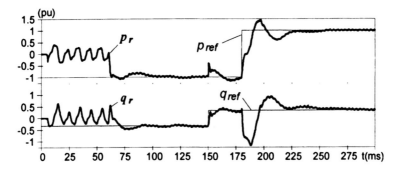

Figure 6-93. Instantaneous powers of the controlled transmission line (right side of the UPLC).

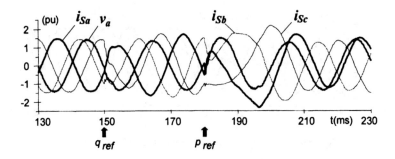

Figure 6-94. Currents of the controlled transmission line.

voltages at both sides of the UPLC, considering only the fundamental positive-sequence component in the calculation of $\|v'_\Sigma\|$ and $\|v'_{s\Sigma}\|$. It shows that $\|v'_\Sigma\|$ (right side of the UPLC) reaches 1 p.u. before $t = 240$ ms.

6.3.4. General Aspects

Although no detailed investigation of PWM converters for the UPLC has been performed in the previous sections, some additional considerations can be noted to orient future designs of UPLCs.

The power ratings of PWM converters that are employed in a UPLC may be key to making it work. The first step that should be taken to reduce power ratings is to eliminate zero-sequence current compensation. This might seem reasonable, since a UPLC should be applied to transmission systems in which no neutral conductor exists. Only this procedure can widely reduce the capacitance at the dc link of the UPLC, which would lead to a great reduction in costs.

Another aspect that should be carefully investigated is to of optimize the PWM control to reduce its switching frequency, without losing active filtering capabilities, at least for low-order harmonics, up to 600 or 700 Hz. Switching losses decrease significantly if the IGBTs are switched at no more than 6 kHz. In this case, PWM converters based on presently available power semiconductor devices may make feasible the implementation of a UPLC. Another solution may be a further reduction in the frequency range of harmonics to be actively filtered.

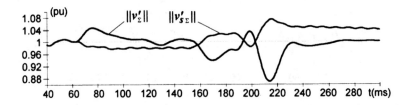

Figure 6-95. Aggregate values of the fundamental positive-sequence voltage v and v_s.

6.4. SUMMARY

The control techniques for active filters have been successfully merged with a FACTS device, the unified power flow controller (UPFC). This allowed the improvement of the dynamic response, and additionally provided active filtering capabilities for the new FACTS device—the universal active power line conditioner (UPLC).

A three-phase, four-wire power system has been considered in the analysis, which allowed the evaluation of the performance of the UPLC under all kinds of unbalances and harmonic pollutions that can arise in three-phase systems. The control algorithm is so flexible that a simplification can be easily made to adapt the equipment for other applications in three-phase systems, with or without a neutral conductor. Active filtering capabilities for the shunt and/or series converter of the UPLC may also be considered.

The UPLC incorporates almost all compensation characteristics of active power line conditioners that can be implemented with PWM converters into a single device. Most problems of active filtering and power flow control, as well as ac bus voltage regulation, can be solved with the UPLC, although it might be interesting, from an economical point of view, to select some compensation characteristics and to leave others to be executed by equipment that does not require high-power PWM converters. The unique power quality problem that is not covered in the presented UPLC approach is the compensation of voltage sags and swells. Although not covered, the authors believe this feature can also be incorporated by adding extra control circuits still based on the *p-q* Theory to the UPLC controller without affecting the other compensation characteristics.

Two important points have been left for future work: the dimensioning of the dc capacitors of the UPLC and the analysis of stability under subsynchronous resonances. These points require a more complex model of the power generating system and a longer time interval of simulation. For this purpose, it might be interesting to develop a model of the UPLC in a real-time simulator.

REFERENCES

[1] B. K. Bose, *Power Electronics and AC Drives*, ISBN 0-13-686882-7, Prentice-Hall, 1986, pp. 130–140.

[2] N. G. Hingorani, "High Power Electronics and Flexible AC Transmission System," *IEEE Power Engineering Review*, July 1988.

[3] P. Wood et al., "Study of Improved Load-Tap-Changing for Transformers and Phase-Angle Regulators," in *EPRI*, Final Report EL-6079, 1988.

[4] W. J. Lyman, "Controlling Power Flow with Phase Shifting Equipment," *A.I.E.E. Transactions*, vol. 49, July 1930, pp. 825–831.

[5] C. P. Arnold, R. M. Duke, and J. Arrillaga, "Transient Stability Improvement Using Thyristor Controlled Quadrature Voltage Injection," *IEEE Transactions on Power Applications and Systems*, vol. PAS-100, no. 3, Mar. 1981, pp. 1382–1388.

[6] R. Baker, G. Güth, W. Egli, and P. Eglin, "Control Algorithm for a Static Phase Shifting Transformer to Enhance Transient and Dynamic Stability of Large Power Systems," *IEEE Transactions on Power Applications and Systems,* vol. PAS-101, no. 9, Sep. 1982, pp. 3532–3542.

[7] A. E. Hammad and J. Dobsa, "Application of a Thyristor Controlled Phase Angle Regulating Transformer for Damping Subsynchronous Oscillations," in *IEEE-PESC'83—Power Electronics Special Conference,* 1983, pp. 111–121.

[8] G. Güth and M. Häusler, "Simulator Investigations of Thyristor Controlled Static Phase Shifter and Advanced Static VAr Compensator," EPRI TR-100504, in *Proceedings of FACTS Conference I—The Future in High-Voltage Transmission,* Nov. 1990.

[9] M. R. Iravani and D. Maratukulam, "Review of Semiconductor-Controlled (Static) Phase Shifters for Power System Applications," *IEEE Transactions on Power Systems,* vol. 9, no. 4, Nov. 1994, pp. 1833–1839.

[10] L. Gyugyi, "Reactive Power Generation and Control by Thyristor Circuits," *IEEE Transactions on Industrial Applications,* vol. 15, no. 5, Sep./Oct. 1979, pp. 521–532.

[11] M. Chamia and L. Ångquist, "Thyristor-Controlled Series Capacitor Design and Field Tests," in *Proceedings of EPRI Flexible AC Transmission System (FACTS) Conference,* Boston, USA, May 1992.

[12] N. Christl et al., "Advanced Series Compensation with Variable Impedance," EPRI TR-100504, in *Proceedings of FACTS Conference I—The Future in High-Voltage Transmission,* Nov. 1990.

[13] M. P. Bottino, B. Delfino, G. B. Denegri, M. Invernizzi, and A. Morini, "Enhancement of Power System Performance over Different Time Frames Using Advanced Series Compensation," in *IPEC'95—International Power Electronics Conference,* Yokohama, Japan, Apr. 1995, pp. 394–399.

[14] L. Gyugyi, "Unified Power Flow Control Concept for Flexible AC Transmission Systems," *IEE Proceedings-C,* vol. 139, no. 4, July 1992, pp. 323–331.

[15] L. Gyugyi, "Solid-State Control of ac Power Transmission," EPRI TR-100504, in *Proceedings of FACTS Conference I—The Future in High-Voltage Transmission,* Nov. 1990.

[16] Å. Ekström, "Theoretical Analysis and Simulation of Force-Commutated Voltage-Source Converters for FACTS Application," EPRI TR-100504, in *Proceedings of FACTS Conference I—The Future in High-Voltage Transmission,* Nov. 1990.

[17] B. T. Ooi, S. Z. Dai, and F. D. Galiana, "A Solid-State PWM Phase-Shifter," *IEEE Transactions on Power Delivery,* vol. 8, no. 2, Apr. 1993, pp. 573–579.

[18] Z. Zhang, J. Kuang, X. Wang, and B. T. Ooi, "Force Commutated HVDC and SVC Based on Phase-Shifted Multi-Converter Modules," *IEEE Transactions on Power Delivery,* vol. 8, no. 2, Apr. 1993, pp. 712–718.

[19] C. Schauder, M. Gernhardt, E. Stacey, T. Lemak, L. Gyugyi, T. W. Cease, and A. Edris, "Development of a ± 100 Mvar Static Condenser for Voltage Control of Transmission Systems," *IEEE Transactions on Power Delivery,* vol. 10, no. 3, July 1995, pp. 1486–1493.

[20] G. Reed, S. Jochi, T. Snow, J. Paserba, N. Morishima, A. Abed, T. Croasdaile, M. Takeda, R. Westover, T. Sugiyama, and Y. Hamasaki, "SDG&E Talega STATCOM Project—System Analysis, Design, and Configuration," in *IEEE/PES Transm. and Distrib. Conference and Exibition 2002: Asia Pacific,* vol. 2, 2002, pp. 1393–1398.

[21] A.-A. Edris, S. Zelingher, L. Gyugyi, and L. J. Kovalsky, "Squeezing More Power from the Grid," *IEEE Power Engineering Review,* June 2002, pp. 4– 6.

[22] S. Mori, K. Matsuno, T. Hasegawa, S. Ohnishi, M. Takeda, M. Seto, S. Murakami, and F. Ishiguro, "Development of a Large Static Var Generator Using Self-Commutated Inverters for Improving Power System Stability," *IEEE Transactions on Power System,* vol. 8, no. 1, Feb. 1993, pp. 371– 377.

[23] L. Gyugyi, C. D. Schauder, S. L. Williams, T. R. Rietman, D. R. Torgerson, and A. Edris, "The Unified Power Flow Controller: A New Approach to Power Transmission Control," *IEEE Transactions on Power Delivery,* vol. 10, no. 2, Apr. 1995, pp. 1085– 1093.

[24] L. Gyugyi, "Dynamic Compensation of AC Transmission Lines by Solid-State Synchronous Voltage Sources," *IEEE Transactions on Power Delivery,* vol. 9, no. 2, Apr. 1994, pp. 904– 911.

[25] R. J. Nelson, J. Bian, and S. L. Williams, "Transmission Series Power Flow Control," *IEEE Transactions on Power Delivery,* vol. 10, no. 1, Jan. 1995, pp. 504– 510.

[26] Q. Yu, S. D. Round, L. E. Norum, and T. M. Undeland, "A New Control Strategy for a Unified Power Flow Controller," in *EPE'95—Eur. Conference Power Electronics Applications,* vol. 2, Sevilla, Spain, Sep. 1995, pp. 2.901– 2.906.

[27] A. Nabae, I. Takahashi, and H. Akagi, "A new neutral-point-clamped PWM inverter," *IEEE Transactions on Ind. Applicat.,* vol. IA-17, no. 5, 1981, pp. 518–523.

[28] H. Fujita, Y. Watanabe, and H. Akagi, "Control and Analysis of a Unified Power Flow Controller," *IEEE Transactions on Power Electronics,* vol. 14, no. 6, Nov. 1999, pp. 1021– 1027.

[29] M. Aredes, J. Häfner, and K. Heumann, "A Combined Series and Shunt Active Power Filter," *IEEE/KTH Stockholm Power Tech Conference,* vol. Power Electronics, Sweden, June 1995, pp. 237– 242.

[30] M. Aredes, *Active Power Line Conditioners,* Dr.-Ing. Thesis (magna cum laude), Technische Universität Berlin, Germany, March 1996.

[31] H. Akagi and H. Fujita, "A New Power Line Conditioner for Harmonic Compensation in Power Systems," *IEEE Transactions on Power Delivery,* vol. 10, no. 3, July 1995, pp. 1570– 1575.

[32] H. Akagi, "New Trends in Active Filters," in *EPE'95—Eur. Conference Power Electronics Applications,* Sevilla, Spain, September 1995, pp. 0.017– 0.026.

[33] H. Fujita and H. Akagi, "The Unified Power Quality Conditioner: The Integration of Series- and Shunt-Active Filters," *IEEE Transactions on Power Electronics,* vol. 13, no. 2, pp. 315– 322, March 1998.

[34] N. G. Hingorani, "Introducing Custom Power," *IEEE Spectrum,* pp. 41– 48, June 1995.

[35] M. Boost and P. D. Ziogas, "State of the Art PWM Techniques: A Critical Evaluation," in *IEEE-PESC'86—Power Electronics Special Conference,* 1986, pp. 425– 433.

[36] F. Z. Peng, H. Akagi, and A. Nabae, "A New Approach to Harmonic Compensation in Power Systems—A Combined System of Shunt Passive and Series Active Filters," *IEEE Transactions on Industrial Applications,* vol. 26, no. 6, Nov./Dec. 1990, pp. 983– 990.

[37] M. Aredes, *Novos Conceitos de Potência e Aplicações em Filtros Ativos,"* M.Sc. Thesis, COPPE— Federal University of Rio de Janeiro, Brazil, Nov. 1991.

[38] H. Akagi, Y. Kanazawa, and A. Nabae, "Generalized Theory of the Instantaneous Reactive Power in Three-Phase Circuits," in *IPEC'83—International Power Electronics Conference*, Tokyo, Japan, 1983, pp. 1375–1386.

[39] T. Tanaka, K. Wada, and H. Akagi, "A New Control Scheme of Series Active Filters," in *IPEC'95—International Power Electronics Conference*, Yokohama, Japan, Apr. 1995, pp. 376–381.

[40] C. L. Fortescue, "Method of Symmetrical Co-ordinates Applied to the Solution of Polyphase Networks," *A.I.E.E. Transactions*, vol. 37, June 1918, pp. 1027–1140.

INDEX

abc Theory, 90
Active and nonactive current calculation by means of a minimization method, 89
Active current, 30
Active filter controller, 111, 113
 based on voltage detection, 152
Active filter dc voltage regulator, 186
Active filter for harmonic damping, 157
 power distribution line, 158
Active filter for damping of harmonic propagation, 159
 adjustment of the active filter gain, 168
 distributed-constant model, 169
Active filter for constant power compensation, 118
Active filter for current minimization, 145
Active filter for harmonic damping, 150
Active filter for sinusoidal current control, 134
Active impedance, 225, 226
Active power, 21, 26, 30
Active power factor, 30
Active voltage, 30
Akagi, H, 1
Alternating current (ac) transmission and distribution power systems, 2
Analog/digital signal processing, 1
Antiresonance phenomena, 225
Apparent power, 2, 21, 22, 25, 30
Arc furnace system, 212

Balanced three-phase system, 34
Balanced voltages and capacitive loads, 54
Budeanu, C. I., 2, 25, 28
Butterworth filters, 126, 127, 128

Calculation of voltage and current vectors when zero-sequence components are excluded, 45
Clarke transformation, 43
 avoiding, 80
 inverse, 44
Combined series active filter and shunt passive filter, 223
Combined series and shunt power conditioners, 265
Comparisons between hybrid and pure active filters, 253
 through simulation results, 259
Comparisons between the *p-q* Theory and the *abc* Theory, 98
Compensated currents, 100
 line, 133
Complex impedance, 22
Complex power, 24
Constant instantaneous power control strategy, 189
Control of active and reactive power in multipulse converters, 288
Control algorithm for imaginary current compensation, 99

376 INDEX

Control method for shunt current compensation, 86
Control theory, 1
Controller design for the UPFC, 274
Current control methods, 178
Current harmonic spectra, 216
Current phasor, 23
Custom Power concept, 14

Deadbeat control, 180
Design of a switching-ripple filter, 248
  effect on system stability, 250
  experimental results, 252
  experimental testing, 251
  principle, 248
Digital controller, 173
  operating principle of PLL and PWM units, 175
  sampling operation in the A/D unit, 177
  specification, 175
  system configuration, 174
Displacement factor, 29
Distortion factor, 28, 29
Distortion power, 26
Distortions and imbalances in voltages and currents, 75
Dual p-q Theory, 68
Dynamic hysteresis-band current controller, 184

Electric power definitions, 19
Electric power theory, 2
Emanuel-Eigeles, A., 3
Energy transfer, 52
Erlicki, M. S., 3

FACTS (flexible AC transmission system), 5
FACTS and UPFC principles, 268
  control of active and reactive power in multipulse converters, 288
  controller design for the UPFC, 274
  power flow control principle, 270
  quasi 24-pulse converter, 286
  shunt multipulse converter controller, 290
  six-pulse converter, 282
  UPFC approach using a shunt multipulse converter, 281

voltage regulation principle, 269
Fifth harmonic in load current, 64
Filter to suppress switching ripples, 233
Frequency response of current control, 181
Fryze, S., 2, 30
Fryze currents minimization method, 94
Fujiwara, K, 3
Fukao, T., 3

Gate-turn-off thyristors (GTOs), 267
Gyugyi, L., 3

Harashima, F., 3
Harmonic compensation, 11, 116
  basic principles, 11
Harmonic current and voltage sources, 8
Harmonic currents, 109
Harmonic damping, 1
Harmonic filtering, 1
Harmonic isolation, 1
Harmonic pollution, 4, 8
Harmonic regulations or guidelines, 8
Harmonic termination, 1
Harmonic voltages in power systems, 5
Hidden currents, 64
Hybrid filters, 237
Hybrid and series active filters, 221
Hysteresis-based current control, 184

Identified and unidentified harmonic-producing loads, 7
Iida, H., 3
Imaginary power, 52, 80, 123
Inaba, H, 3
Instantaneous abc Theory, 87
Instantaneous active and reactive power theory (see p-q Theory), 1
Instantaneous active current, 50
Instantaneous active power, 51
Instantaneous imaginary power, 52
Instantaneous power, 20, 50
Instantaneous power theory, 41
Instantaneous reactive current, 50
Instantaneous reactive power, 3, 51, 52, 87
Instantaneous real power, 52
Instantaneous zero-sequence power, 52
Integration of series and shunt active filter controllers, 305
Interference with communication systems, 5

INDEX 377

Kanazawa, Y., 1

Load balancing, 1
Low-voltage transformerless hybrid active filter, 255
Low-voltage, transformerless, pure shunt active filter, 258

Machida, T., 3
Minimized currents, 91
Miyairi, S., 3
Modeling of digital current control, 178
Modified p-q Theory, 82
 control method for shunt current compensation, 86

Nabae, A., 1, 4
Negative-sequence components, 74
Nonsinusoidal conditions, 25

Operating principle of PLL and PWM units, 175
Overheating of capacitors for power-factor correction, 5
Overheating of transformers and electrical motors, 4

Phase-locked-loop (PLL) circuit, 141
Phasor notation, 22
Physical meanings of instantaneous real, imaginary, and zero-sequence powers, 79
Positive-sequence voltage detector, 138
 main circuit, 138
 phase-locked-loop (PLL) circuit, 141
 simulation results, 145
Power conditioning, 1
Power conversion circuits,, 1
Power definitions under sinusoidal conditions, 20
Power distribution line, 158
Power electronics technology, 1
Power factor, 24, 29
Power flow control, 1, 14
Power flow control principle, 270
Power in three-phase unbalanced systems, 36
Power semiconductor devices, 1
Power tetrahedron, 28

p-q Theory, 1
 applications to power electronics equipment, 4
 avoiding the Clarke transformation, 80
 basis of, 42
 comparisons with the conventional theory, 53
 for three-phase, three-wire systems, 49
 historical background, 42
 instantaneous powers, 48
 modified, 82
Proportional control, 179
PWM converter, 111, 112
PWM voltage control with minor feedback control loop, 300

Quasi 24-pulse converter, 286

Reactive current, 30
Reactive power, 2, 21, 22, 26
Reactive power factor, 30
Reactive voltage, 30
Reactive-power control for power factor correction, 1
Real power, 52
Relation between phasors and sinusoidal time functions, 23

Sasaki, H., 3
Selection of power components to be compensated, 101
Selective filter basic cell (SFBC), 209
Series active filter, 221
 basic, 221
 compensation principle, 228
 control circuit, 231
 experimental results, 234
 experimental system, 226
 filter to suppress switching ripples, 233
 filtering characteristics, 230
 harmonic current flowing from the load to the source, 230
 harmonic current flowing from the source to the shunt passive cilter, 231
 hybrid filters, 237
 output voltage, 229
 shunt passive filter harmonic voltage, 229
 source harmonic current, 228
Series active filter controller, 301

Series active filter integrated with a double-series diode rectifier, 238
  characteristics comparisons, 247
  circuit configuration and delay time, 241
  first-generation control circuit, 241
  second-generation control circuit, 244
  stability analysis and characteristics comparison, 246
  stability of the active filter, 242
  transfer function of the control circuits, 246
Series voltage compensation, 70, 71
Seventh harmonic in load current, 67
Shunt active filter, 109
  based on voltage detection, 151
  general description of, 111
  three-phase, four-wire, 182
Shunt current compensation, 59, 110
Shunt multipulse converter controller, 290
Shunt passive filter harmonic voltage, 229
Shunt selective harmonic compensation, 208
Single-phase, sinusoidal systems, 20
Sinusoidal balanced voltage and nonlinear load, 55
Sinusoidal time function, 22
Sinusoidal voltage sources, 2
Sinusoidal voltages and currents, 53
Sinusoidal waveform, 23
Six-pulse converter, 282
Source harmonic current, 228
Stand-alone series active filter, 13
Stand-alone shunt active filter, 11
Static synchronous compensator (STATCOM), 16, 267
Static var compensator (SVC), 267
Strycula, E. C., 3
Synchronous sampling, 178

Takahashi, I., 3
Three-phase complex power, 35
Three-phase diode rectifier, 9, 10
Three-phase instantaneous active power, 47
  in terms of Clarke components, 47
Three-phase instantaneous reactive power, 88
Three-phase reactive power, 36
Three-phase systems, 31
  balanced, 34

classifications of, 31
power in, 36
Three-phase, four-wire shunt active filters, 182
  constant instantaneous power control strategy, 189
  converter topologies for, 183
  dynamic response, 196
  economical aspects, 201
  experimental results, 203
  influence of the system parameters, 195
  optimal power flow conditions, 187
  performance analysis and parameter optimization, 195
  sinusoidal current control strategy, 192
Three-phase, four-wire systems, 71
Three-phase, three-wire shunt active filters, 116
Three-phase, three-wire systems, 49
Thyristor bridge, 125
Thyristor-controlled reactor (TCR), 267
Thyristor-controlled series capacitor (TCSC), 267
Total harmonic distortion (THD), 5
Total instantaneous energy flow per time unit, 80
Triangle of powers, 24
Tsuboi, K., 3

Unified power flow controller (UPFC), 16, 266, 267
Unified power quality conditioner (UPQC), 293
  analysis of the UPQC dynamic, 308
  controller, 299
  general aspects, 307
  general description, 294
  power circuit, 297
  three-phase, four-wire UPQC, 297
Universal active power line conditioner (UPLC), 343
  controller, 347
  controller for configuration #2, 355
  general aspects, 370
  general description, 344
  normalized system parameters, 355
  performance, 355
  simulation results of configuration #1, 360

simulation results of configuration #2, 368
UPQC combined with passive filters (the hybrid UPQC), 326
   controller, 331
   experimental results, 337
UPQC controller, 299
UPFC approach using a shunt multipulse converter, 281

Voltage and current phasors, 22
Voltage flicker, 5
Voltage phasor, 23
Voltage regulation, 1
Voltage regulation principle, 269
Voltage waveform distortion, 5
Voltage/current sensors, 1
Voltage-flicker reduction, 1

Zero-sequence components, 80
zero-sequence power, 52
   in a three-phase sinusoidal voltage source, 72

Lightning Source UK Ltd.
Milton Keynes UK
UKOW04n1510101114

241400UK00001B/26/P